Implementing Cisco IP Telephony and Video, Part 2 (CIPTV2) Foundation Learning Guide (CCNP Collaboration Exam 300-075 CIPTV2)

William Alexander Hannah CCIE #25853

Akhil Behl CCIE #19564

Cisco Press

800 East 96th Street

Indianapolis, IN 46240

Implementing Cisco IP Telephony and Video, Part 2 (CIPTV2) Foundation Learning Guide (CCNP Collaboration Exam 300-075 CIPTV2)

William Alexander Hannah CCIE #25853

Akhil Behl CCIE #19564

Copyright© 2016 Cisco Systems, Inc.

Published by:
Cisco Press
800 East 96th Street
Indianapolis, IN 46240 USA

Printed in the United States of America

First Printing March 2016

Library of Congress Control Number: 2015961048

ISBN-13: 978-1-58714-455-4

ISBN-10: 1-58714-455-7

Warning and Disclaimer

This book is designed to provide information about Cisco Unified IP Telephony and Video administration and to provide test preparation for the CIPTV Part 2 Version 10.5 exam (CCNP Collaboration CIPTV2 300-075), which is part of the CCNP Collaboration certification. Every effort has been made to make this book as complete and accurate as possible, but no warranty or fitness is implied.

The information is provided on an "as is" basis. The author, Cisco Press, and Cisco Systems, Inc., shall have neither liability nor responsibility to any person or entity with respect to any loss or damages arising from the information contained in this book or from the use of the discs or programs that may accompany it.

The opinions expressed in this book belong to the author and are not necessarily those of Cisco Systems, Inc.

Trademark Acknowledgments

All terms mentioned in this book that are known to be trademarks or service marks have been appropriately capitalized. Cisco Press or Cisco Systems, Inc. cannot attest to the accuracy of this information. Use of a term in this book should not be regarded as affecting the validity of any trademark or service mark.

Special Sales

For information about buying this title in bulk quantities, or for special sales opportunities (which may include electronic versions; custom cover designs; and content particular to your business, training goals, marketing focus, or branding interests), please contact our corporate sales department at corpsales@pearsoned.com or (800) 382-3419.

For government sales inquiries, please contact governmentsales@pearsoned.com.

For questions about sales outside the U.S., please contact intlcs@pearson.com.

Feedback Information

At Cisco Press, our goal is to create in-depth technical books of the highest quality and value. Each book is crafted with care and precision, undergoing rigorous development that involves the unique expertise of members from the professional technical community.

Readers' feedback is a natural continuation of this process. If you have any comments regarding how we could improve the quality of this book, or otherwise alter it to better suit your needs, you can contact us through e-mail at feedback@ciscopress.com. Please make sure to include the book title and ISBN in your message.

We greatly appreciate your assistance.

Publisher: Paul Boger

Associate Publisher: Dave Dusthimer

Business Operation Manager, Cisco Press: Jan Cornelssen

Executive Editor: Brett Bartow

Managing Editor: Sandra Schroeder

Development Editor: Marianne Bartow

Senior Project Editor: Mandie Frank

Copy Editor: Keith Cline

Technical Editors: Steve Foy

Editorial Assistant: Vanessa Evans

Designer: Mark Shirar

Composition: codeMantra

Indexer: Lisa Stumpf

Proofreader: Debbie Williams

Americas Headquarters	Asia Pacific Headquarters	Europe Headquarters
Cisco Systems, Inc.	Cisco Systems (USA) Pte. Ltd.	Cisco Systems International BV
San Jose, CA	Singapore	Amsterdam, The Netherlands

Cisco has more than 200 offices worldwide. Addresses, phone numbers, and fax numbers are listed on the Cisco Website at **www.cisco.com/go/offices.**

CCDE, CCENT, Cisco Eos, Cisco HealthPresence, the Cisco logo, Cisco Lumin, Cisco Nexus, Cisco StadiumVision, Cisco TelePresence, Cisco WebEx, DCE, and Welcome to the Human Network are trademarks; Changing the Way We Work, Live, Play, and Learn and Cisco Store are service marks; and Access Registrar, Aironet, AsyncOS, Bringing the Meeting To You, Catalyst, CCDA, CCDP, CCIE, CCIP, CCNA, CCNP, CCSP, CCVP, Cisco, the Cisco Certified Internetwork Expert logo, Cisco IOS, Cisco Press, Cisco Systems, Cisco Systems Capital, the Cisco Systems logo, Cisco Unity, Collaboration Without Limitation, EtherFast, EtherSwitch, Event Center, Fast Step, Follow Me Browsing, FormShare, GigaDrive, HomeLink, Internet Quotient, IOS, iPhone, iQuick Study, IronPort, the IronPort logo, LightStream, Linksys, MediaTone, MeetingPlace, MeetingPlace Chime Sound, MGX, Networkers, Networking Academy, Network Registrar, PCNow, PIX, PowerPanels, ProConnect, ScriptShare, SenderBase, SMARTnet, Spectrum Expert, StackWise, The Fastest Way to Increase Your Internet Quotient, TransPath, WebEx, and the WebEx logo are registered trademarks of Cisco Systems, Inc. and/or its affiliates in the United States and certain other countries.

All other trademarks mentioned in this document or website are the property of their respective owners. The use of the word partner does not imply a partnership relationship between Cisco and any other company. (0812R)

About the Authors

William Alexander Hannah, CCIE Collaboration #25853, CCSI #32072, along with numerous other Cisco Unified Communications and data center specializations, and VMware certifications, has been an independent IT and telephony consultant, author, and technical editor for more than 12 years. He has been a technical trainer for more than 8 years and has taught more than 20 different courses for Cisco. Alex is a Senior Courseware Developer and Subject Matter Expert for Global Knowledge, designing all CCNP Collaboration courseware, labs, and infrastructure. He has done a wide array of IT and telephony consulting for many different companies along the eastern portion of the United States. A former Senior Architect and Senior Presales Engineer for two Cisco Gold Partners in the Southern Virginia area, Alex is now the principal owner of Hannah Technologies LLC, an IT consulting and training firm based in Midlothian, Virginia. Alex has implemented advanced IP telephony and video installations in his area for more than 12 years. When he is not working, he can be found on a boat, wakeboarding with friends and family. He can be reached at alex@hannahtechnologies.com.

Akhil Behl is a Pre-Sales Manager with a leading service provider. His charter involves an overarching technology portfolio encompassing IoT, collaboration, security, infrastructure, service management, cloud, and data center. He has 12+ years of experience working in leadership, advisory, business development, and consulting positions with various organizations; leading global accounts, driving toward business innovation and excellence. Previously, he was in a leadership role with Cisco Systems.

Akhil has a Bachelor of Technology degree in electronics and telecommunications from IP University, India, and a Master's degree in business administration from Symbiosis Institute, India. Akhil holds dual CCIE in Collaboration and Security, PMP, ITIL, VCP, TOGAF, CEH, ISO/IEC 27002, and many other industry certifications.

He has published several research papers in national and international journals, including IEEE, and has been a speaker at prominent industry forums such as Interop, Enterprise Connect, Cloud Connect, Cloud Summit, Cisco Sec-Con, IT Expo, Computer Society of India, Singapore Computer Society, and Cisco Networkers.

Akhil is the author of the following Cisco Press books:

- *CCIE Collaboration Quick Reference*
- *Securing Cisco IP Telephony Networks*
- *Implementing Cisco IP Telephony and Video (Part 1)*

He is a technical editor for Cisco Press and other publications. Akhil can be reached at akbehl@technologist.com.

About the Technical Reviewer

Steve Foy, CCSI #96106, is an IT professional and certified CCNP in Collaboration. Steve is employed by Global Knowledge, and teaches and develops classes and labs supporting Cisco Collaboration courses, in addition to customized courses for clients. He has been a Certified Cisco Systems Instructor (CCSI) since 1995. Steve has experience with Cisco Communications and Collaboration products dating back to 1999, and has been in the IT/data communications industry since 1979. He has worked for Paradyne and AT&T in previous employments. Steve is co-author of the Cisco Press publication *Cisco Voice over Frame Relay, ATM, and IP* (ISBN-10: 1578702275). Steve is married to Charlene (Chaz), and has four children and five grandchildren. He lives in the Tampa Bay area of Florida.

Dedications

William Alexander Hannah:

This book is dedicated to several people who have been major influences in my life and career. First and foremost, I would like to dedicate this book to my father. He blessed me with the wisdom, drive, and determination to push through obstacles and always strive to achieve my personal best. Dad, you were a tremendous role model for me growing up. I hope that as you look down on me from heaven you will know that I have strived every day to make you proud and earn your respect. To my mother, Sheila, thank you for being patient with me and showing me unconditional love no matter what the circumstances. To Kim, I love you and Kendall very much and I am so lucky to have you both in my life; thank you for supporting me through this process, you are truly the love of my life. To my sister, Kristol, and brother, Brandon, keep it real. I love you both very much. To my best friends, Jon and Ricki, thanks for putting up with me and being there for me to vent and for being a shoulder for support! To all my extended family and friends, thank you for the support and love during my journey.

Akhil Behl:

I would like to dedicate this book first to my family, my wonderful and beautiful wife, Kanika, and my lovely children, Shivansh and Shaurya, for their love, patience, sacrifice, and support while working on this project. They have been very kind and supportive, as always, during my journey to write yet another book. Moreover, my loving wife Kanika has been pivotal while writing the book as she reviewed my work and suggested amendments and improvements.

To my parents, Vijay Behl and Ravi Behl, for their continuous love, encouragement, guidance, and wisdom. To my brothers, Nikhil Behl and Ankit Behl, who have always been there to support me in all my endeavors. And I would like to thank God for all his blessings in my life.

Acknowledgments

William Alexander Hannah:

I cannot thank the staff at Cisco Press enough for this opportunity. It has truly been a lifelong dream to be published in an industry that I have great passion for and love. Brett and team, thank you for being patient, great motivators, and educators during my journey. I would like to thank the team at Global Knowledge (Lisa, Lia, Lori, Tyler, Rick, Pam, and Stuart) for allowing me the platform to train thousands of engineers and students over the past 8 years. It has truly been the highlight of my career to give back to individuals and see them achieve their dreams. To my fellow Cisco instructors (Steve, Ted, Joel, and Dennis), thank you for putting up with me all these years. I would like to thank my mentors, former employers, and engineering peers: Patrick, David, Alan, Jim, Adash, Sue, Jose, Duane, Tom, Travis, Greg, Shawn, Will, Larry, Hunter, Tres, Heather, and Trent. You all provided me a tremendous platform to learn and excel in my craft. It was truly an honor to work with each of you and learn from the best group of guys and gals in the world. It is amazing that in a small area like southern Virginia our drive and passion created more than ten CCIE Collaboration certified engineers. I know our paths do not cross as often as they should, but I cannot thank you all enough from the bottom of my heart. Each of you has played a vital role in shaping me and grooming me for the journey that lies ahead.

Akhil Behl:

I would like to thank the following amazing people and teams for helping me write this book.

The Cisco Press editorial team: Brett Bartow, the Executive Editor, for seeing the value and vision in the proposed title and providing me the opportunity to write this book; and Marianne Bartow, Development Editor, and Ellie Bru, Development Editor, and Vanessa Evans, Editorial Assistant, for their support and guidance all throughout the writing of this book. It is my sincere hope to work again with them in the near future. And my special thanks to everyone else in the Cisco Press production team, for their support and commitment.

I would like to thank my mentors and my peers who have guided and stood by me all of these years. Thank you to all my managers and peers from Cisco who have been supportive of what I wanted to do and helped me achieve it.

Contents at a Glance

Introduction   xxi

Chapter 1   Cisco Collaboration Solution Multisite Deployment Considerations   1

Chapter 2   Understanding Multisite Deployment Solutions   33

Chapter 3   Overview of PSTN and Intersite Connectivity Options   69

Chapter 4   URI-Based Dial Plan for Multisite Deployments   119

Chapter 5   Remote Site Telephony and Branch Redundancy Options   141

Chapter 6   Cisco Collaboration Solution Bandwidth Management   159

Chapter 7   Call Admission Control (CAC) Implementation   183

Chapter 8   Implementing Cisco Device Mobility   209

Chapter 9   Cisco Extension Mobility   241

Chapter 10   Implementing Cisco Unified Mobility   261

Chapter 11   Cisco Video Communication Server and Expressway Deployment   287

Chapter 12   Deploying Users and Endpoints in Cisco VCS Control   311

Chapter 13   Interconnecting Cisco Unified Communications Manager and Cisco Video
             Control Server   333

Chapter 14   Cisco Unified Communications Mobile and Remote Access   349

Chapter 15   Cisco Inter-Cluster Lookup Service (ILS) and Global Dial Plan Replication
             (GDPR)   377

Chapter 16   Cisco Service Advertisement Framework (SAF) and Call Control
             Discovery (CCD)   397

Appendix A   Answers Appendix   423

Glossary   429

Index   441

Reader Services

Register your copy at www.ciscopress.com/title/ISBN for convenient access to downloads, updates, and corrections as they become available. To start the registration process, go to www.ciscopress.com/register and log in or create an account*. Enter the product ISBN 9781587144554 and click Submit. Once the process is complete, you will find any available bonus content under Registered Products.

*Be sure to check the box that you would like to hear from us to receive exclusive discounts on future editions of this product.

Contents

Introduction   xxi

Chapter 1   Cisco Collaboration Solution Multisite Deployment Considerations   1

Multisite Deployment Issues Overview   2

Voice and Video Call Quality Issues   5

Bandwidth Challenges   7

Availability Challenges   10

Dial Plan Challenges   12

Overlapping Numbers   12

Nonconsecutive Numbers   13

Variable-Length Numbering   13

Direct Inward Dialing (DID) Ranges and E.164 Addressing   14

Optimized Call Routing   15

Various PSTN Requirements   16

Scalability   17

Fixed Versus Variable-Length Numbering Plans   17

Detection of End of Dialing in Variable-Length Numbering Plans   20

Optimized Call Routing and PSTN Backup   22

PSTN Requirements   23

Issues Caused by Different Methods of PSTN Dialing   24

Dial Plan Scalability Issues   26

NAT and Security Issues   27

Summary   29

References   30

Review Questions   30

Chapter 2   Understanding Multisite Deployment Solutions   33

Multisite Deployment Solution Overview   34

Quality of Service   36

QoS Advantages   37

Overview of Solutions for Bandwidth Challenges   39

Low-Bandwidth Codecs and RTP Header Compression   41

Codec Configuration in CUCM   42

Disabling the Annunciator for Remote Branches   43

Local Versus Remote Conference Bridges   44

Transcoders   44

Leading Practices for Transcoder Design   45

Mixed Conference Bridge   46

Multicast MOH from the Branch Router Flash   47

An Example of Multicast MOH from the Branch Router Flash   49

An Example of Multicast MOH from the Branch Router Flash Cisco IOS Configuration   51

Alternatives to Multicast MOH from Remote Site Router Flash   52

Preventing Too Many Calls by CUCM Call Admission Control   52

Availability   53

PSTN Backup   55

MGCP Fallback   55

Fallback for IP Phones: SRST, CME SRST, or SIP SRST   56

Using CFUR to Reach Remote Site Cisco IP Phones During WAN Failure   58

Using CFUR to Reach Users of Unregistered Software IP Phones on Other Devices   58

AAR and CFNB   59

Mobility Solutions   60

Overview of Dial Plan Solutions   61

NAT and Security Solutions   62

CUBE in Flow-Through Mode   62

Cisco Expressway C and Cisco Expressway E As a Solution to NAT and Security Issues in a Multisite Environment   63

Summary   64

References   65

Review Questions   65

Chapter 3   Overview of PSTN and Intersite Connectivity Options   69

Overview of Multisite Connection Options   70

CUCM Connection Options Overview   71

Cisco IOS Gateway Protocol Functions Review   72

SIP Trunk Characteristics   73

H.323 Trunk Overview   74

Trunk Implementation Overview   76

Gatekeeper-Controlled ICT and H.225 Trunk Configuration   77

Trunk Types Used by Special Applications   78

Dial Plan Requirements for Multisite Deployments with Distributed Call Processing   79

Implementing Site Codes for On-Net Calls   81

Digit-Manipulation Requirements When Using Access and Site Codes   82

Access and Site Code Requirements for Centralized
Call-Processing Deployments   83

Implementing PSTN Access in Cisco IOS Gateways   84

PSTN Access Example   85

Transformation of Incoming Calls Using ISDN TON   85

ISDN TON Example: Calling Number Transformation of Incoming
Call   87

Implementing Selective PSTN Breakout   88

Configuring IP Phones to Use Local PSTN Gateway   88

Implementing PSTN Backup for On-Net Intersite Calls   90

Digit-Manipulation Requirements for PSTN Backup of On-Net Intersite
Calls   90

Implementing TEHO   92

TEHO Example Without Local Route Groups   93

TEHO Example with Local Route Groups   95

Implementing Globalized Call Routing   96

Globalized Call Routing: Number Formats   98

Normalization of Localized Call Ingress on Gateways   102

Normalization of Localized Call Ingress from Phones   104

Localized Call Egress at Gateways   105

Localized Call Egress at Phones   107

Globalized Call Routing Example: Emergency Dialing   109

Considering Globalized Call Routing Interdependencies   112

Globalized Call Routing and TEHO Advantages   113

Globalized Call Routing TEHO Example   113

Summary   115

References   116

Review Questions   116

Chapter 4   **URI-Based Dial Plan for Multisite Deployments   119**

URI Dialing Overview   120

URI Endpoint Addressing Overview   123

URI Partitions and Calling Search Spaces   125

URI Call Sources Overview   126

Blended Addressing   127

FQDNs in Directory URIs   128

URI Call Routing   129

Non-Numeric URI Call Routing Process   132

Numeric URI Call Routing Process   134

   Routing URI Calls over SIP Trunks   134

Summary   136

References   137

Review Questions   137

Chapter 5   **Remote Site Telephony and Branch Redundancy Options   141**

Cisco Unified Communications Manager Express   141

Cisco Business Edition   143

Survivable Remote Site Telephony   144

SRST and E-SRST Configuration   146

SRST IOS Dial Plan   148

CUCM SRST Configuration   149

Multicast Music on Hold in SRST   150

MGCP Fallback   153

Cisco Call Forward Unregistered   154

Summary   156

References   156

Review Questions   156

Chapter 6   **Cisco Collaboration Solution Bandwidth Management   159**

Bandwidth Management Options   159

Voice and Video Codecs   161

Codec Selection   162

   Media Resource Group and Media Resource Group List   166

Multicast Music on Hold   168

   Multicast MOH IP Address and Port Considerations   172

Local Conference Bridge   172

Transcoder   176

Summary   179

References   180

Review Questions   180

Chapter 7   **Call Admission Control (CAC) Implementation   183**

Call Admission Control Characteristics   184

CUCM Call Admission Control   184

Location-Based CAC   185

Location Bandwidth Manager   187

Enhanced Location-Based CAC   189

Resource Reservation Protocol   196

RSVP Configuration   198

RSVP SIP Preconditions   199

Automated Alternate Routing   202

IOS Call Admission Control   204

Local CAC   204

Reservation-Based CAC   205

Measurement-Based CAC   206

Summary   206

References   206

Review Questions   207

Chapter 8   Implementing Cisco Device Mobility   209

Device Roaming Overview   210

Issues with Roaming Devices   210

Using Device Mobility to Solve Roaming Device Issues   212

Device Mobility Overview   213

Device Mobility: Dynamic Phone Configuration Parameters   213

Device Mobility Dynamic Configuration by Location-Dependent
Device Pools   216

Device Mobility Configuration Elements   217

Relationship Between Device Mobility Configuration Elements   218

Device Mobility Operation   220

Device Mobility Operation Flowchart   221

Device Mobility Considerations   224

Review of Line and Device CSSs   225

Device Mobility and CSSs   225

Examples of Different Call-Routing Paths Based on
Device Mobility Groups and Tail-End Hop-Off   226

Device Mobility Interaction with Globalized Call Routing   228

Advantages of Using Local Route Groups and Globalized Call
Routing   229

An Example of Globalized Call Routing That Is *Not*
Configured with a Different Device Mobility Group   230

An Example of Globalized Call Routing That Is *Not*
Configured with the Same Device Mobility Group   231

An Example of Globalized Call Routing   232

Device Mobility Configuration   233

Summary   236

References   237

Review Questions   237

Chapter 9   Cisco Extension Mobility   241

Overview of Roaming Between Sites   241

    Challenges with Roaming Users   242

CUCM Extension Mobility Overview and Characteristics   243

    Extension Mobility: Dynamic Phone Configuration Parameters   244

    Extension Mobility with Dynamic Phone Configuration by Device
       Profiles   245

CUCM Extension Mobility Operation   245

    Cisco Extension Mobility and CSSs   247

CUCM Extension Mobility Device Profile Overview   248

    Relationship Between Extension Mobility Configuration Elements   249

    Default Device Profile and Feature Safe   251

CUCM Extension Mobility Configuration   252

Summary   257

References   257

Review Questions   257

Chapter 10   Implementing Cisco Unified Mobility   261

Cisco Unified Mobility Overview   262

    Mobile Connect and Mobile Voice Access Characteristics   263

Cisco Unified Mobility Call Flow   264

    Mobile Connect Call Flow   264

    Mobile Voice Access Call Flow   266

Cisco Unified Mobility Implementation Requirements   267

    Cisco Unified Mobility Configuration Elements   268

Cisco Unified Mobility MGCP or SCCP Gateway PSTN Access   271

    MVA Call Flow with MGCP or SCCP PSTN Gateway Access   272

Calling Search Space Handling in Cisco Unified Mobility   273

    CSS Handling in Mobile Voice Access   273

Cisco Unified Mobility Access List Functions   274

    Operation of Time-of-Day Access Control   274

Cisco Unified Mobility Configuration   275

    Configuring Mobile Connect   275

Configuring Mobile Voice Access   281

Summary   284

References   285

Review Questions   285

Chapter 11   Cisco Video Communication Server and Expressway Deployment   287

Cisco VCS and Expressway Series Overview   288

CUCM with Cisco Expressway Series   289

Cisco VCS Control   289

Cisco VCS-C with Cisco VCS Expressway   290

CUCM and Cisco VCS-C (Combined Solution)   290

Common Terminology for Cisco Video and Legacy Video   290

Cisco VCS and Cisco Expressway Series Deployment Options   292

Cisco VCS Deployment   292

Cisco Expressway Series Deployment   293

CUCM and Cisco VCS-C Interconnection   295

Cisco VCS and Cisco Expressway Series Platforms, Licenses, and
Features   296

Cisco VCS and Cisco Expressway Licensing   297

Cisco VCS and Cisco Expressway Feature Comparison   297

Cisco VCS and Cisco Expressway Clustering   298

Clustering Considerations   299

Cluster Deployment Overview   300

Cisco VCS and Cisco Expressway Series Initial Configuration   301

Summary   306

References   306

Review Questions   307

Chapter 12   Deploying Users and Endpoints in Cisco VCS Control   311

Cisco VCS User Authentication Options   312

LDAP Authentication Configuration Example   313

Endpoint Registration   314

Endpoint Authentication   316

Cisco VCS Authentication Methods   317

Registration Restriction Policy   318

Cisco TMS Provisioning   319

Deploying Cisco Jabber Video for TelePresence   320

Cisco VCS Zones   320

Local Zone   321

Default Subzone   322

Subzone   323

Traversal Subzone   323

Links   324

Zone Bandwidth Restrictions: Within   325

Zone Bandwidth Restrictions: In&Out   325

Zone Bandwidth Restrictions: Total   326

Pipes   327

Pipe Bandwidth Restrictions   328

Summary   329

References   330

Review Questions   330

Chapter 13   **Interconnecting Cisco Unified Communications Manager and Cisco Video Control Server   333**

Cisco Unified Communications Manager and Cisco VCS Interconnection Overview   334

Call Flow Between CUCM and Cisco VCS   335

Cisco VCS Dial Plan Components   337

Transforms   338

Admin Policy   338

FindMe Feature   339

Search Rules   340

Configuration of CUCM and Cisco VCS Interconnections   340

FindMe Configuration Procedure   341

Summary   344

References   345

Review Questions   345

Chapter 14   **Cisco Unified Communications Mobile and Remote Access   349**

Cisco Mobile Remote Access Overview   349

Cisco Mobile Remote Access Components   351

Cisco Mobile Remote Access Operation   352

Cisco Mobile Remote Access Firewall Traversal   352

HTTPS Reverse Proxy   354

DNS SRV Setup   354

Registering Remote Jabber Client with CUCM   355

Cisco Unified Communications Mobile and Remote Access
    Configuration   357

    CUCM Configuration for Cisco Unified Communications MRA   358

    IM&P Configuration for Cisco Unified Communications MRA   363

    Cisco Expressway (Expressway-C and Expressway-E) Configuration for
        Cisco Unified Communications MRA   366

    Troubleshooting Cisco MRA   373

Summary   373

References   374

Review Questions   374

Chapter 15   Cisco Inter-Cluster Lookup Service (ILS) and Global Dial Plan
              Replication (GDPR)   377

Inter-Cluster Lookup Service Overview   378

ILS Networking Overview   378

ILS Networking Configuration   380

    ILS-Based SIP URI Dialing/Routing   381

    ILS Calls Via SIP Trunk and Cisco Unified Border Element   383

Directory URI, Enterprise Alternate, and
    +E.164 Alternate Number Exchange   385

Global Dial Plan Replication Overview   386

GDPR Configuration   388

Global Dial Plan Catalogs   391

Summary   393

References   393

Review Questions   394

Chapter 16   Cisco Service Advertisement Framework (SAF) and Call Control
              Discovery (CCD)   397

Complex Dial Plan Implementation Challenges   397

Cisco Service Advertisement Framework Overview   399

    SAF Architecture   399

SAF Characteristics and Operation   402

    SAF Clients   402

    SAF Client Protocol   403

    SAF Forwarders (SAF Forwarding Nodes)   403

    SAF Forwarder Protocol   405

    SAF Message   406

Call Control Discovery Service Overview   406

Call Control Discovery Schema   408

CCD Characteristics and Operation   408

Use Case 1: Normal Calls via SAF-Enabled Network
to Remote Call Control   410

Use Case 2: Calls via PSTN When the SAF Forwarder Is Down   411

Use Case 3: Normal Calls via SAF-Enabled Network to CUBE   411

SAF and CCD Configuration   412

SAF Client Configuration   412

SAF Forwarder Configuration   417

Summary   419

References   420

Review Questions   420

Appendix A   Answers Appendix   423

Glossary   429

Index   441

Command Syntax Conventions

The conventions used to present command syntax in this book are the same conventions used in the IOS Command Reference. The Command Reference describes these conventions as follows:

- **Boldface** indicates commands and keywords that are entered literally as shown. In actual configuration examples and output (not general command syntax), boldface indicates commands that are manually input by the user (such as a **show** command).

- *Italic* indicates arguments for which you supply actual values.

- Vertical bars (|) separate alternative, mutually exclusive elements.

- Square brackets ([ ]) indicate an optional element.

- Braces ({ }) indicate a required choice.

- Braces within brackets ([{ }]) indicate a required choice within an optional element.

Introduction

Professional career certifications have been an important part of the IT industry for many years and will continue to become more important. Many reasons exist for these certifications, but the most popularly cited reason is that of credibility and the knowledge to get the job done. All other considerations held equal, a certified employee/consultant/job candidate is considered more valuable than one who is not. CIPTV2 sets the stage with the above objective in mind and helps you learn and comprehend the topics for the exam and at the same time prepares you for real-world configuration of Cisco's audio and video technology.

Goals and Methods

The most important goal of this book is to provide you with knowledge and skills in Cisco Collaboration solution, with a focus on deploying the Cisco Unified Communications Manager (CUCM), Cisco TelePresence Video Communications Server (VCS), Cisco Expressway Series Solution, and associated Cisco Collaboration solution features. Subsequently, the obvious goal of this book is to help you with the Cisco IP Telephony and Video (CIPTV) Part 2 exam, which is part of the Cisco Certified Network Professional Voice (CCNP) Collaboration certification. This book provides questions at the end of each chapter to reinforce the chapter's content.

The organization of this book helps you discover the exam topics that you need to review in more depth, fully understand and remember those details, and test the knowledge you have retained on those topics. This book does not try to help you pass by memorization, but to truly learn and understand the topics. The knowledge contained in this book is vitally important for you to consider yourself a truly skilled Cisco Collaboration professional. This book helps you pass the Cisco IP Telephony and Video Part 2 exam by using the following methods:

- Helps you discover which test topics you have not mastered

- Provides explanations and information to fill in your knowledge gaps

- Provides practice exercises on the topics and the testing process via test questions at the end of each chapter

Who Should Read This Book?

This book is designed as a foundation for Cisco Collaboration and a certification-preparation book. It provides you with the knowledge required to pass the CCNP Voice Cisco IP Telephony and Video exam in the CCNP Collaboration exams series.

In today's world, technology is evolving at a rapid rate and you need to remain progressive.

How This Book Is Organized

The book covers the following topics:

Chapter 1, "Cisco Collaboration Solution Multisite Deployment Considerations," sets the stage for this book by identifying all the relevant challenges and considerations in multisite deployments requiring Cisco Collaboration solutions.

Chapter 2, "Understanding Multisite Deployment Solutions," provides insight to the solutions of challenges identified in Chapter 1 that are described in this book.

Chapter 3, "Overview of PSTN and Intersite Connectivity Options," provides an overview of the various mechanisms that the sites in an organization can be interconnected to in order to reap the benefits of Cisco Collaboration network.

Chapter 4, "URI-Based Dial Plan for Multisite Deployments," provides insight to SIP URI for both audio and video calls.

Chapter 5, "Remote Site Telephony and Branch Redundancy Options," introduces and explains the various options for deploying remote site telephony solutions and providing a level of redundancy for remote branches/offices including Cisco Business Edition, Cisco Unified Communications Manager Express, SRST, and other mechanisms.

Chapter 6, "Cisco Collaboration Solution Bandwidth Management," describes the various mechanisms to manage bandwidth, which is one of the most expensive and precious resources in a Cisco Collaboration solution.

Chapter 7, "Call Admission Control (CAC) Implementation," provides insight to bandwidth control using various mechanisms, including CAC, E-LCAC, RSVP, to ensure that the call quality is maintained and calls are served as per the organization's policies.

Chapter 8, "Implementing Cisco Device Mobility," provides insight to deployment of Cisco's device mobility solution to support roaming users.

Chapter 9, "Cisco Extension Mobility," explains the concept and implementation of hot desking or hoteling in a Cisco Collaboration environment.

Chapter 10, "Implementing Cisco Unified Mobility," discusses the concept and implementation of Cisco's Mobility solution components (SNR and MVA), which are features that are most commonly used when an organization has a mobile or remote workforce.

Chapter 11, "Cisco Video Communication Server and Expressway Deployment," describes the deployment models pertinent to Cisco VCS and Expressway solution. Cisco VCS and Expressway solutions are state-of-the-art solutions that offer video, audio, and conferencing to enterprise users and mobile users.

Chapter 12, "Deploying Users and Endpoints in Cisco VCS Control," expands on the previous chapter and discusses the deployment of users and endpoints on a Cisco VCS solution.

Chapter 13, "Interconnecting Cisco Unified Communications Manager and Cisco Video Control Server," discusses the integration of Cisco VCS with CUCM and explains the leading practices of performing such integration.

Chapter 14, "Cisco Unified Communications Mobile and Remote Access," describes the concept and configuration of the Cisco MRA feature for remote users leveraging Jabber client on the go for voice, video, presence, and other collaboration features and functions.

Chapter 15, "Cisco Inter-Cluster Lookup Service (ILS) and Global Dial Plan Replication (GDPR)," dives deep into concepts and deployment of automated dial plan replication and leveraging URI-based dial plan across multiple clusters in an organization.

Chapter 16, "Cisco Service Advertisement Framework (SAF) and Call Control Discovery (CCD)," explains the use of Cisco's Service Advertisement Framework and associated services (such as Call Control Discovery) for automated replication of dial plans using the underlying network in a Cisco Collaboration solution.

Answers Appendix allows you to check the validity of your answers at the end of each chapter as you review the questions.

Chapter 1

Cisco Collaboration Solution Multisite Deployment Considerations

When deploying Cisco Unified Communications (UC) in a multisite environment, some unique aspects and design considerations need to be addressed and considered. Any information technology (IT) professional would likely admit that *complexity* is the last word anyone wants to hear in his or her data center and IT environment. Unified Communications can be broken down into components or building blocks. A multisite deployment implementation can be achieved if you properly plan and follow best practices.

Upon completing this chapter, you will be able to explain issues pertaining to a UC multisite deployment. Once you understand the issues, possible solutions are provided. Where applicable, best practices are mentioned according to Cisco Solutions Reference Network Designs (SRNDs) as well as Cisco Validated Designs (CVDs). Each recommended architecture is explained in greater detail throughout the remainder of this book.

Deploying a multisite UC environment requires a deep understanding of how to craft a proper multisite dial plan that allows for scalability, proper planning for bandwidth allocation for not only IP phones but also video endpoints, quality of service (QoS) design and implementation, and a highly available wide-area network (WAN) and local-area network (LAN) architecture, including survivable remote site telephony (SRST). This chapter helps identify issues that arise in multisite UC deployments.

Upon completing this chapter, you will be able to meet these objectives:

■ Describe aspects that pertain to multisite deployment

■ Describe QoS aspects in a multisite deployment

■ Describe bandwidth aspects in a multisite deployment

■ Describe availability in a multisite deployment

■ Describe dial plan aspects in a multisite deployment

■ Describe fixed-length versus variable-length numbering plans

- Describe how to optimize call routing and implement PSTN backup solutions

- Describe overlapping and nonconsecutive numbering plans

- Describe various PSTN requirements

- Describe how to create a scalable dial plan architecture

- Describe NAT and possible security issues in modern unified communications

Multisite Deployment Issues Overview

The goal of any successful business is to grow; usually this entails expansion and possibly adding sites or locations. In today's modern IT environments, the pace of expansion and the pressure of delivering new technologies to business units can be overwhelming at times. This is only compounded by a different type of end user coming into the workforce (the bring your own device [BYOD] end users who want to connect their personal devices to the corporate network and work in a manner that is efficient and effective for them). Figure 1-1 illustrates several issues with multisite deployments, including availability, quality, and bandwidth concerns, dial plan issues, and security concerns.

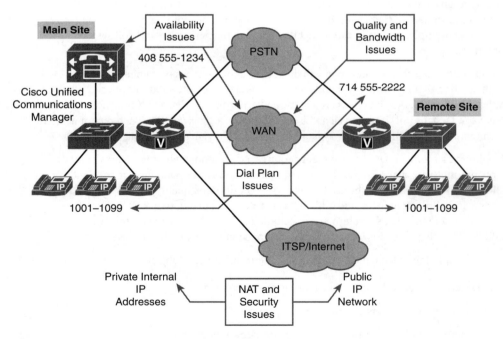

Figure 1-1   *Common Issues in Multiple-Site Deployments*

To provide workers with unified communications, video-capable devices, instant messaging (IM), voicemail, call centers, and enterprise-grade communication features, IT shops are challenged to support multiple systems and oftentimes multiple sites. In a multisite deployment, several issues can occur if not properly planned for:

- Quality issues
- Bandwidth
- Availability
- Dial plan
- NAT and security

Voice and video communications are considered real-time communications; they utilize the User Datagram Protocol (UDP), and more specifically the Real-time Transport Protocol (RTP). Think of RTP as a "fire and forget" protocol; if a packet is transmitted, it needs to be prioritized over a packet-switching network. It will not tolerate delay and will not be retransmitted. All traffic is treated equally by default in routers and switches. Due to voice and video being delay-sensitive packets, they must be given priority over other network traffic. QoS is a network and Unified Communications engineer's best friend when it comes to mitigating call or video quality issues. QoS allows you to prioritize voice and video over other types of network traffic. Cisco has a best practice architecture for QoS deemed Medianet. The Enterprise Medianet Quality of Service design principles are beyond the scope of this text; however, QoS is an important topic in Unified Communications and a properly planned collaboration solution.

Cisco Unified Communications (UC) can include voice and video RTP streams, signaling traffic, management traffic, and application traffic (such as rich media conferencing). The additional bandwidth that is required when deploying a Cisco UC solution has to be calculated and provisioned for to ensure that data applications and Cisco UC applications do not overload the available bandwidth. Bandwidth reservations can be made at a network level through proper QoS deployment and technologies such as Resource Reservation Protocol (RSVP). Bandwidth reservations can also be made at the Unified Communications application level by implementing Call Admission Control (CAC) and selecting the proper codec for voice and video calls. As of Unified Communications Manager 9x./10.x/11.x, newer technologies such as Enhanced Locations CAC, Intracluster Enhanced Locations CAC, and Intercluster Enhanced Locations CAC are available. These newer technologies are discussed in Chapter 7, "Call Admission Control (CAC) Implementation."

When deploying Cisco Unified Communications Manager (CUCM) with centralized call processing (servers in a main or headquarters site or data center with multiple branch or remote sites without local call processing servers), IP phones register with CUCM over the IP LAN and potentially over the WAN. If voice gateways such as Integrated Service Routers (ISRs) or Aggregation Services Routers (ASRs) in remote sites are using Media

Gateway Control Protocol (MGCP) as a signaling protocol, they also depend on the availability of CUCM acting as an MGCP call agent. Certain analog voice cards such as voice interface cards (VICs), which provide plain old telephone system (POTS) capability as well as high density analog devices (such as VG 350s), can register to CUCM using Skinny Client Control Protocol (SCCP), which is dependent on the communication path to CUCM. It is important to implement fallback solutions for IP phones and gateways in scenarios in which the connection to the CUCM servers is broken because of IP WAN failure. One common technique is to implement a highly available WAN as well as provide a feature on the ISR/ASR routers called survivable remote site telephony (SRST). SRST allows a gateway at a remote site to become the call-processing engine in the event of a WAN failure. The ISR/ASR router provides registration and call-processing capabilities to Cisco IP phones as well as certain virtual interface cards (VICs) and HD analog gateways. Fallback solutions also apply to H.323 or Session Initiation Protocol (SIP) gateways but require the correct dial peers to support this functionality. Each failover technology is examined in later chapters.

The goal of a properly designed Unified Communications dial plan is to limit "dial plan overlap," meaning users typically have unique extensions or directory numbers (DNs). There are techniques in CUCM in which the same extension can exist inside the same partition. In the event this occurs, the DN is considered a shared line. Unified Communications engineers typically design a single site in which each user has a unique DN inside a common partition for that site. When you design a multisite deployment or global deployment, DNs can overlap across multiple sites, the difference being these DNs are often in separate partitions and are separated out logically in the CUCM dial plan and database. A partition is a logical container (think of it as a padlock over a container) in which DNs, route patterns, meet-me numbers, voicemail ports, and so on can be placed. A calling search space (CSS) is the key by which IP phones and video endpoints are granted permission that allows them to dial certain numbers or unlock those partitions. A design challenge arises in multisite deployments regarding overlapping DNs, variable-length dial plans, various public switched telephone network (PSTN) access codes, and nonconsecutive numbers. Each of these challenges can be solved by designing a robust multisite dial plan. Some techniques used to mitigate these issues include site access codes, a properly planned extensions length, translation patterns, and proper route patterns. Each technique is examined in later sections. In general, avoid overlapping numbers across sites whenever possible for an efficient design.

Cisco Unified Communications and Unified IP Phones/Video endpoints use IP and private IP addresses primarily to communicate within the enterprise. One issue arises in a multiple-site deployment when the various UC systems need to interact and communicate with devices or businesses on the public Internet. Some UC examples include instant messaging (IM) in the form of Cisco Jabber and video business-to-business (B2B) communication in the form of Cisco Expressways or Video Communications Servers (VCS). Last but not least is the Internet telephony service providers (ITSPs), which rely on SIP trunks versus primary rate interface (PRI) or POTS telephone lines to provide communication paths into modern IT environments. SIP trunks are likely terminated onto Cisco Unified Border Elements (CUBEs), which can be a demarcation point between the private and public networks. Security and firewall concerns have become paramount recently with spikes in global hacking. To provide secure communications, the private IP addresses within the enterprise must be translated into public IP addresses. Public IP addresses make the IP phones and video endpoints visible from the Internet and therefore subject to attacks. Network Address Translation (NAT) is one of the preferred technologies of allowing public devices and connections through the firewall and security policies to communicate with internal IP phones and video endpoints. NAT challenges and design considerations are discussed in depth in later sections.

Note The challenge of NAT and security is not limited to multisite deployments. Voice over IP (VoIP) and communications protocols such as Media Gateway Control Protocol (MGCP), Skinny Client Control Protocol (SCCP), H.323, and Session Initiation Protocol (SIP) all require design considerations any time their traffic is subjected to NAT and their traffic traverses through a CUBE or firewall. In addition, some larger environments may invoke security in the data center in the form of virtual firewalls to segment traffic from the network to various sections of the data center. Special design considerations are required for voice and RTP any time a NAT translation occurs. Video devices are especially problematic to NAT traversal and translation, and separate techniques are addressed for video devices.

Voice and Video Call Quality Issues

IP networks were not originally designed to carry real-time traffic. Instead, they were designed for resiliency and fault tolerance. Transmission Control Protocol (TCP) is a great example of this; if a packet fails to be delivered, we simply retransmit. This technique does not work with User Datagram Protocol (UDP) and Real-time Transport Protocol (RTP) protocols, which carry voice and video over IP. It makes no sense to receive the same word over and over again in a conversation just because it was delayed or, worst case, dropped. Each packet is processed separately by a router or Layer 3 switch in an IP network, sometimes causing different packets in a communications stream or word to take different paths to the destination. Imagine a scenario where a branch office has redundant MPLS providers back to a main site, using various router load-balancing protocols and high availability. It is entirely possible the traffic would take different paths to the same destination. The different paths in the network may have a different amount of packet loss, delay, and delay variation (jitter) because of bandwidth, distance, and congestion differences. The destination must be able to receive packets out of order and sequence them. This challenge is solved by the use of RTP sequence numbers, ensuring proper reassembly and playout to the application. When possible, it is best to not rely solely on these RTP mechanisms. Proper network design, using Cisco router Cisco Express Forwarding (CEF) switch cache technology, performs per-destination load sharing by default. Per-destination load sharing is not a perfect load-balancing paradigm, but it ensures that each IP flow (voice call) takes the same path.

Another common design consideration is that bandwidth is shared by multiple users and applications; the amount of bandwidth required for an individual IP flow varies significantly during short lapses of time. Most data applications are bursty by nature, whereas Cisco real-time audio communications with RTP use the same continuous-bandwidth stream. The bandwidth available for any application, including CUCM and voice-bearer traffic, is unpredictable. During peak periods, packets need to be buffered in queues waiting to be processed because of network congestion. *Queuing* is a term that anyone who has ever experienced air flight is familiar with. When you arrive at the airport, you must get in a line (queue) because the number of ticket agents (bandwidth) available to check you in is less than the flow of traffic arriving at the ticket counters (incoming IP traffic). If congestion occurs for too long, the queue (packet buffers) gets filled up, and passengers are annoyed. (Packets are dropped.) Higher queuing delays and packet drops are more likely on highly loaded, slow-speed links such as WAN links used between sites in a multisite environment. Quality challenges are common on these types of links, and you need to handle them by implementing QoS. Without the use of QoS, voice packets experience delay, jitter, and packet loss, impacting voice quality. It is critical to properly configure Cisco QoS mechanisms end to end throughout the network for proper audio and video performance.

During peak periods, packets cannot be sent immediately because of interface congestion. Instead, the packets are temporarily stored in a queue, waiting to be processed. The amount of time the packet waits in the queue, called the queuing delay, can vary greatly based on network conditions and traffic arrival rates. If the queue is full, newly received packets cannot be buffered anymore and get dropped (tail drop). Figure 1-2 illustrates tail drop. Packets are processed on a first-in, first-out (FIFO) model in the hardware queue of all router interfaces. Voice conversations are predictable and constant (sampling is every 20 milliseconds by default), but data applications are bursty and greedy. Voice, therefore, without any special QoS or queuing mechanism, is subject to degradation of quality because of delay, jitter, and packet loss.

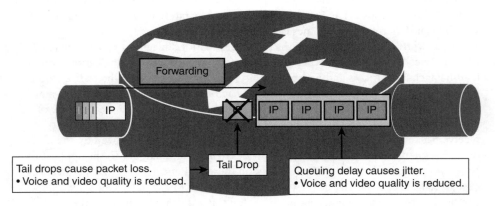

Figure 1-2   *Quality of Service Issues Example: Jitter and Packet Drop*

Bandwidth Challenges

Each site in a multisite deployment is usually interconnected by an IP WAN, or occasionally by a metropolitan-area network (MAN), such as Metro Ethernet. Within the past 10 to 15 years, various WAN technologies have emerged such as MPLS, SONET, Frame Relay, ATM, T1, and satellite, to name a few. Bandwidth on WAN links is limited and relatively expensive. The goal is to use the available bandwidth as efficiently as possible. Unnecessary traffic should be removed from the IP WAN links through content filtering, firewalls, and access control lists (ACLs). IP WAN acceleration methods for bandwidth optimization should be considered as well, such as Cisco Wide Area Application Services (WAAS), Cisco Intelligent WAN (IWAN) technologies, and perhaps caching technologies such as Akamai. Because available bandwidth on the WAN can become scarce, any period of congestion could result in service degradation unless QoS is deployed throughout the network. Figure 1-3 demonstrates the Cisco WAAS solution.

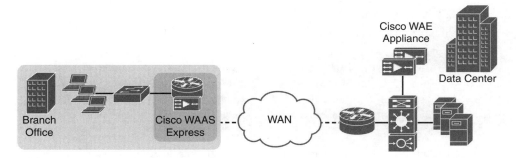

Figure 1-3   *Cisco WAAS Example*

Voice RTP streams produced by Cisco IP phones and video endpoints are a constant and predictable packet size. They are small in size but sent at a very high frequency rate (that is, a high number of small sized packets going across the wire or network link). In bandwidth-challenged locations or slow-speed WAN links, voice streams can be considered wasteful if the wrong voice codec is selected. G.711 uses a consistent 64 kbps for the payload size plus Layer 2 overhead. G.729, however, uses an 8-kbps payload size plus Layer 2 overhead. The Layer 2 overhead of packetization, the encapsulation of digitized voice into an RTP, UDP, IP, and Layer 2 header, is extremely high compared to the payload size. The more voice packets that are sent, the more headers are added to the RTP payload, and thus the more bandwidth required on the link.

The G.711 audio codec requires 64 kbps for the payload or RTP stream, whereas packetizing the G.711 voice sample in an IP/UDP/RTP header every 20 ms requires an additional 16 kbps for overhead. The overhead consists of a 12-byte RTP header, an 8-byte UDP header, and a 20-byte IP header. The header to payload ratio is 1:4; the bandwidth required is 80 kbps. This metric only considers the encapsulation to IP packets, the actual Layer 2 encapsulation (which varies based on WAN technologies) is not considered. For example, the header size of a generic routing encapsulation (GRE) tunnel or IPsec virtual private network (VPN) across a Layer 2 transport is much higher.

The G.729 codec is used across the WAN in situations when bandwidth is a concern due to the smaller packet size. G.729 uses 8 kbps for the payload size and a sampling rate of every 20 ms yields 16 kbps plus Layer 2 overhead for the header; the bandwidth required is 24 kbps. In addition, G.729 has a 2:1 header to payload ratio as compared to G.711.

Note   Sampling rate determines the bandwidth required per codec.

You may wonder how the 16-kbps value for the header bandwidth was calculated. The 40 bytes of header information must be converted to bits to figure out the packet rate of the overhead. Because a byte has 8 bits, 40 bytes * 8 bits in a byte = 320 bits. The 320 bits are sent 50 times per second based on the 20-ms rate (1 millisecond is 1/1000 of a second, and 20/1000 = .02). So:

.02 * 50 = 1 second

320 bits * 50 = 16,000 bits/sec, or 16 kbps

Note   This calculation does not take Layer 2 encapsulation into consideration. For additional information, refer to *QoS Solution Reference Network Design (SRND)* (http://www.cisco.com/go/srnd) or *Cisco QoS Exam Certification Guide*, Second Edition (Cisco Press, 2004). For more information on QoS, go to http://www.cisco.com/go/qos.

Voice packets are benign compared to the bandwidth consumed by data applications. Data applications can fill the entire maximum transmission unit (MTU) of an Ethernet frame (1518 bytes or 9216 bytes if jumbo Ethernet frames have been enabled). In comparison to data application packets, voice packets are small (approximately 60 bytes for G.729 and 200 bytes for G.711 with the default 20-ms sampling rate).

Because of the inefficiency of voice packets, all unnecessary voice streams should be kept away from the IP WAN. A great example of this is media resources, in particular music on hold (MOH), conference bridges (CFB), and annunciators. Each of the types of resources requires additional bandwidth across the IP WAN. These media resources can be optimized in such a way that they do not have to traverse the IP WAN all the time, thereby saving bandwidth. You can achieve this optimization by placing local media resources at the remote sites where applicable.

In Figure 1-4, a conference bridge has been deployed at the main site. No conference bridge exists at the remote site. If three IP phones at a remote site join a conference, their RTP streams are sent across the WAN to the conference bridge. The conference bridge, whether using software or hardware resources, mixes the received audio streams and sends back three unique unicast audio streams to the IP phones over the IP WAN. The conference bridge removes the receiver's voice from his unique RTP stream so that the user does not experience echo because of the delay of traversing the WAN link and mixing RTP audio streams in the conference bridge.

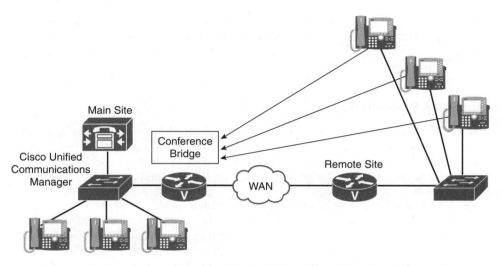

Figure 1-4   *Bandwidth Issue Example: Centralized Media Resources and Bandwidth*

Centralized conference resources cause bandwidth, delay, and capacity challenges in the voice network. Each G.711 RTP stream requires 80 kbps (plus the Layer 2 overhead), resulting in 240 kbps of IP WAN bandwidth consumption by this voice conference. If the conference bridge were not located on the other side of the IP WAN, this traffic would not need to traverse the WAN link, resulting in less delay and bandwidth consumption. If the remote site had a CUCM region configuration that resulted in calls with the G.729 codec back to the main site, the software conferencing resources of CUCM would not be able to mix the audio conversations. Software-based conferencing on CUCM can only handle the G.711 codec. Hardware conferencing or hardware transcoder media resources in a voice gateway are required to accommodate G.729 audio conferencing. Local hardware conference resources eliminate this need. All centrally located media resources (MOH, annunciator, conference bridges, videoconferencing, and media termination points) suffer similar bandwidth, delay, and resource-exhaustion challenges.

Cisco has a best practice architecture for media resources, which is available in the Cisco Validated Designs (CVDs) and Solutions Reference Network Designs (SRNDs). Chapter 6, "Cisco Collaboration Solution Bandwidth Management," is devoted to media resource and bandwidth management coverage. A general concept when planning media resources is *conference remotely* and *transcode centrally*. This is achieved by conferencing remotely using packet voice data modules (PVDMs), which are router/hardware-based conference resources at remote branches. Various Cisco applications such as Unity Connection and Unified Contact Center Express (UCCX) can only accept G.711 streams depending on how they are installed. These applications require a transcoding resource to convert a WAN G.729 codec into G.711. In a centralized call processing architecture, these applications are usually located in a headquarters or data center. You need PVDMs or digital signal processors (DSPs) located near these servers to perform transcoding of calls coming from remote branches into these applications; the idea being to transcode centrally at main sites or data centers. In certain hybrid layouts, a mixture of local and remote transcoders and conference bridges are used to achieve the desired result.

Availability Challenges

When deploying CUCM in multisite environments, CUCM-based services are accessed over the IP WAN. Availability of the IP network, especially of the IP WAN that interconnects sites, is critical for several services and protocols. Protocols and services that are affected in the event of a WAN failure include the following:

- **Signaling in CUCM multisite deployments with centralized call processing:** Remote Cisco IP phones and video endpoints register with a centralized CUCM server. Remote MGCP gateways are controlled by a centralized CUCM server that acts as an MGCP call agent. VIC cards or high-density analog gateways provide POTS capabilities at remote branches and can register with a centralized CUCM server that acts as a SCCP call agent. SIP and H.323 protocols are peer-to-peer technologies and can survive in the event the WAN goes down provided proper dial peers and SRST configurations are in place.

- **Signaling in CUCM multisite deployments with distributed call processing:** In such environments, sites are connected via H.323 (non-gatekeeper-controlled, gatekeeper-controlled, or H.225), SIP trunks, or intercluster trunks (ICTs). In the event of a WAN failure, these connection types stop processing the signaling traffic between clusters.

- **Media exchange:** RTP streams sent between endpoints that are located at different sites rely on the IP WAN to be stable and available. In the event of an IP WAN failure, the audio paths for RTP stop functioning. This can be detrimental to any active calls across the WAN and to future calls placed between sites until function-ality is restored.

- **Other services:** UC has a host of auxiliary services and protocols that all rely on the IP WAN. These include Cisco IP phone Extensible Markup Language (XML) services and access to applications such as attendant console, CUCM Assistant Cisco IP Manager Assistant (IPMA), VCS, or Expressway cluster signaling, media resources that register with CUCM using SCCP, centralized video conferencing using TelePresence Conductor and TelePresence Server, and centralized voicemail using Cisco Unity Connection. Scheduling of video resources such as meeting room reservations that rely on TelePresence Management Suite (TMS) to communicate with the endpoint and Microsoft Exchange are included in this category.

If the IP WAN connection is broken, these services are not accessible. The unavailability might be acceptable for some services, but strategic applications such as UC, voicemail,

video, and auxiliary services should be made available during WAN failure via ba
mechanisms.

Figure 1-5 shows a UC network in which the main site is connected to a remote site via
a centralized call-processing environment. The main site is also connected to a remote
cluster through an ICT, representing a distributed call-processing environment. The
combination of both centralized and distributed call processing represents a hybrid
call-processing model in which small sites use the CUCM resources of the main site,
but large remote offices have their own CUCM cluster. The bottom left of Figure 1-5
shows a SIP trunk terminated on a CUBE, which is typically implemented over a WAN
connection such as MPLS to an ITSP. The benefit of the SIP trunk is that the ITSP
provides the gateways to the public switched telephone network (PSTN) instead of you
needing to provide gateways at the main site.

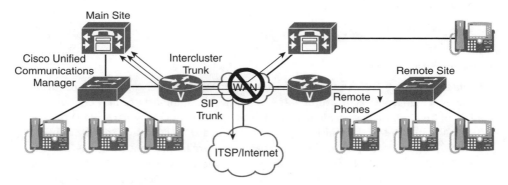

Figure 1-5 *Availability Issues Example: IP WAN Failure*

An IP WAN outage in Figure 1-5 will cause an outage of call-processing services for
the remote site connected in a centralized fashion. The remote cluster will not suffer
a call-processing outage, but the remote cluster will not be able to dial the main site
over the IP WAN during the outage via the ICT. Mission-critical voice applications
(voicemail, interactive voice response [IVR], and so on) located at the main site will be
unavailable to any of the other sites during the WAN outage.

If the ITSP is using the same links that allow IP WAN connectivity, all calls to and from
the PSTN are also unavailable.

Note A deployment like the one shown in Figure 1-5 is considered a bad design
because of the lack of IP WAN fault tolerance and PSTN backup. A high availability
(HA) design would include multiple redundant WAN links, HA routing protocols,
multiple ICT trunks, and redundant CUBEs. The Cisco CVDs have detailed sections on
providing fault tolerance and HA solutions in a UC environment.

yment, with a single or multiple CUCM cluster, dial plan design
on of several issues that do not exist in single-site deployments,

rs

ibers

variable-length numbering

- Direct inward dialing (DID) ranges and E.164 addressing

- Optimized call routing

- Various PSTN requirements

- Scalability

Overlapping Numbers

Users located at different sites can have the same DNs assigned. An example of this is
a user in Virginia with a DID of 804-424-1601; the UC administrator may configure
a DN of 1601 on that user's phone. Another user in Colorado may have a DID of
303-860-1601; the UC administrator may configure a DN of 1601 on that user's phone
as well. This can occur provided the two extensions are in different partitions inside
the CUCM. Because DNs usually are unique only within a site, a multisite deployment
requires a solution for overlapping numbers. In this example, how could the Virginia-
based DN of 1601 dial a Denver-based DN of 1601? They are the same number in
separate partitions. The solution is creative site codes or creative use of CSS design.

> **Note** The solutions to the problems listed in this chapter are discussed in more detail in
> Chapter 2, "Understanding Multisite Deployment Solutions."

In Figure 1-6, Cisco IP phones at the main site use DNs 1001 to 1099, 2000 to 2157, and
2365 to 2999. At the remote site, 1001 to 1099 and 2158 to 2364 are used. These DNs
have two issues. First, 1001 to 1099 overlap; these DNs exist at both sites, so they are
not unique throughout the complete deployment. This causes a problem: If a user in the
remote site were to dial only the four digits 1001, which phone would ring? This issue
of overlapping dial plans needs to be addressed by digit manipulation. In addition, the
nonconsecutive use of the range 2000 to 2999 (with some duplicate numbers at the two
sites) requires a significant number of additional entries in call-routing tables because the
ranges can hardly be summarized by one entry (or a few entries).

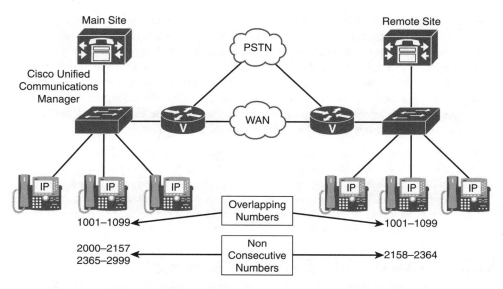

Figure 1-6 *Dial Plan Challenges: Overlapping and Nonconsecutive Numbers*

Nonconsecutive Numbers

Contiguous ranges of numbers are important to summarize call-routing information, analogous to contiguous IP address ranges for route summarization. For example, a remote branch may have a PSTN DID range of 757-466-1XXX, thus providing that branch with 1000 DNs from extension 1000 through 1999 (assuming a four-digit dial plan). In CUCM, you can summarize these patterns and do not need to enter all 1000 entries into the routing table/dial plan as simply 1XXX. Such blocks of extensions can be represented by one entry in the call-routing table, such as route patterns and translation patterns (in CUCM), dial peer destination patterns (in IOS), and voice translation rules (in IOS), which keep the routing table short and simple. If each endpoint requires its own entry in the call-routing table, the table gets too big, lots of memory is required, and lookups take more time. Therefore, nonconsecutive numbers at any site are not optimal for efficient call routing. A nonoptimal design is to skip ranges of numbers for this remote site. Imagine what the routing table would look like if it only had DN range 1000 to 1050 then 1190 to 1300 followed by 1550 to 1600. These "gaps" would require multiple routing entries in the CUCM database.

Variable-Length Numbering

Some countries, such as the United States and Canada, have fixed-length numbering plans for PSTN numbers. North America uses the North American Numbering Plan (NANP). This dictates that PSTN phone numbers are ten digits in the format XXX-XXX-XXXX, a three-digit number plan area code (NAA), followed by a three-digit exchange (NXX), followed by a four-digit subscriber code.

Others, such as Mexico and England, have variable-length numbering plans; their PSTN numbers vary in length. A problem with variable-length numbers is that the complete length of the number dialed can be determined in CUCM only by waiting for the interdigit timeout. Interdigit timeout refers to the time CUCM waits to determine you are done dialing a number. Waiting for the interdigit timeout, known as the T.302 timer, adds to the post-dial delay, which may annoy users. Further, the T.302 timer is a service parameter in CUCM that needs to be set on every server in the cluster; the default is 15 seconds, which for many people is far too long.

Throughout my consulting tenor, I have found that lowering the T.302 timer to around 7 to 8 seconds is the best solution for many organizations; lower and you may disconnect users mid-dial, any longer would annoy users who have completed dialing and are waiting on CUCM to connect the call. Ways to mitigate this issue are discussed later in this chapter. You can allow users to specify a terminating digit to represent they are done dialing a variable-length pattern.

Direct Inward Dialing (DID) Ranges and E.164 Addressing

When considering integration with the PSTN, internally used DNs have to be related to external PSTN numbers (public DIDs and E.164 addressing). In layman terms, it is how you coordinate the mapping of external DIDs to an internal DN scheme. A misconception among many junior UC engineers is that CUCM contains a screen or mechanism to track DID to DN mappings; this simply does not exist, and proper planning and a few Excel spreadsheets are often used. You can create translation patterns for every DID and translate them into DNs if you wanted to go to an extreme and where warranted, but that is adding complexity. Depending on the numbering plan (fixed or variable) and services provided by the PSTN, the following solutions are common:

■ Each internal DN relates to a fixed-length PSTN number.

■ Another solution is to not reuse any digits of the PSTN number, but to simply map each internally used DN to any PSTN number assigned to the company.

Each internal DN has its own dedicated PSTN number. The DN can, but does not have to, match the least-significant digits of the PSTN number. In countries with a fixed numbering plan, such as the NANP, this usually means that the four-digit office or subscriber codes are used as internal DNs. If these are not unique, office codes or administratively assigned site codes might be added to the number, resulting in five or more digits being used for internal DNs.

An example to provide clarity is a remote branch in California with a DID range of 415-586-7200 through 7299 may choose to assign internal DNs or DNs to phones using a four-digit extension from 7200 to 7299. Assume there is an additional remote branch in Chicago with DID range 312-733-7200 to 7299. You could create DNs on the phones in Chicago with extensions 7200 to 7299 and place the DNs in separate partitions (logically separating them in CUCM). How does one dial between sites now? One common technique is to use site codes; in the dial plan for San Francisco, users would append a site code to the four-digit extension if they were trying to reach Chicago phones. For example, a user in San Francisco may dial 557200 to reach an extension of 7200 in Chicago. The site code would be 55, representing all phones in Chicago. CUCM can uniquely route the call to a site based on the site code and using a translation pattern or digit-stripping mechanisms in CUCM.

Another solution is to not reuse any digits of the PSTN number, but to simply map each internally used DN to any PSTN number assigned to the company. In this case, the internal and external numbers do not have anything in common. If the internally used DN matches the least-significant digits of its corresponding PSTN number, significant digits can be set at the gateway or trunk. Also, general external phone number masks, transformation masks, or prefixes can be configured. This is true because all internal DNs are changed to fully qualified PSTN numbers in the same way.

An example of this technique is a UC dial plan in which sites have contiguous blocks of DNs, a site in New York may receive extensions 1000–1999, a site in San Diego may receive blocks 2000–2999, and so on. The internal numbering scheme has nothing to do with the DID ranges from the PSTN. New York DIDs may be 212-618-6750 through 212-618-6799. To map the public DID to an internal DN, you need to invoke digit manipulation in the form of translation patterns, transforms, significant digits, or other techniques in CUCM to mask and change the DID to fit the internal scheme. This approach can be laborious because a one-for-one translation is required.

What if a remote site has no DIDs in fixed-length numbering plans? To avoid the requirement of having one DID number per internal DN when using a fixed-length numbering plan, it is common in some organizations to disallow DIDs to internal extensions. Instead, the PSTN trunk has a single number, and all PSTN calls routed to that number are sent to an attendant, an auto-attendant, a receptionist, or a secretary. From there, the calls are *transferred* to the appropriate internal extension.

Internal DNs are part of a variable-length number. In countries with variable-length numbering plans, a typically shorter "subscriber" number is assigned to the PSTN trunk, but the PSTN routes all calls *starting* with this number to the trunk. The caller can add digits to identify the extension. There is no fixed number of additional digits or total digits. However, there is a maximum, usually 32 digits, which provides the freedom to select the length of DNs. This maximum length can be less.

For example, in E.164 the maximum number is 15 digits, not including the country code. A caller simply adds the appropriate extension to the company's (short) PSTN number when placing a call to a specific user. If only the short PSTN number without an extension is dialed, the call is routed to an attendant within the company. Residential PSTN numbers are usually longer and do not allow additional digits to be added; the feature just described is available only on trunks.

Optimized Call Routing

Having an IP WAN between sites with local PSTN access at all sites allows for PSTN toll bypass by sending calls between sites over the IP WAN instead of using the PSTN. In such scenarios, the PSTN should be used as a backup path only in case of WAN failure.

Another solution, which extends the idea of toll bypass and can potentially reduce toll charges, is to also use the IP WAN for PSTN calls. With tail-end hop-off (TEHO), the IP WAN is used as much as possible, and the gateway that is closest to the dialed PSTN destination is used for the PSTN breakout. An example is a New York-based IP phone dialing a San Diego PSTN number. Provided the enterprise has an IP WAN between the sites and a local gateway in San Diego with a PSTN circuit (POTS, PRI, or SIP trunk), you could in fact route that call across the WAN to go out the San Diego gateway as a local call, thus bypassing costly long-distance charges if the call were sent out the New York gateways as a long-distance call. In certain areas of the world, TEHO or toll bypass is illegal, because the telephone companies are often government regulated. Check the telephone laws in your specific country or locality to determine any legal issues that could arise from this technique.

Various PSTN Requirements

Various countries and sometimes even several PSTN providers within the same country can have numerous requirements regarding the PSTN dial rules. This situation can cause issues when calls can be routed via multiple gateways. If the requirements of a primary gateway are different from the requirements of a backup gateway, numbers must be transformed accordingly. In the United States, smaller cities and localities often will allow the use of a seven-digit local dialing plan, meaning you can dial seven digits for local PSTN calls within that area. In larger metropolitan areas, this is often expanded and mandated that ten-digit dialing is used to represent a local call. Imagine if you have multiple paths for a call to go out with redundant gateways, one being in a small locality and the backup gateway being in a major city. Digit manipulation would be required to expand a local call from seven to ten digits if it routed out the backup gateway in a site where ten-digit dialing is mandated.

Additional PSTN considerations surrounding automatic number identification (ANI) also need to be addressed. The ANI of calls that are being received from the PSTN can be represented in various ways: as a seven-digit subscriber number, as a ten-digit number including the area code, or in international format with the country code in front of the area code. To standardize the calling number for all calls (which can be displayed in the call logs or display screens on IP phones and video endpoints), the format that is used must be known, and the number must be transformed accordingly. In countries where PSTN numbers do not have fixed lengths, it is impossible to detect the type of number (local, national, or international) by looking at only the length of the number. In such cases, the type of number must be specified in signaling messages (for example, by the ISDN type of number [TON]). *Type of number* is an ISDN term in which the calling number or ANI of the call contains additional information elements (IE) describing the number as a local, national, or international call. Gateways can read this information and append or strip the appropriate number of digits and pass a uniform ANI length to IP phones and video endpoints.

Note   In the United States, TON is mainly used on PRIs; SIP trunks may not support TON depending on the carrier.

Scalability

In large or very large deployments, dial plan scalability issues arise. When interconnecting multiple CUCM clusters or CUCM Express routers via trunks, it is difficult to implement a dial plan on an any-to-any basis where each device or cluster needs to know the numbers or prefixes that are found at every other system. In addition to the need to enter almost the same dial plan at each system, a static configuration does not reflect true reachability. If there are any changes, the dial plans at each system must be updated. Although there are solutions that allow centralized dial plan configuration (for example, Cisco Session Management Edition [SME] or H.323 gatekeepers), in very large deployments a dynamic discovery of DN ranges and prefixes simplifies the implementation and provides a more scalable solution.

Fixed Versus Variable-Length Numbering Plans

A fixed numbering plan features fixed-length area codes and local numbers. The United States utilizes fixed-length dial plans. An open numbering plan or variable-length numbering plan features variance in length of area code or local number, or both, within the country. Figure 1-7 illustrates an international deployment with various numbering schemes in place including fixed and variable length. Take a moment to familiarize yourself with the dialing habits in both schemes.

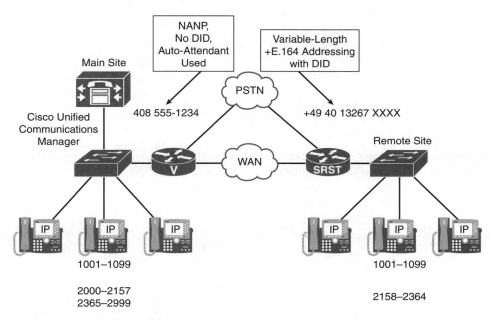

Figure 1-7 *Variable-Length Numbering, +E.164 Addressing, and DID Example*

Figure 1-7 features a main site in the United States. The NANP PSTN number is 408 555-1234. DIDs are not used. All calls placed to the main site are managed by an attendant. There is a remote site in Germany with a PSTN number of +49 404 13267.

The German location uses four-digit extensions, and DID is allowed, since digits can be added to the PSTN number. When calling the German office attendant (not knowing a specific extension), users in the United States dial +9 011 49 404 13267. Note that the + is replaced by the international prefix 011 and the access code 9. If they know that they want to contact extension 1001 directly, they dial +9 011 49 404 13267 1001.

Table 1-1 contrasts the NANP and a variable-length numbering plan (Germany's numbering plan).

Table 1-1 *Fixed- Versus Variable-Length Numbering Plans*

Component	Description	Fixed-Length Numbering Plan (NANP)	Variable-Length Numbering Plan (Germany)
Country code	A code of one to 3 digits is used to reach the particular telephone system for each nation or special service. Obtain the E.164 standard from http://itu.org to see all international country codes.	1	49
Area code	Used within many nations to route calls to a particular city, region, or special service. Depending on the nation or region, it may also be called a numbering plan area, subscriber trunk dialing code, national destination code, or routing code.	3 digits	3 to 5 digits
Subscriber number	Represents the specific telephone number to be dialed, but it does not include the country code, area code (if applicable), international prefix, or trunk prefix.	3-digit exchange code plus a 4-digit station code	3 or more digits
Trunk prefix	The initial digits to be dialed in a domestic call, before the area code and the subscriber number.	1	0
Access code	A number that is traditionally dialed first "to get out to the PSTN," used in private branch exchanges (PBXs) and VoIP systems.	9	0

Table 1-1 *continued*

Component	Description	Fixed-Length Numbering Plan (NANP)	Variable-Length Numbering Plan (Germany)
International prefix	The code dialed before an international number (country code, area code if any, and then subscriber number).	011	00 or + (+ is used by cell phones)

An area code is used within many countries to route calls to a particular city, region, or special service. Depending on the country or region, it may also be referred to as one of the following:

- Numbering plan area (NPA)
- Subscriber destination code
- National destination code
- International Country code

The subscriber number represents the specific telephone number to be dialed, but does not include the country code, area code (if applicable), international prefix, or trunk prefix.

A trunk prefix refers to the initial digits that are dialed in a call within the United States, preceding the area code and the subscriber number.

An international prefix is the code that is dialed before an international number (the country code, the area code if any, and then the subscriber number).

The table contrasts the NANP and a variable-length numbering plan (the German numbering plan, in this example).

Some examples include the following:

- **Within the U.S.:** 9-1-408-555-1234 or 408-555-1234 (within the same area code)
- **U.S. to Germany:** 9-011-49-404-132670
- **Within Germany:** 0-0-404-132670 or 0-132670 (within the same city code)
- **Germany to the U.S.:** 0-00-1-408-555-1234 (Note: The 1 in 00-1-408 is the U.S. country code, not the trunk prefix.)

Note In the examples shown following Table 1-1, dialing out from the United States illustrates the common practice of dialing 9 first as an access code to dial out. This use is common, but optional, in a dial plan. However, if the access code is used, the 9 must be stripped before reaching the PSTN, whereas the other dialed prefixes must be sent to the PSTN for proper call routing.

It is worth noting that the logic of routing calls by CUCM over the WAN or through the PSTN is appropriately transparent to the phone user. The phone user has no idea if the call is being routed via a PSTN circuit or the IP WAN. Appropriate digit manipulation needs to occur in either scenario.

Detection of End of Dialing in Variable-Length Numbering Plans

There are three ways to detect end of dialing in variable-length numbering plans:

- Interdigit timeout
    - Simple to configure
    - Least convenient
- Use of # key
    - Different implementation in Cisco IOS Software (simple) versus CUCM (complex)
    - Convenient
    - Requires users to be aware of this option
- Use of overlap sending and receiving
    - Convenient
    - Must be supported by PSTN
    - Complex implementation

As the preceding list shows, one issue that can arise is how to detect that a user is done dialing a number when using a variable-length dial plan. You must ensure that you give ample time for IP phone users to complete their call without disconnecting them prematurely. Also confirm that there is a mechanism in place to allow the users to signify they are done dialing and that CUCM should start routing the call immediately.

From an implementation perspective, the simplest way to detect end of dialing is to wait for an interdigit timeout to expire. This approach, however, provides the least comfort to the end user because it adds post-dial delay. In an environment with only a few numbers of variable length (for example, NANP, where only international calls are of variable length), waiting for the interdigit timeout might be acceptable. However, even in such an environment, it might make sense to at least reduce the value of the timer, because the default value in CUCM is high (15 seconds).

Note   In CUCM, the interdigit timer is set by the cluster-wide Cisco Call Manager service parameter T302 timer that is found in CUCM Administration by navigating to **System > Service Parameters** under the Cisco Call Manager Service.

In Cisco IOS Software, the default for the interdigit timeout is 10 seconds. You can modify this value using the voice-port **timeouts interdigit** command.

Another solution for detecting end of dialing on variable-length numbers is the use of the # key. An end user can press the # key to indicate that dialing has finished. The implementation of the # key is different in CUCM versus Cisco IOS Software. In Cisco IOS gateways, the # is seen as an instruction to stop digit collection. It is not seen as part of the dialed string. Therefore, the # is not part of the configured destination pattern. In CUCM, the # is considered to be part of the dialed number and, therefore, its usage has to be explicitly permitted by the administrator by creating patterns that include the #. If a pattern includes the #, the # has to be used; if a pattern does not include the #, the pattern is not matched if the user presses the # key. Therefore, it is common in CUCM to create a variable-length pattern twice: once with the # at the end, and once without the #. An example is the following two route patterns inside CUCM: 9.011! and 9.011!#. Note that the "9." would be discarded with the strip PreDot discard digit command at the route pattern level. You can additionally specify the discard PreDot and trailing pound instruction for the 9.011!# route pattern; thereby allowing both patterns inside CUCM—one for variable length and the other including a terminating digit of pound.

An alternative way to configure such patterns is to end the pattern with ![0–9#]. In this case, a single pattern supports both ways of dialing (with and without the #). However, be aware that the use of such patterns can introduce other issues. For example, this can be a concern when using discard digits instructions that include trailing-# (for example, PreDot-trailing-#). This discard digit instruction will have an effect only when there is a trailing # in the dialed number. If the # is not used, the discard digit instruction is ignored. Therefore, the PreDot component of the discard digit instruction is also not performed. PreDot is a form of digit manipulation in CUCM that strips off all digits before the dot.

Allowing the use of the # to indicate end of dialing provides more comfort to end users than having them wait for the interdigit timeout. However, this possibility has to be communicated to the end users, and it should be consistently implemented. As previously mentioned, it is automatically permitted in Cisco IOS Software, but not in CUCM.

The third way to indicate end of dialing is the use of overlap send and overlap receive. If overlap is supported end to end, the digits that are dialed by the end user are sent one by one over the signaling path. Then, the receiving end system can inform the calling device after it receives enough digits to route the call (number complete). Overlap send and receive is common in some European countries, such as Germany and Austria. From a dial plan implementation perspective, overlap send and receive is difficult to implement when different PSTN calling privileges are desired. In this case, you have to collect enough digits locally (for example, in CUCM or Cisco IOS Software) to be able to decide to permit or deny the call. Only then can you start passing digits on to the PSTN one by one using overlap. For the end user, however, overlap send and receive is comfortable because each call is processed as soon as enough digits have been dialed. The number of digits that are sufficient varies per dialed PSTN number. For example, one local PSTN destination might be reachable by a seven-digit number, whereas another local number might be uniquely identified only after receiving nine digits.

Optimized Call Routing and PSTN Backup

Using an IP WAN enables savings on the cost of long-distance or international PSTN calls in a multisite environment. There are two ways to save costs on long-distance or international PSTN calls in a multisite deployment:

■ **Toll bypass:** Calls between sites that use the IP WAN instead of the PSTN are toll-bypass calls. The PSTN is used only when calls over the IP WAN are not possible (either because of a WAN failure or because the call is not admitted by CAC). An example is dialing between IP phones at two sites; the call traverses the IP WAN for RTP and signaling versus going across the PSTN.

■ **TEHO:** TEHO extends the concept of toll bypass by also using the IP WAN for calls to the remote destinations in the PSTN. With TEHO, the IP WAN is used as much as possible and PSTN breakout occurs at the gateway that is located closest to the dialed PSTN destination. Local PSTN breakout is used as a backup in case of an IP WAN or CAC failure. An example is dialing a San Diego number from New York. Provided you have an IP WAN connecting both sites, the call can flow across the IP WAN and exit a voice gateway or CUBE in San Diego as a local call, thereby saving costly PSTN charges.

Note   Some countries do not allow the use of TEHO. When implementing TEHO, ensure that the deployment complies with legal requirements.

When using the IP WAN to reach remote PSTN destinations or internal DNs at a different site, it is important to consider backup paths. When the IP WAN is down or when not enough bandwidth is available for an additional voice call, calls should be routed via the local PSTN gateways as a backup path.

In the example shown in Figure 1-8, a call from Chicago to San Jose would be routed as shown in the following steps:

Step 1.   A Chicago user dials 9 1 408 555-6666, the number for a PSTN phone that is located in San Jose.

Step 2.   The call is routed from the CUCM Express in Chicago to the CUCM cluster in San Jose over the IP WAN.

Step 3.   The CUCM in San Jose routes the call to the San Jose gateway, which breaks out to the PSTN with a (now) local call to the San Jose PSTN.

Step 4.   The San Jose PSTN phone rings.

If the WAN were unavailable for any reason before the call, the Chicago gateway would have to be properly configured to route the call with the appropriate digit manipulation through the PSTN at a potentially higher toll cost to the San Jose PSTN phone.

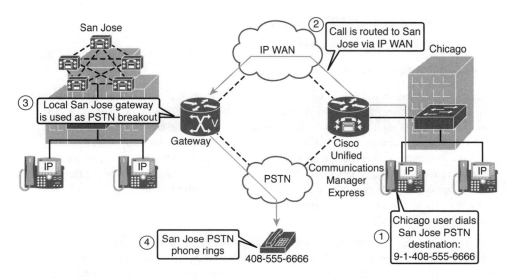

Figure 1-8 *TEHO Example*

> **Note** The primary purpose of implementing TEHO is because of a reduction of operating costs from calling through the PSTN. In the TEHO example shown in Figure 1-8, there would be potential cost savings. However, costs savings are typically considerably higher when the remote location and destination call are international.

PSTN Requirements

Various countries can have different PSTN dialing requirements, which makes it difficult to implement dial plans in international multisite deployments.

There are several design challenges regarding PSTN access, including the following:

- Dial rules for the called-party number on outbound PSTN calls
  - Dial PSTN access code
  - Dial national access code
  - Dial international access code
- Dial presentation of called- and calling-party numbers on inbound and outbound calls
  - Dial length of number and its components
  - Dial ISDN number types

- Dial overlap send and overlap receive

- Dial + prefix on E.164 numbers

■ Dial emergency dialing

One of the issues in international deployments is various PSTN dial rules. For example, in the United States, the PSTN access code is 9, whereas in most countries in Europe 0 is used as the PSTN access code. The national access code in the United States is 1, whereas 0 is commonly used in Europe. The international access code is 011 in the United States, and 00 is used in many European countries. Some PSTN provider networks require the use of the ISDN TON, but others do not support it. Some networks allow national or international access codes to be combined with ISDN TON. Others require you to send the actual number only (that is, without any access codes) when setting the ISDN TON.

The same principle applies to the calling-party number. As mentioned earlier, in variable-length numbering plans, the TON cannot be detected by its length. Therefore, the only way to determine whether the received call is a local, national, or international call is by relying on the availability of the TON information in the received signaling message.

Some countries that have variable-length numbering plans use overlap send and overlap receive. With overlap send, a number that is dialed by an end user is passed on to the PSTN digit by digit. Then, the PSTN indicates when it has received enough numbers to route the call. Overlap receive describes the same concept in the opposite direction: When a call is received from the PSTN in overlap mode, the dialed number is delivered digit by digit, and not en bloc. Some providers that use overlap send toward their customers do not send the prefix that is configured for the customer trunk, but only the additional digits that are dialed by the user who initiates the call.

When dialing PSTN numbers in E.164 format (that is, numbers that start with the country code), the + sign is commonly prefixed to indicate that the number is in E.164 format. The advantage of using the + sign as a prefix for international numbers is that it is commonly accepted as a symbol for internationally formatted telephone numbers around the world. In contrast, PSTN access codes such as 011 (used in the NANP) or 00 (often used in Europe) are known only in the respective countries.

Finally, emergency dialing can be an issue in international deployments. Because various countries have several emergency numbers and numerous ways to place emergency calls, users are not sure how to dial the emergency number when roaming to other countries. An international deployment should allow roaming users to employ their home dialing rules when placing emergency calls. The system should then modify the called number as required at the respective site.

Issues Caused by Different Methods of PSTN Dialing

Different local PSTN dial rules can cause several issues, especially in international deployments. Imagine an executive flies from New York to Paris. Typically, the executive has not been properly trained on how to dial a local call according to the Paris

PSTN dialing requirements. Vice versa, a European executive travels to the United States. He or she may not know the proper format of calls in the NANP dialing plan, which is fixed length.

The following list outlines how to store PSTN contacts so that they can be used from any site:

- Different ways to store or configure PSTN destinations:
  - Dial speed dials
  - Dial fast dials
  - Dial address book entries
  - Dial call lists
  - Dial AAR targets
  - Dial call-forward destinations
- Stored number can be used at multiple sites (countries) because of roaming users using local PSTN gateways:
  - Dial Cisco Extension Mobility
  - Dial Cisco Device Mobility
  - Dial PSTN backup
  - Dial TEHO and LCR

The main problem that needs to be solved in international environments is how to store contacts' telephone numbers. Address book entries, speed and fast dials, call list entries, redial capability, and other numbers should be in a format that allows them to be used at any site, regardless of the local dial rules that apply to the site where the user is currently located. Call-forwarding destinations should also be in a universal format that allows the configured number to be used at any site.

The main reason for a universal format is that a multisite deployment has several features that make it difficult to predict which gateway will be used for the call. For example, a roaming user might use Cisco Extension Mobility or Device Mobility. Both features allow an end user to use local PSTN gateways while roaming. If no universal format is used to store speed dials or address book entries, it will be difficult for the end user to place a PSTN call to a number that was stored according to the NANP dial rules while in countries that require different dial rules. Even when not roaming, the end user can use TEHO or least cost routing (LCR), so that calls break out to the PSTN at a remote gateway, not at the local gateway. If the IP WAN link to the remote gateway is down, the local gateway is usually used as a backup. How should the number that is used for call routing look in such an environment? It is clearly entered according to local dial rules by the end user, but ideally it is changed to a universal format before call routing is performed. After the call is routed and the egress gateway is selected, the number could then be changed as required by the egress gateway.

Dial Plan Scalability Issues

In large CUCM deployments, it can be difficult to implement dial plans, especially when using features such as TEHO with local PSTN backup. Dial plans are difficult to implement in large Cisco UC deployments, and the following list outlines several scalability issues to take into consideration:

- Dial static configuration for multiple sites or domains is very complex because of any-to-any call-routing requirements.

- Dial centralized H.323 gatekeepers or SIP network services offer dial plan simplification.

   - Dial less configuration because of any-to-one call-routing topology

   - Dial static configuration nevertheless (no dynamic recognition of routes, no automatic PSTN rerouting)

   - Dial no built-in redundancy

- Dial while, an optimal solution, is desirable for large deployments. Services such as Global Dial Plan Replication (GDPR) and Service Advertisement Framework (SAD) and Call-Control Discovery (CCD) allow dynamic learning of dial plans in large networks. These concepts and their implementation are discussed in later chapters.

In large CUCM deployments, it can be difficult to implement scalable and easy-to-use dial plans, especially when using features such as TEHO with local PSTN backup or globalized +E.164 dial plans.

The main scalability issue of large deployments is that each call-routing domain (for example, a CUCM cluster or a CUCM Express router) must be aware of how to get to all other domains. Suppose that you have three CUCM clusters spread across the globe handling global communications; you have users traveling between sites who would like to retain their local dialing habits, features, and functionality.

Such a dial plan can become very large and complex, especially when multiple paths (for example, a backup path for TEHO) must be made available. Because each call routing domain must be aware of the complete dial plan, a static configuration does not scale. For example, any changes in the dial plan must be applied individually at each call-routing domain.

Centralized H.323 gatekeepers or SIP network services can be used to simplify the implementation of such dial plans, because there is no need to implement the complete dial plan at each call-routing domain. Instead of an any-to-any dial plan configuration, only the centralized component must be aware of where to find each number. This approach, however, means that you rely on a centralized service. If the individual call-routing entities have no connectivity to the centralized call-routing intelligence, all calls will fail. Further, the configuration is still static. Any changes at one call-routing domain (for example, new PSTN prefixes because of changing the PSTN provider) must also be implemented at the central call-routing component.

In addition, these centralized call-routing services do not have built-in redundancy. Redundancy can be provided, but requires additional hardware, additional configuration, and so on. Redundancy is not an integrated part of the solution.

The ideal solution for a large deployment is to allow an automatic recognition of routes. Internal as well as external (for PSTN backup) numbers should be advertised and learned by call-routing entities. A dynamic routing protocol for call-routing targets addresses scalability issues in large deployments. A new technique and technology has emerged with the advent of CUCM Version 10.x called Global Dial Plan Replication (GDPR). GDPR is a feature that is based on the concepts in previous CUCM releases. In Version 9.x, Cisco introduced the Intercluster Lookup Service (ILS) and Call Control Discovery (CCD). Think of ILS and CCD as a mechanism in which one CUCM cluster can advertise its DNs to another CUCM cluster via the IP network. It does this by broadcasting its dial plan using Service Advertisement Framework (SAF). GDPR, ILS, CCD, and Cisco SAF are explained in more detail in Chapter 16, "Cisco Service Advertisement Framework (SAF) and Call Control Discovery (CCD)."

NAT and Security Issues

In single-site deployments, CUCM servers and IP phones usually use private IP addresses because there is no need to communicate with the outside IP world. NAT is not configured for the phone subnets, and attacks from the outside are impossible as they are behind the corporate firewall. In modern multisite environments, this requires a paradigm shift as users have multiple devices in multiple sites all requiring communication paths that may transmit across the public Internet. A great example of this is instant messaging utilizing Cisco Jabber and the IM and Presence Server. How do you allow IMs on devices that roam outside of the LAN? Another example is video devices such as a Cisco DX80 endpoint on an executive's desk in his home office. How would you get that voice and video traffic back into the LAN and through a firewall? These newer technologies raise interesting security and protocol issues that are outlined in the list that follows:

- In single-site deployments, CUCM and IP phones do not require access to public IP networks:

  - NAT is not required.

  - Not reachable from the outside.

  - Not subject to attacks from outside (except from ITSP environment).

- In multisite deployments, private links or VPN tunnels can be used:

  - Requires gateway configuration at each site

  - Allows only intersite communication

  - Blocks access to and from outside (unless traffic is tunneled)

- Access to public IP networks is required in some situations

  - Connections to ITSPs or destinations on the Internet.

  - NAT required; CUCM and IP phones are exposed to the outside.

  - CUCM and IP phones are subject to attacks.

As the preceding list shows, if you focus on multisite deployments, IPsec VPN tunnels can be used between sites. The VPN tunnels allow only intersite communication; access to the protected internal networks is not possible from the outside (only from the other site through the tunnel). Therefore, attacks from the outside are blocked at the gateway. To configure IPsec VPNs, the VPN tunnel must be configured to terminate on the two gateways in the different sites. Sometimes this is not possible. For instance, the two sites may be under different administration, or perhaps security policies do not allow the configuration of IPsec VPNs.

In these cases, or when connecting to a public service such as an ITSP, you must configure NAT for CUCM servers and IP phones. When CUCM servers and IP phones are reachable with public IP addresses, they are subject to attacks from the outside world, which introduces potential security issues.

In such a case, or when connecting to a public service such as an ITSP, NAT has to be configured for CUCM servers and IP phones. Cisco UC accomplishes this NAT traversal by utilizing a third-party session border controller or a Cisco CUBE solution.

In Figure 1-9, Company A and Company B both use IP network 10.0.0.0/8 internally. For the companies to communicate over the Internet, the private addresses are translated to public IP addresses. Company A uses public IP network A, and Company B uses public IP network B. All CUCM servers and IP phones are reachable from the Internet and communicate with each other.

As soon as CUCM servers and IP phones can be reached with public IP addresses, they are subject to attacks from the outside world, introducing potential security issues.

Recently released UC technologies from Cisco have attempted to address security concerns with NAT and UC. Cisco has CUBEs, which are specialized Integrated Services Routers (ISRs) or Aggregation Services Routers (ASRs) that terminate ITSP circuits across Multiprotocol Label Switching (MPLS) or the open Internet. These devices provide a demarcation point and firewall features such as NAT and ACLs to limit which public entities are allowed to communicate through these devices. In the United States since the mid 2000s, the use of session border controllers and SIP technologies is outpacing that of traditional PSTN technologies such as PRIs or POTS telephone lines. Recent trends have shown that by 2020 SIP trunks will surpass the number of PRIs in the United States. Many telcos are moving their backbones to an all IP-based network; this transition only makes sense for the customer-facing offerings. SIP trunks are an all IP solution with many benefits, which are discussed throughout this book.

Company A Private IP	Company A Public IP	Company B Public IP	Company B Private IP
10.0.0.0/8	Public IP A	Public IP B	10.0.0.0/8

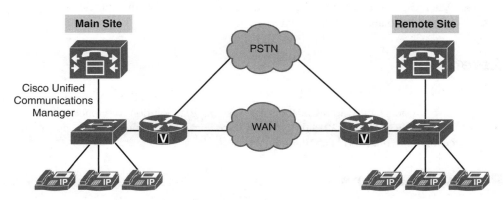

Figure 1-9 *Network Address Translation Security Issues for CUCM and IP Phones*

In addition, Cisco has launched an Expressway and VCS platform line that is capable of NAT traversal for video and IM devices. The technology uses a pair of servers (one on the LAN and another in the demilitarized zone [DMZ], which is a special network in the firewall for communications with external devices). These devices utilize a traversal zone or technology that establishes a communication path from the outside work into the DMZ, and a trust relationship is set up from the DMZ to the internal network. This allows for Cisco Jabber and video endpoints to register and communicate with internal devices without the use of a VPN client or VPN technology. Additional information about Cisco VCS and Expressway is covered in later chapters of this book.

Summary

Implementing a large communications network has never been a trivial task. Yet, with the right level of planning, leveraging expertise, and best practices, it is possible. Sticking to a flexible yet robust infrastructure and dial plan is the key to achieve enterprise-grade communications. The following key points were discussed in this chapter:

- Multisite deployment introduces issues and complexity, including call and video quality, bandwidth concerns, high availability, dial plan design, and NAT security.

- During congestion, voice and video packets have to be buffered or queued; otherwise, they may get dropped.

- Bandwidth in the IP WAN is limited and should be used as efficiently as possible.

- A multisite deployment has several protocols and services that depend on the availability of the IP WAN.

- A multisite dial plan has to address overlapping and nonconsecutive numbers, variable-length numbering plans, DID ranges, and ISDN TON and should minimize PSTN costs.

- When CUCM servers and IP phones need to be exposed to the outside, they can be subject to attacks from the Internet.

References

For additional information, refer to the following:

Cisco Systems, Inc. Cisco Collaboration Systems 10.x Solution Reference Network Designs (SRND), May 2014. http://www.cisco.com/c/en/us/td/docs/voice_ip_comm/cucm/srnd/collab10/collab10.html.

Review Questions

Use these questions to review what you have learned in this chapter. The answers appear in Appendix A, "Answers Appendix."

1. **Which of the following best describes DID?**

   a. E.164 international dialing

   b. External dialing from an IP phone to the PSTN

   c. VoIP security for phone dialing

   d. The ability of an outside user to directly dial into an internal phone

2. **Which of the following statements is the least accurate about IP networks?**

   a. IP packets can be delivered in the incorrect order.

   b. Buffering results in variable delays.

   c. Tail drops result in constant delays.

   d. Bandwidth is shared by multiple streams.

3. **Which statement most accurately describes overhead for packetized voice?**

   a. VoIP packets are large compared to data packets and are sent at a high rate.

   b. The Layer 3 overhead of a voice packet is insignificant and can be ignored in payload calculations.

  c. Voice packets have a small payload size relative to the packet headers and are sent at high packet rates.

  d. Packetized voice has the same overhead as circuit-switching voice technologies.

4. **What does the + symbol refer to in E.164?**

  a. Country code

  b. Area code

  c. International access code

  d. User's phone number

5. **Which two of the following are dial plan issues requiring a CUCM solution in multisite distributed deployments?**

  a. Overlapping directory numbers

  b. Overlapping E.164 numbers

  c. Variable-length addressing

  d. Centralized call processing

  e. Centralized phone configuration

6. **What is a requirement for performing NAT for Cisco IP phones between different sites through the Internet?**

  a. Use DHCP instead of fixed IP addresses

  b. Exchange RTP media streams with the outside world

  c. Use DNS instead of hostnames in CUCM

  d. Exchange signaling information with the outside world

7. **Which is the most accurate description of E.164?**

  a. An international standard for phone numbers including country codes and area codes

  b. An international standard for local phone numbers

  c. An international standard for dialing only local numbers to the PSTN

  d. An international standard for phone numbers for DID

8. **Which of the following is the most accurate description of TEHO?**

  a. Using the PSTN for cost reduction

  b. Using the IP WAN link for cost reduction

    c. Using the IP WAN link for cost reduction with remote routing over the WAN, and then transferring into a local PSTN call at the remote gateway

    d. Using the PSTN for cost reduction with minimal IP WAN usage

9. **What is the greatest benefit of toll bypass?**

    a. It increases the security of VoIP.

    b. It creates an effective implementation of Unified Communications.

    c. It reduces operating costs by routing internal calls over WAN links as opposed to the PSTN.

    d. It implements NAT to allow the routing of calls across the public Internet.

Chapter 2

Understanding Multisite Deployment Solutions

A multisite deployment involves several considerations that usually do not apply to single-site deployments. When implementing Cisco Unified Communications Manager (CUCM) in a multisite environment, you must address these concerns. This chapter provides information on how to leverage Cisco Collaboration technology to ensure that multisite deployments are successful.

Upon completing this chapter, you will be able to explain issues pertaining to a Unified Communications (UC) multisite deployment. Once you understand these issues, possible solutions are provided and where applicable best practices are mentioned according to Cisco Solutions Reference Network Designs (SRNDs) as well as Cisco Validated Designs (CVDs). Each recommended architecture is explained in greater detail throughout the remainder of this book.

Deploying a multisite Unified Communications environment requires a deep understanding of how to craft a proper multisite dial plan that allows for scalability, proper planning for bandwidth allocation for not only IP phones but also video endpoints, quality of service (QoS) design and implementation, and a highly available wide-area network (WAN) architecture, including survivable remote site telephony (SRST). This chapter will aid you in identifying issues that arise in multisite UC deployments.

Upon completing this chapter, you will be able to meet these objectives:

- Describe solutions that pertain to multisite deployments

- Describe how QoS solves voice and video quality issues in a multisite deployment

- Describe solutions to bandwidth issues in a multisite deployment

- Describe how low-bandwidth codecs and RTP-header compression can reduce IP WAN bandwidth consumption

- Describe how local conference bridges can reduce IP WAN bandwidth consumption

- Describe how transcoders can reduce IP WAN bandwidth consumption

- Describe how mixed conference bridges can reduce IP WAN bandwidth consumption

- Describe how multicast MOH from a branch ISR or ASR router flash can reduce IP WAN bandwidth consumption

- Describe the purpose of CAC and how it can reduce IP WAN bandwidth

- Describe survivability and availability features in a multisite deployment

- Describe how to use PSTN as a backup path when the IP WAN is not available

- Describe how to provide survivability for remote MGCP gateways

- Describe how SRST can provide survivability to remote phones

- Describe the rule of CFUR in a SRST environment

- Describe how to use the PSTN as a backup path when calls over the IP WAN are not admitted due to congestion or CAC

- Describe Cisco Unified Communications Manager mobility solutions

- Describe dial plan solutions for multisite deployments

- Describe how Cisco Expressway and Cisco Unified Border Element can provide NAT and security solutions

Multisite Deployment Solution Overview

The goal of any multisite deployment is to provide end users with a consistent and user-friendly experience with UC and video technologies while maintaining maximum availability. Note the "maximum availability" portion of the previous statement; this is where project scope and complexity come into the picture. Multisite deployments typically add complexity as they introduce several components, such as the following:

- Multiple Integrated Services Routers (ISRs) or Aggregation Services Routers (ASRs) gateways with various protocols

- Quality of service (QoS) across wide-area network (WAN) links

- Cisco Unified Border Elements (CUBEs) for public switched telephone network (PSTN) access

- Dial plan constraints and considerations

- Access codes or site codes for multisite dialing habits

- Survivable remote site telephony (SRST)

- Media resource placement considerations

Figure 2-1 demonstrates some techniques that can be incorporated into a multisite solution.

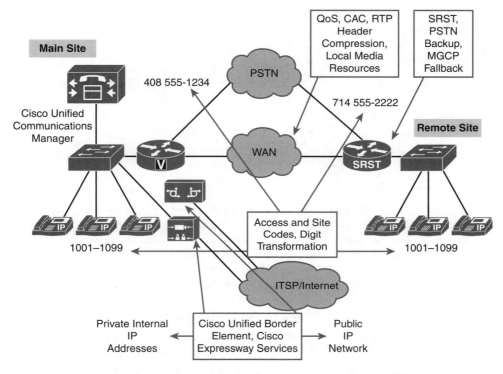

Figure 2-1   *Scenarios and Considerations for Multisite Deployments*

Availability issues can arise should the IP WAN fail or become saturated with various types of network traffic. With centralized call processing shown in Figure 2-1, the Cisco Unified Communications Manager (CUCM) servers or cluster may be physically located at a centralized data center or the headquarters site along with other data servers. The centralization of IT resources into a data center has many benefits in terms of IT, but can be problematic when WAN issues arise; simply put, services stop working. When connectivity fails across the WAN, Cisco Unified survivable remote site telephony (Cisco Unified SRST) can help alleviate the situation. SRST allows an ISR or ASR gateway at a remote site to become the call processing engine in the event of a WAN failure. The ISR/ASR gateways provide registration and call-processing capabilities to Cisco IP phones and video endpoints as well as certain voice interface cards (VICs) cards and high-density (HD) analog gateways. SRST can be configured to leverage any local public switched telephone network (PSTN) trunks or circuits a remote site may have. You can configure H.323 or Session Initiation Protocol (SIP) protocols along with dial peers to allow for PSTN dialing even if the main circuits are centralized at the data center. These local circuits can be used to allow emergency dialing capabilities or provide business continuity during WAN failure as calls continue to process from the outside world into IP phones.

Quality and bandwidth issues are solved by quality of service (QoS), call admission control (CAC), Real-time Transport Protocol (RTP) header compression, and local media resources. Each of these technologies and techniques is discussed in this chapter.

Dial plan implementation and planning for multisite deployments may include access and site codes, as well as digit manipulation. Digit manipulation can occur at various places inside the CUCM server or even at the ISR/ASR router level using dial peers and voice translation profiles and rules.

Network Address Translation (NAT) and security issues are solved by the deployment of a Cisco Unified Border Element (CUBE) or Cisco Expressway C and E servers. These technologies provide demarcation capabilities and advanced security feature sets between the LAN and WAN or Internet for incoming and outgoing voice and video calls.

Quality of Service

QoS refers to the capability of a network to provide better service to selected network traffic at the direct expense of other traffic. The primary goal of QoS is to offer improved service, including dedicated bandwidth, controlled jitter and latency (required by some real-time and interactive traffic), and better loss characteristics by giving priority to certain communication flows. QoS can be thought of as "managed unfairness" because whenever one type of traffic is given a higher priority, another is implicitly given a lower priority. The QoS designer must assess the level of each type of traffic and prioritize them best to suit the business needs of each organization.

Fundamentally, QoS enables you to provide better service to certain flows. This is done by either raising the priority of a flow or limiting the priority of another flow. When using congestion-management tools, you try to raise the priority of a flow by queuing and servicing queues in different ways. The queue-management tool used for congestion avoidance raises priority by dropping lower-priority flows before higher-priority flows. Policing and shaping provide priority to a flow by limiting the throughput of other flows. Link efficiency tools prevent large sized packet flows (such as file transfers) from severely degrading small sized packet flows, such as voice.

The following are the leading practices associated with QoS implementation:

1. Identify various types of traffic (voice, video, signaling, mission critical, enterprise resource planning [ERP], data, and so on).

2. Divide traffic into classes (voice and video real-time traffic, mission-critical traffic, signaling, less-important traffic, and so on).

3. Create a QoS policy per class, which is typically done on an aggregation WAN link or egress port by connecting a high-speed network to a slow-speed network. This QoS policy applies a differential treatment to the various traffic classes created and identified.

Figure 2-2 shows packets flowing into a router and the concept of identification, traffic class creation, and QoS policies being applied, reordering and prioritizing the traffic based on the classes.

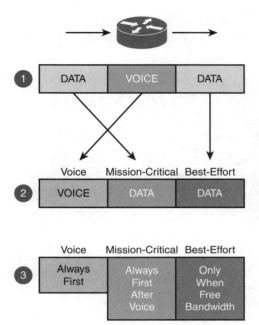

Figure 2-2 *Quality of Service Review*

QoS Advantages

Many engineers ask, "What is the advantage of QoS? Why can't we simply upgrade our WAN links to Gigabit+ speeds?" Well, first is the obvious answer: cost! Imagine the capital expenditure (CAPEX) and operational expenditure (OPEX) expenses required for upgrading not only the WAN links but also possibly upgrading legacy devices to support the increased speeds. Many organizations face network refreshes every 3 to 5 years simply to keep pace with industry trends. Second, just because you provide more bandwidth on the WAN does not necessarily mean it is a good thing. Imagine malware breaks out on data network; it could in theory consume an entire gigabit link. Therefore, QoS and creating specific queues and prioritizing traffic is preferred.

QoS can improve the quality of voice and video calls when bandwidth utilization is high by giving priority to RTP packets and video frames. Figure 2-3 demonstrates how voice (audio) traffic is given absolute priority over all other traffic with a technique called a priority queue (PQ) or low-latency queuing (LLQ). In some documentation and manuscripts, the LLQ is referred to as a PQ, which is reserved for real-time traffic such as voice. In addition to the PQ, LLQ also employs a class-based weighted fair queue (CBWFQ) for lower classes of traffic such as transactional, network, traditional data, and scavenger classes. LLQ is a technique in which a dedicated queue is set up on a router and only services real-time traffic. The PQ is serviced first before all other types of traffic in times of congestion. One important item to discuss is that the queuing mechanisms on Cisco Catalyst switches and ISR/ASR routers only use the queue if there is network congestion and the transmit ring or TX ring is full. LLQ reduces jitter, which is caused by variable queuing delays, and lost voice packets, which are caused by tail drops that occur when buffers are full.

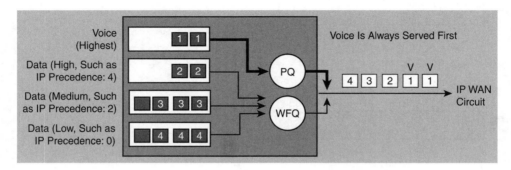

Figure 2-3   *Quality of Service VoIP Packet Prioritization*

With QoS configured, voice traffic has absolute priority over other traffic and provides the following guarantees for voice traffic on the network:

- Prevents jitter caused by variable queuing delays
- Ensures bandwidth for signaling
- Prevents packet loss caused by tail drops in queues

To avoid a possible scenario where voice RTP packets could potentially start blocking of other traffic (remember the voice queue or PQ is serviced first), voice bandwidth should be limited by defining the maximum bandwidth to be used by the maximum number of calls with the priority command within the LLQ configuration. The number of voice calls should also be limited by a CAC mechanism. The two techniques are often used side by side. You can use a LLQ to prioritize voice RTP on the wire, while employing a locations CAC inside the CUCM application. The two techniques combined ensure that additional calls will not try to further saturate the WAN link. As an analogy, QoS builds a carpool lane for prioritized drivers with LLQ. CAC is the mechanism that limits the maximum number of cars that can be in the carpool lane simultaneously. One lesser-used technique is to configure automated alternate routing (AAR) inside CUCM. AAR is used to route the calls through the PSTN in the event the WAN is saturated or unavailable.

Note   CAC and AAR mechanisms are discussed in Chapter 7, "Call Admission Control (CAC) Implementation."

One final consideration that should be given for QoS and voice calls is to configure QoS to guarantee a certain level of bandwidth for signaling traffic with Class-based weighted fair queuing (CBWFQ). Signaling traffic is often referred to as the signals being sent to and from the IP phones or video endpoints to the CUCM servers. Typically with signaling traffic you are speaking about Skinny Client Control Protocol (SCCP) or Session Initiation Protocol (SIP) traffic from the endpoint to the CUCM. However, signaling traffic can encompass gateway traffic using H.323 or Media Gateway Control Protocol (MGCP). Signaling traffic uses a very nominal amount of bandwidth in terms of any other network traffic. Therefore, it is not necessary to configure large queues or bandwidth reservations. However, failure to configure a queue for signaling could result in voice or video calls not being properly set up due to the signaling messages not making it between devices in a timely manner.

Overview of Solutions for Bandwidth Challenges

Bandwidth on the IP WAN is a finite resource that needs to be conserved when possible. Not everyone has the luxury of possessing OC-192 connections between every site. It is entirely possible that during your Cisco UC career you may be asked on an interview, by your boss, or even peers which techniques can be implemented to conserve WAN bandwidth. Some available techniques for WAN conservation with Cisco UC include the following:

- Using low-bandwidth codecs

- Using RTP header compression

- Deploying local annunciators or disabling remote annunciators

- Deploying local conference bridges

- Deploying local media termination points (MTPs)

- Deploying transcoders or mixed conference bridges

- Deploying local music on hold (MOH) servers (requires a local CUCM server) or using multicast MOH from branch router flash

- Limiting the number of voice calls using CAC or possibly Resource Reservation Protocol (RSVP)

If you use low-bandwidth (compressed) codecs, such as G.729, the required bandwidth for digitized voice is 8 kbps for the payload, or approximately 24 kbps in a full RTP packet compared to the 64 kbps for the payload, or approximately 80 kbps in a full RTP packet, that is required by G.711. In recent years, Cisco has embraced newer lower-bandwidth codecs such as Internet Low Bitrate Codec (iLBC), which are open source. One of the main issues with G.729 is it is a proprietary codec, meaning certain devices cannot play back recordings or perhaps voicemail greetings that have been encoded with G.729. Looking into the future, Cisco has plans in CUCM Version 11.x to start supporting the Opus codec, which is capable of adjusting the bandwidth used for encoding dynamically.

When you use RTP header compression (compressed RTP or cRTP), the IP, User Datagram Protocol (UDP), and RTP header can be compressed to 2 or 4 bytes (depending on whether the UDP checksum is preserved), compared to the 40 bytes that is required by these headers if cRTP is not used. cRTP is enabled per link on both ends of a point-to-point WAN link. It should be selectively used on a slow WAN link, typically less than 768 kbps. It does not need to be enabled end to end across all faster WAN links.

Annunciators are an easy way for CUCM to play back error messages in a human voice, providing the end user with an error greeting or informative message rather than a simple reorder tone. If spoken announcements are not required, the use of annunciators can be disabled for Cisco IP phones that do not have a local annunciator. Otherwise, local annunciators could be deployed. Annunciators in a UC environment always exist and are configured on the CUCM servers. There is a particular service on the CUCM server that is called the *IP Voice Streaming Application service*, which supplies these information messages to end users. Therefore, local annunciators can be implemented only if a local CUCM cluster is deployed or if clustering over the IP WAN is being used and a CUCM subscriber exists in the same LAN as the local phones.

Implementing a local conference bridge is a technique where IP phones send their RTP streams to a local conference resource that is configured on an ISR or ASR gateway. This technique saves bandwidth by preventing RTP streams traversing the WAN destined for a centralized conference bridge. Cisco IP phones thereby use the local conference bridge. Hardware-based conference bridges are achieved by configuring packet voice data modules (PVDMs) on the router. In some texts and documents, the term *packet voice data module* is used interchangeably with digital signal processor (DSP). If there is a CUCM Subscriber local to the LAN or branch site, software-based G.711 conferences could be achieved by using the local Subscriber as a conference resource. It is important to note that PVDMs and hardware-based conferences can support multiple mixed codecs, whereas software-based conferencing can support G.711 only by default.

Local media termination points (MTPs) can be required in situations where you need to convert G.711uLaw to G.711aLaw (like codecs) or in situations when H.323 slow-start to H.323 fast-start networking is necessary. More recently, with SIP trunks taking the markets by storm, SIP early-offer to SIP delay-offer support may require an MTP. Local MTPs are deployed using a mixture of CUCM Subscribers or hardware-based DSPs at remote sites. Locally deployed MTPs prevent traffic from traversing the IP WAN and using centralized MTPs.

If low-bandwidth codecs are not supported by all endpoints, transcoders can be used so that low-bandwidth codecs can be used across the IP WAN. IP phones and video endpoints can natively be configured to utilize G.729 using the regions concept in CUCM. If traffic leaving a remote branch is configured to use G.729 across the WAN, when it reaches a central site it can then be converted back to G.711 before arriving at Unity Connection or Contact Center Express server, which may be G.711 only. Transcoders allow a voice stream to be converted from G.729 to G.711 and vice versa. It is important to note that transcoding is only supported in hardware-based PVDMs and DSPs; CUCM cannot transcode calls. Throughout this book, G.711 and G.729 are used as example codecs. However, with hardware-based PVDM resources, you can configure a mixture of codecs for transcoding, MTP, or conferences such as G.722 and newer codecs available in CUCM 9.x and later.

Deploying a local MOH server means that CUCM servers have to be present at each site. In centralized call-processing models in which CUCM servers are not present at remote sites, it is recommended that you use multicast MOH from branch router flash. This approach eliminates the need to stream MOH over the IP WAN. If this is not an option, use multicast MOH instead of unicast MOH to reduce the number of MOH streams that have to traverse the IP WAN. Multicast MOH requires multicast routing to be enabled in the routed IP network.

Use software-based CAC with CUCM or a gatekeeper to avoid oversubscription of WAN bandwidth from too many voice calls. CAC can be used alongside LLQ on routers and switches to achieve a network, which prevents oversubscription, and queue traffic spikes. If absolute guaranteed bandwidth is required for mission-critical immersive telepresence, video, or voice calls, implement RSVP in the environment. RSVP is a protocol that runs on IOS devices to guarantee bandwidth on a per-flow basis to applications requiring bandwidth guarantees. It is very difficult in many environments to implement RSVP because it requires manually touching and configuring it on each hop or network segment.

Low-Bandwidth Codecs and RTP Header Compression

The use of a low-bandwidth codec and compressed Real-Time Transfer Protocol (cRTP) reduces the bandwidth requirements of a call on a WAN link. In Figure 2-4, a voice packet for a call using the G.711 codec and a 20-ms packetization period is being passed along a Frame Relay WAN link. The RTP frame has a total size of 206 bytes, composed of 6 bytes of Frame Relay header, 20 bytes of IP header, 8 of bytes UDP header, 12 bytes of RTP header, and a 160-byte payload of digitized voice. The packet rate is 50 packets per second (pps), resulting in a bandwidth need of 82.4 kbps. Note that when cRTP is used in conjunction with a lower-bandwidth codec such as G.729, the bandwidth can be considerably reduced to 11.2 or 12 kbps for the same call.

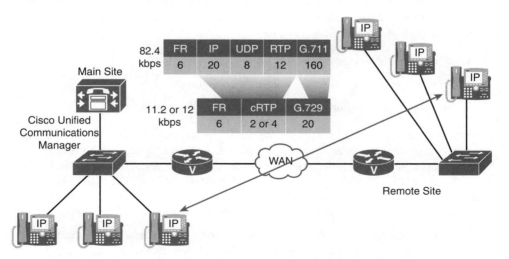

Figure 2-4 *Low-Bandwidth Codecs and RTP Header Compression*

Note   The traditional default codec with CUCM is G.711, which utilizes a 160-byte sample size and a 20-ms packet interval. In newer releases of CUCM, a new codec G.722 was introduced. G.722 provides improved speech quality due to a wider speech bandwidth of 50–7000 Hz compared to G.711, which was optimized for a plain old telephone system (POTS) wireline quality of 300–3400 Hz. G.722 samples audio data at a rate of 16 kHz (using 14 bits) double that of traditional telephony interfaces. This results in superior audio quality and clarity that uses the same bandwidth as G.711.

When you use cRTP and change the codec to G.729 using CUCM regions, the required bandwidth changes as follows: The frame now has a total size of 28 or 30 bytes per frame, composed of 6 bytes of Frame Relay header, 2 or 4 bytes of cRTP header (depending on whether the UDP checksum is preserved), and a 20-byte payload of digitized, compressed voice. The packet rate is still 50 pps (because the packetization period was not changed), resulting in bandwidth needs of 11.2 or 12 kbps.

Seven G.729 calls with cRTP enabled require less bandwidth than one G.711 call without cRTP (assuming that cRTP is used without preserving the UDP checksum).

Note   While the audio codec configuration preserves the codec end to end, cRTP is configured on a link-by-link basis. RTP header compression is configured on a per-link basis on both routers in a point-to-point WAN connection.

Codec Configuration in CUCM

The codec that is used for a call is determined by the region configuration in CUCM. Each region in CUCM is configured with the maximum audio bandwidth requirements or codecs to be used per call. Regions in CUCM can be configured to utilize specific codecs in each of the following scenarios:

- Within the configured region

- Toward a specific other region (manually configured)

- Toward all other regions (not manually configured)

Regions are assigned to device pools (one region per device pool), and a device pool is assigned to each device, such as a Cisco IP phone. A Cisco IP phone can be configured with a maximum of one device pool at any one time. The codec actually used depends on the capabilities of the two devices that are involved in the call. The assigned codec is the first codec of the applicable codec preference list that is supported by both devices and does not exceed the permitted bandwidth of that region.

In newer releases of CUCM, you can not only specify a codec maximum bit rate or highest quality codec but also configure a preferred list of codecs that devices can negotiate. If devices cannot agree on a codec, a transcoder is invoked. If a transcoder were unavailable for use during a call with different codecs configured, the audio call would fail. Typically, this throws most engineers off; the behavior of a codec negotiation failure still results in both IP phones ringing because the signaling is separate from audio. However, when one side of the call answers or goes off hook, the call fails due to a codec negotiation failure by both devices.

Disabling the Annunciator for Remote Branches

The annunciator is a CUCM feature that sends one-way audio of prerecorded messages over RTP to IP phones. An example is the replacement to the fast busy reorder tone with the recorded message "Your call cannot be completed as dialed; please . . .". CUCM comes preconfigured with several different annunciator messages, including messages for unallocated numbers, call rejected, number changed, invalid number format, and precedence level exceeded. One could do a quick browse of the Trivial File Transfer Protocol (TFTP) server on CUCM, and you can download and listen to these messages. If announcements should not be sent over a saturated IP WAN link, Media Resource Group Lists (MRGL) can be used so that remote phones do not have access to the annunciator media resource, which can be implemented in a design like the one shown in Figure 2-5.

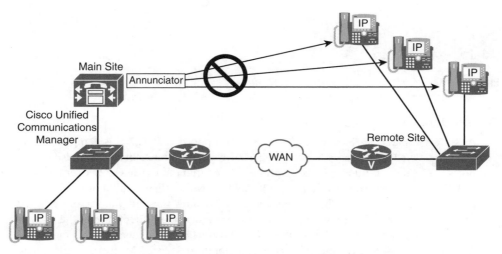

Figure 2-5 *Disabling the Annunciator in CUCM to Conserve WAN Bandwidth*

Note   Not every call requires annunciator messages, and because the messages usually are rather short, the bandwidth that may be preserved by disabling the annunciator is minimal.

Local Versus Remote Conference Bridges

As shown in Figure 2-6, if a local conference bridge is deployed at the remote site using the remote site's gateway DSPs or PVDMs, it keeps voice streams off the IP WAN for conferences in which all members are physically located at the remote site. The same solution can be implemented for MTPs. MRGLs specify which conference bridge (or MTP) should be used and by which IP phone.

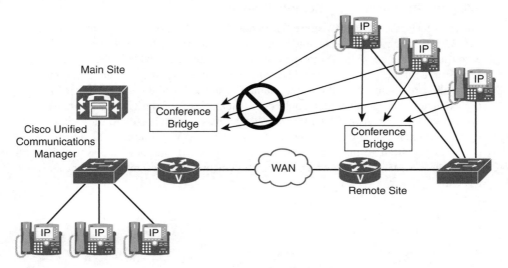

Figure 2-6   *Utilizing Local Conference Bridges*

Transcoders

As shown in Figure 2-7, a voicemail system that supports only G.711 is deployed at the main site. One CUCM server is providing a software conference bridge that also supports G.711 only. If remote Cisco IP phones are configured to use G.729 over the IP WAN with CUCM regions to conserve WAN bandwidth, they would not be able to join conferences or access the voicemail system. To allow these IP phones to use G.729 and to access the G.711-only services, a hardware transcoder is deployed at the main site in the gateway using DSP resources. This illustrates the "Golden Rule" of media resources discussed in Chapter 1, "Cisco Collaboration Solution Multisite Deployment Considerations," where resources are allocated to optimize RTP traffic. The Golden Rule describes allocation of transcoders centrally in the environment and conference bridges at remote sites.

Remote Cisco IP phones now send G.729 voice streams to the transcoder over the IP WAN, which saves bandwidth. The transcoder changes the stream to G.711 and passes it on to the conference bridge or voicemail system, allowing the audio connection to work.

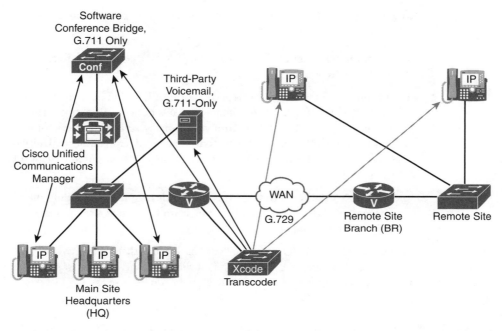

Figure 2-7 *Transcoder Implementation for G.711-Only Devices Such As Voicemail*

Leading Practices for Transcoder Design

When designing transcoders to allow G.711-only devices to communicate with remote IP phones using G.729, proceed with the following steps:

Step 1.   Implement an IOS-based transcoding resource. Because CUCM does not support software transcoding resources, the only option is to use a hardware-based transcoding resource by first configuring the transcoder at the Cisco IOS router and then adding the transcoder to CUCM. Remember that all IOS-based resources register to CUCM using SCCP.

Step 2.   Implement CUCM-based regions so that *only* G.729 is permitted on the IP WAN, and an MRGL-based transcoder is used in which the IP phone has access to via a device pool or hard coded as a setting to the phone itself. To do so, all IP phones and G.711-only devices, such as third-party voicemail systems or software conference bridges that are located in the headquarters site, are placed in a region (such as Headquarters region) and remote IP phones are placed in another region (such as Branch 1 region). The transcoding resource is put into a third region (such as XCODER region).

Step 3.   Now using the CUCM regions relationships, set the maximum codec for calls within and between regions:

- **Within Brach 1 set to G.711:** This allows local calls between remote IP phones to use G.711.

- **Within HQ set to G.711:** This allows local calls within headquarters to use G.711. These calls are not limited to calls between IP phones. This also includes calls to the G.711-only third-party voicemail system or calls that use the G.711-only software conference bridge. In addition, Unified Contact Center Express or Enterprise may also be configured to support G.711 only.

- **Within XCODER set to G.711:** Because this region includes only the transcoder media resource, this setting is irrelevant because there are no calls within this region.

- **Between BR and HQ set to G.729:** This ensures that calls between remote IP phones and headquarters devices, such as IP phones, software conference bridges, and voicemail systems, do not use the G.711 codec for calls that traverse the IP WAN.

Note   Calls between IP phones at headquarters and remote IP phones do not require a transcoder. They simply use the best allowed codec that is supported on both ends from the CUCM region settings (ideally, G.729). One important note: Cisco IP phones and many video endpoints support *both* G.711 and G.729 codecs; they are essentially built in to the phone as part of its firmware and capabilities. The only time a transcoder needs to be invoked is for devices that *only* support G.711, such as Cisco Unified Contact Center Express (UCCX), Cisco Unified Contact Center Express (UCCE), Unity Connection, and so on, depending on how these devices were installed.

A transcoder is invoked only when the two endpoints of a call cannot find a common codec that is permitted by region configuration. This is the case in this example. The remote IP phones (which support G.711 and G.729) are not allowed to use G.711 over the IP WAN, and the headquarters voicemail system and software conference bridge do not support G.729. CUCM detects this problem based on its region configurations, and the capability negotiation performed during call setup signaling identifies the need for a transcoder.

Mixed Conference Bridge

As illustrated in Figure 2-8, a hardware conference bridge is deployed at the main site gateway. The hardware conference bridge is configured to support mixed conferences, in which members use different codecs. Headquarters IP phones that join the conference can use G.711, whereas remote IP phones can join the conference using a low-bandwidth codec such as G.729. The end result is a minimum WAN utilization with relatively high voice quality.

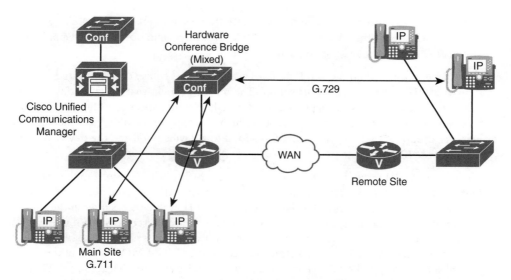

Figure 2-8 *Mixed (Hardware) Conference Bridges Enable Low-Bandwidth Codecs to Be Used by Remote Conference Participants*

One new feature in the later versions of CUCM (Version 9.x and later) is the addition of homogeneous and heterogeneous video conferencing using IOS-based PVDM3 resources. PVDM3s are available on the ISR and ASR Generation 2 routers such as the 2911 and 3945.

A homogeneous video conference is a video conference in which all participants connect using the same video format attributes. All the video phones support the same video format, and the conference bridge sends the same data stream format to all the video participants. In a heterogeneous video conference, all the conference participants connect to the conference bridge with phones that use different video format attributes. In heterogeneous conferences, transcoding and transrating features are required from the DSP to convert the signal between the various formats. Additional information on both conference types is provided in later chapters.

Multicast MOH from the Branch Router Flash

Multicast MOH from the branch routers flash is a technique for multisite deployments that use centralized call processing along with a centralized MOH server streaming multicast MOH. A branch ISR or ASR router is also configured to play back the same MOH file and spoofs the multicast IP address used by the centralized CUCM MOH process.

The multicast MOH feature only works with multicast MOH and is based on MOH capabilities of Cisco Unified SRST. The Cisco IOS SRST gateway is configured for multicast MOH and continuously sends a multicast MOH stream, regardless of whether the ISR or ASR router is in SRST mode (standby or fallback mode).

Neither CUCM nor the remote IP phones are aware that the SRST gateway is involved. To them it appears as though a multicast MOH stream has been generated by the CUCM MOH server and received by the remote IP phones (when behind the scenes CUCM is configured to stop broadcasting multicast MOH after a set number of hops). The SRST gateway then spoofs the CUCM multicast MOH IP address and port, thereby supplying multicast to the IP phone at the branch.

To achieve this, the remote Cisco IP phones are configured to use the centralized CUCM MOH server as their MOH source. The building blocks that make up this design include a multicast MOH server, configured in Communications Manager, which is added to a media resource group (MRG), which is subsequently added to a media resource group list (MRGL). The MRGL is added to a device pool, and the device pool is added to the IP phone.

The CUCM MOH server is configured for multicast MOH (mandatory), and the max-hops value in the MOH server configuration is set to 1 for the affected audio sources. The max-hops parameter specifies the Time To Live (TTL) value that is used in the IP header of the RTP packets. The CUCM MOH server and the Cisco IOS SRST gateway located at the remote site have to use the same multicast address and port number for their streams. This way, MOH packets generated by the CUCM MOH server at the central site are dropped by the central-site aggregation layer router because the TTL has been exceeded. As a consequence, the MOH packets do not cross the IP WAN. The SRST gateway permanently generates a multicast MOH stream with an identical multi-cast IP address and port number so that the Cisco IP phones simply listen to this stream as it appears to be coming from the CUCM MOH server.

As an alternate solution, instead of setting the max-hops parameter for MOH packets to 1, use one of the following methods:

■ **Configure an access control list (ACL) on the WAN interface at the central site:** This prevents packets that are destined for the multicast group address or addresses from being sent out the interface onto the IP WAN.

■ **Disable multicast routing on the WAN interface:** Do not configure multicast routing on the WAN interface to ensure that multicast streams are not forwarded into the WAN.

Note   Depending on the configuration of MOH in CUCM, a separate MOH stream for each enabled codec is sent for each multicast MOH audio source. The streams are incremented either based on IP addresses or based on port numbers. (The recommendation and default is per IP address.) Assuming that one multicast MOH audio source and G.711 a-law, G.711 u-law, G.729, and the wideband codec are enabled, there will be four multicast streams. Make sure that all of them are included in the ACL to prevent MOH packets from being sent to the IP WAN. Cisco recommends using the end of the multicast Class D IPv4 range, or 239.1.1.1. The Class D range is from 224.0.0.0 to 239.255.255.255. Most applications on the network that require multicast traffic would use the beginning portion of the Class D space; if we use the latter half, there is less chance for a collision or misconfiguration. Each stream will consume one IP address. Therefore, G.711-ulaw would use 239.1.1.1, G.711-alaw would use 239.1.1.2, and so on. Each new MOH stream would use every fourth IP address; MOH stream 2 would start using 239.1.1.4.

When using multicast MOH from a branch router's flash, G.711 has to be enabled between the CUCM MOH server and the remote Cisco IP phones via the region configuration. This is necessary because the branch SRST MOH feature supports only the G.711 codec. Therefore, the stream that is set up by CUCM in the signaling messages also has to be G.711. Because the packets are not sent across the WAN, configuring the high-bandwidth G.711 codec is no problem as long as it is enabled only for MOH. All other audio streams (such as calls between phones) that are sent over the WAN should use the low-bandwidth G.729 codec. A recommended best practice involves creating a multicast MOH region in CUCM and setting its relationship to every other region to G.711.

An Example of Multicast MOH from the Branch Router Flash

Multicast MOH implementation is illustrated in Figure 2-9.

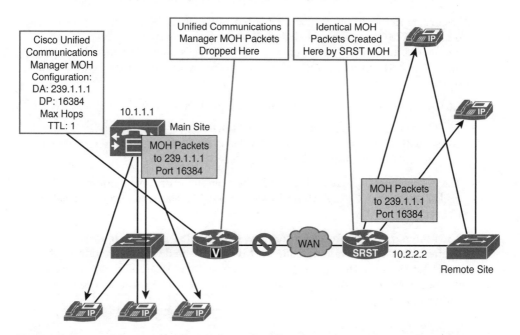

Figure 2-9 *Multicast MOH from Branch Router Flash Operation*

As illustrated in Figure 2-9, the CUCM MOH server is configured for multicast MOH with a destination Class D multicast group address of 239.1.1, a destination port 16384, and a max-hops TTL value of 1. Cisco recommends using an IP address in the range reserved for administratively controlled applications on private networks of 239.0.0.0 to 239.255.255.255.

The SRST ISR or ASR gateway located at the remote site is configured with the same multicast MOH IP address and port number as the CUCM MOH server. IP phones at the main site can be configured to receive either unicast or multicast MOH packets from a CUCM server. Typically, you configure the main headquarters site to have a MRGL and subsequent MRG that contains a unicast MOH server. The implicit benefit of this is MOH that starts at the beginning of the recording for each caller. IP phones at the remote site then receive multicast MOH packets that are generated by the SRST gateway. At no time do multicast MOH packets cross the WAN.

Note   Multicast traffic must be allowed on your enterprise networks, which in some organizations can violate security measures or auditing measures. This can lead to contention among IT teams in a segmented environment. Should you run into this, we recommend using the ACL method or disallowing multicast traffic across the WAN on the ISR or ASR router.

When a remote phone at a branch location is put on hold, the following happens:

1. According to the device pool and inherited MRGL and MRG of the remote phone, the CUCM MOH server is used as the MOH media resource. This server can be configured to use multicast or unicast MOH. There is nothing wrong with having multiple MOH servers configured in your environment, some using multicast and others serving unicast.

2. CUCM signals the IP phone to receive MOH on IP address 239.1.1.1, port 16384.

3. The CUCM MOH server sends multicast MOH packets to IP address 239.1.1.1, port 16384, with a TTL value of 1. The TTL value is extremely important here; it limits the hops that the multicast packet can traverse.

4. The aggregation layer router located at the main site drops the multicast MOH packet sent by the CUCM MOH server because TTL has been exceeded inside the IP header.

5. The router at the remote site is configured as an SRST gateway. In its Cisco Unified SRST configuration, multicast MOH is enabled with an IP address of 239.1.1.1 and port 16384. The SRST gateway streams MOH all the time, even if it is not in fallback mode. One important note: The MOH file must reside in the router's flash!

Note   You can also stream MOH from a USB port and a USB thumb drive using the command **moh usbflash0:{*MOH Filename*}.wav**. This is beneficial for seasonal MOH messages or a MOH that changes frequently. Rather than touching each individual configuration, merely remove the USB thumb drive and insert the new MOH thumb drive.

6. The IP phones listen to the multicast MOH stream that was sent from the SRST gateway to IP address 239.1.1.1, port 16384, and play the received MOH stream.

7. Whether the remote gateway is in SRST mode or not, MOH packets never cross the IP WAN, and the ISR or ASR router is continuously streaming the MOH file.

> **Note** If an MOH file is used in router flash, only up to five MOH files can be configured to play at a time. This is unlike CUCM, where many different MOH files can be configured.

An Example of Multicast MOH from the Branch Router Flash Cisco IOS Configuration

As shown in Figure 2-10, the name of the audio file on the branch router flash is moh-file.au, and the configured multicast address and port number are 239.1.1.1 and 16384, respectively. The optional **route** command can be used to specify a source interface address for the multicast stream. If no route option is specified, the multicast stream is sourced from the configured Cisco Unified SRST default address. This is specified by the **ip source-address** command under the Cisco Unified SRST configuration (10.2.2.2 in this example). Note that you can stream only a single audio file from flash and that you can use only a single multicast address and port number per router.

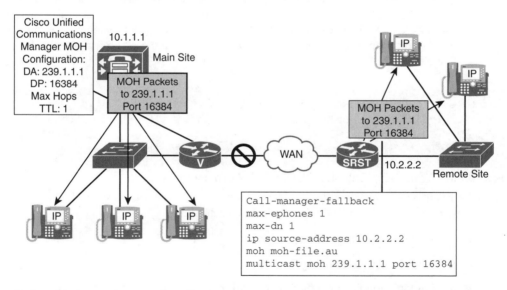

Figure 2-10 *Multicast MOH from the Remote Site Router Flash Cisco IOS Configuration*

A Cisco Unified SRST license is required regardless of whether the SRST functionality will actually be used. The license is required because the configuration for streaming multicast MOH from branch router flash is done in SRST configuration mode. Also, even if SRST functionality will not be used, at least one **max-ephones** for every Cisco IP phone supported and at least one **max-dn** for every directory number (DN) on all phones must be configured. On ISR Gen 2 routers, the Cisco End User License Agreement (EULA) and the SRST feature set and right to use (RUU) licenses must also be accepted.

Alternatives to Multicast MOH from Remote Site Router Flash

Sometimes multicast MOH from remote site router flash cannot be used. For instance, perhaps the remote site router does not support the feature or does not have a Cisco Unified SRST feature license. In that case, consider the following alternatives:

- **Using multicast MOH across the WAN versus unicast across the WAN:** When you use multicast MOH over the IP WAN, the number of required MOH streams can be significantly reduced. Thus, less bandwidth is required compared to multiple unicast MOH streams. The IP network, however, has to support multicast routing for the path from the MOH server to the remote IP phones.

- **Using G.729 for MOH to remote sites:** If multicast MOH is not an option (for instance, because multicast routing cannot be enabled in the network), you may still be able to reduce the bandwidth consumed by MOH. If you change the codec that is used for the MOH streams to G.729 and potentially enable cRTP on the IP WAN, each individual MOH stream requires less bandwidth and hence reduces the load on the WAN link. The bandwidth savings are identical to those that are achieved when using G.729 and cRTP for standard audio streams, which was discussed earlier. To use G.729 for MOH streams, the MOH server and the remote IP phones have to be put into different regions, and the audio codec between these two regions must be limited to G.729.

Note   G.729 codec may not sound as crisp or clear when used as an MOH file. Due to the encapsulation and encoding, G.729 typically does not make for a very friendly-sounding MOH stream.

Preventing Too Many Calls by CUCM Call Admission Control

CUCM allows the number of calls traversing a WAN link from site to site to be limited by the following call admission control (CAC) mechanisms:

- Enhanced location-based CAC (E-LCAC)

- Resource Reservation Protocol (RSVP)-enabled locations

- Session Initiation Protocol (SIP) preconditions

- Gatekeepers

Note   For in-depth details on CAC, see Chapter 7, "Call Admission Control (CAC) Implementation."

CUCM E-LCAC is applicable to calls between two entities that are configured in CUCM as well as intercluster calls between two separate CUCM clusters in the same service domain. A great example is a multinational corporation with a U.S. CUCM cluster and a European CUCM cluster. E-LCAC is based on a network model or topology that you define in CUCM. All bandwidth limits that are configured along the path of the network model are checked during call setup for link oversubscription and available bandwidth. This is a much-needed improvement over traditional location CAC, which was hub-and-spoke topology and could not account for various multisite topologies.

RSVP is a special way to configure locations. When RSVP is configured to be used between a pair of locations, the audio streams flow through two routers, also known as RSVP agents. The call leg between the two RSVP agents is subject to Cisco IOS RSVP CAC. If enough guaranteed bandwidth exists between the network segments, the RSVP router agents signals back to CUCM to let the call proceed. RSVP is typically only applied to a single cluster; so for multicluster topologies, it is recommended to use E-LCAC.

SIP preconditions is a solution similar to RSVP-enabled locations, except that it is designed only for SIP trunks between clusters or to the PSTN. With SIP preconditions, calls processed through a SIP trunk flow through a local Cisco IOS router at each end of the SIP trunk. The ISR or ASR routers then split the call into multiple call legs, just like with RSVP-enabled locations. In this case, however, the call is not within a cluster, but between clusters or to the PSTN. In some manuscripts and documents, you may see this referred to as *intercluster RSVP*.

Gatekeepers are an optional component used within the H.323 protocol and provide address resolution and CAC functions. H.323 gatekeepers can be configured to limit the number of calls between H.323 zones. Gatekeepers and H.323-based communications networks are being progressively replaced with Cisco Unified Border Element (CUBE) and SIP-based communications networks.

Availability

Another topic for discussion is high availability in multisite deployments. How does one ensure that call processing and features remain intact during WAN failure or congestion? High availability can be achieved in several ways:

- **PSTN backup:** Use the PSTN as a backup for on-net intersite calls. IP phones that normally transmit RTP and signaling across the WAN for site to site calling can use the PSTN as an alternate path in the event the WAN goes down. One important note is that digit manipulation needs to occur to expand the dialed number to a fully qualified PSTN number.

- **MGCP fallback:** Configure an MGCP gateway to fall back, and use the locally configured plain old telephone service (POTS), H.323, or SIP dial peers when the connection to its call agent is lost. MGCP uses TCP and UDP ports 2427 and 2428 and the signaling is "backhauled" to CUCM. If the communication path is severed, the signaling path is torn down, causing catastrophe on a PRI D Channel. If MGCP fallback is configured, the gateway uses a locally configured dial plan should the backhaul link go down. During normal operations, the MGCP fallback commands are ignored. If H.323 is deployed as an alternate protocol, the dial peers have the same functionality in or out of SRST mode, and the same configuration can be used for both scenarios, which is another reason many seasoned engineers choose H.323 for its resiliency.

- **Fallback for IP phones with SRST:** Cisco IP phones using either SIP or SCCP must register to a call-processing device for the phones to work. IP phones that register over the IP WAN can have a local Cisco IOS SRST gateway configured as a backup to a CUCM server in their CUCM group configuration. When the connection to the primary CUCM server is lost, they can reregister with the local SRST gateway. Alternatively, SIP SRST or Call Manager Express (CME) can be used, which provides more features than standard Cisco Unified SRST.

- **Call Forward Unregistered (CFUR):** This is a call-forwarding configuration of IP phones that becomes effective when the IP phone is not registered. This setting can be used to forward calls to voicemail or perhaps to an alternate location if the phone is not registered in CUCM.

Note   CFUR was first introduced in Call Manager Version 4.2, and it appeared again in CUCM Version 6 and later versions.

- **Automated alternate routing (AAR) and Call Forward on No Bandwidth (CFNB):** AAR allows calls to be rerouted over the PSTN when calls over the IP WAN are not admitted by CAC. Both of these techniques can be used to reroute calls to alternate destinations in the event that the WAN is oversubscribed.

Note   AAR is configured for calls to and from IP phones within the same cluster. Calls to a different cluster over a SIP or H.323 trunk do not use AAR. They are configured to fail over to the PSTN with CAC by configuring route groups and route lists with path selection within the route pattern.

- **Mobility solutions:** When users or devices roam between sites, they can lose features or have suboptimal configuration because of a change in their actual physical location. CUCM extension mobility and the device mobility feature of CUCM can solve such issues. In addition, Cisco Unified Mobility allows integration of cell phones and home office phones by enabling reachability on any device via a single (office) number.

Note   The high-availability options for remote site are discussed in Chapter 5, "Remote Site Telephony and Branch Redundancy Options."

PSTN Backup

As shown in Figure 2-11, calls to a remote site or branch office within the same cluster are configured to use the IP WAN first. In the event of a WAN failure or in times of WAN congestion, AAR can be implemented to use the PSTN as a backup option. The end result is reduced operating cost with toll bypass over the WAN, and successful delivery of the same calls over the PSTN but potentially at a higher operating cost if the WAN fails.

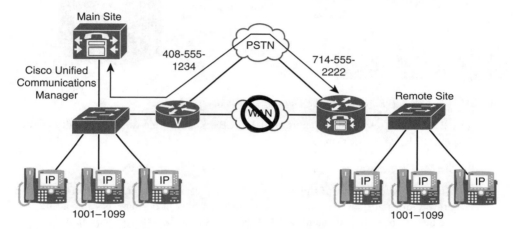

Figure 2-11   *Intersite Calls Are Rerouted Over the PSTN in the Case of an IP WAN Failure*

Note that when failover occurs to the PSTN using AAR, number expansion often occurs, meaning the original four-digit dialed extension needs to be expanded to a fully qualified PSTN number. This is all done seamlessly by CUCM with no end user intervention; the end user dials four digits and is unaware of the rerouting through the PSTN and the number expansion.

MGCP Fallback

MGCP gateway fallback is a feature that improves the availability of remote MGCP gateways.

A WAN link connects the MGCP gateway and backhaul signaling path at a remote site to the CUCM at a main site, which is the MGCP call agent. If the WAN link fails, the fallback feature keeps the gateway working as an H.323 or SIP gateway and rehomes to the MGCP call agent when the WAN link becomes active again. Think of this as failover for gateway protocols.

Figure 2-12 shows how MGCP fallback improves availability in a multisite environment. Note that ISR or ASR routers can run multiple voice protocols simultaneously without issue; it is entirely possible to run SCCP, MGCP, H.323, and SIP all on the same router. The default voice protocol or "application," however, is H.323 unless changed. All routers default to H.323 application mode.

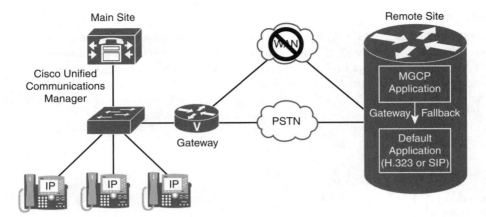

Figure 2-12   *MGCP Fallback to the Default Application*

Figure 2-12 illustrates normal operation of MGCP fallback while the connectivity to the call agent (CUCM) is functional:

■ The MGCP gateway is registered with CUCM over the IP WAN.

■ CUCM is the call agent of the MGCP gateway that is controlling its interfaces. The gateway does not have (or does not use) a local dial plan because all call routing intelligence is at the call agent. MGCP functions in a client/server model, with CUCM as the server and the gateway as the client.

When the MGCP gateway loses the connection to its call agent, as shown in Figure 2-11, it falls back to its default call control application (POTS, H.323, or SIP). The gateway now uses a local dial plan configuration, such as dial peers, voice translation profiles, and so on. Hence, it can operate independently of its MGCP call agent. Without MGCP fallback, the MGCP gateway would be unable to process calls when the connection to its call agent is lost.

Fallback for IP Phones: SRST, CME SRST, or SIP SRST

Fallback for IP phones is provided by the SRST Cisco IOS feature and improves the availability of remote IP phones.

A WAN link connects IP phones at a remote site to the Cisco Communications Manager at a central site via the SIP or SCCP protocol for signaling. If the WAN link fails, Cisco Unified SRST enables the local ISR or ASR gateway to provide call-processing services for IP phones. IP phones register with the gateway (which is listed as a backup CUCM server or SRST reference in the server's group configuration of the IP phones). The actual configuration order is an SRST reference added to CUCM containing the IP address of the branch router. In addition, a CUCM server group is created containing up to three CUCM servers for the IP phone to attempt to register with. Each of the previous components is loaded into the device pool for a particular branch or remote site, which is in turn applied to the IP phone. The IP phone will send a keepalive message to its primary CUCM server and to all backups. In the event that all CUCM servers are unavailable (for example, because of a WAN outage), the IP phone will attempt to register with its local router using the SRST reference.

A gateway functioning in SRST obtains the configuration of the IP phones from the phones themselves and can route calls between the IP phones or out to the PSTN.

Note When a Cisco IP phone is in SRST mode, the configuration on the phone should never be erased, because the phone will not function until the connection to CUCM is restored.

Figure 2-12 in the previous section demonstrated how fallback for IP phones improves availability in a multisite deployment with centralized call processing. This figure illustrated normal operation of Cisco Unified SRST while the connectivity between IP phones and their primary server (CUCM) is functional:

- Remote IP phones are registered with CUCM over the IP WAN.

- CUCM handles call processing for IP phones.

When Cisco IP phones lose contact with all CUCM servers, they register with the local Cisco Unified SRST router to sustain the call-processing capability necessary to place and receive calls. When the WAN connection between a router and the CUCM fails or when connectivity with CUCM is lost for any reason, Cisco Unified IP phones in the remote site become unusable for the duration of the failure. Cisco Unified SRST overcomes this problem and ensures that the Cisco Unified IP phones offer continuous (although minimal) service by providing call-handling support for Cisco Unified IP phones directly from the Cisco Unified SRST router. The system automatically detects a failure and uses Simple Network-Enabled Auto-Provisioning (SNAP) technology to autoconfigure the remote site router to provide call processing for Cisco Unified IP phones that are registered with the router. When the WAN link or connection to the primary CUCM is restored, call handling reverts to the primary CUCM.

One additional item to discuss is that there are three various forms of SRST:

- SCCP SRST or traditional SRST for SCCP-based devices.

- SIP SRST for SIP-based endpoints.

- CME SRST, which allows for advanced configuration in SRST mode. CME SRST adds the features such as hunt pilots and hunt lists (blast groups as they are commonly referred to) to an SRST registered phone.

Using CFUR to Reach Remote Site Cisco IP Phones During WAN Failure

As discussed, IP phones located at remote locations can use an SRST gateway as a backup for CUCM in case of IP WAN failure. The gateway can use its local dial plan to route calls destined for the IP phones in the main site over the PSTN. But how should intersite calls be routed from the main to the remote site while the IP WAN is down?

The problem in this case is that CUCM does not consider any other entries in its dial plan if a dialed number matches a *configured* but *unregistered* DN. Therefore, if users at the main site dial internal extensions during the IP WAN outage, their calls fail (or go to voicemail). To allow remote IP phones to be reached from the IP phones at the headquarters site, configure CFUR for the remote site phones, as shown in Figure 2-13. CFUR should be configured with the PSTN number of the remote site gateway so that internal calls for remote IP phones get forwarded to the appropriate PSTN number.

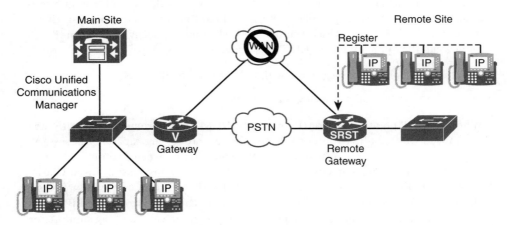

Figure 2-13 *Using CFUR to Reach Remote Site IP Phones Over the PSTN During WAN Failure*

Using CFUR to Reach Users of Unregistered Software IP Phones on Other Devices

If a mobile user has a laptop with a softphone (for instance, Cisco IP Communicator or Cisco Jabber) and the user shuts down the laptop or places it into standby/sleep mode, CFUR can forward calls destined for the unregistered softphone to the user's cell phone, as illustrated in Figure 2-14. The user does not have to set up Call Forward All (CFA) manually before closing the softphone application. If the softphone is not registered, calls are forwarded to the user's cell phone. This is another application of the CFUR feature that improves availability in CUCM deployments.

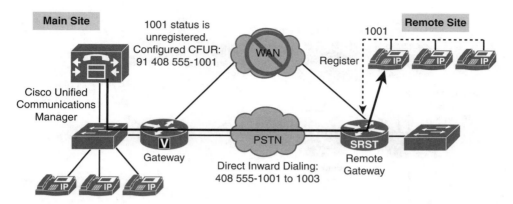

Figure 2-14 *Using CFUR to Reach Users of Unregistered Software IP Phones on Other Devices Such As Cell Phones*

Note This use case is for Call Forward Unregistered (CFUR), which is not related to SRST; the two components can be used together to provide seamless transitions and user experiences in times of WAN congestion or outages. These concepts can also be applied to situations when users with only softphones such as Cisco Jabber are not using their computer and their device is unregistered in CUCM.

AAR and CFNB

Automatic alternate routing (AAR) allows rerouting of calls over the PSTN if there is not enough bandwidth for Voice over IP (VoIP) calls. An alternate destination is derived from the external phone number mask and prefix that is configured per AAR group in CUCM. Individual destinations can be configured per phone, and this is applicable to calls within a single cluster. AAR does not support cross-cluster capabilities.

If a call over the IP WAN to another IP phone in the same cluster is not admitted by CAC, the call can be rerouted over the PSTN using AAR, as shown in Figure 2-15. The AAR feature includes a CFNB (Call Forward No Bandwidth) option that allows the alternative number to be set for each IP phone. In the example, because the remote site does not have PSTN access, the call is not rerouted to the IP phone over the PSTN (instead of over the IP WAN). It is alternatively rerouted to the cell phone of the affected user. AAR and CFNB improve availability in multisite environments by providing the ability to reroute on-net calls that failed CAC over the PSTN.

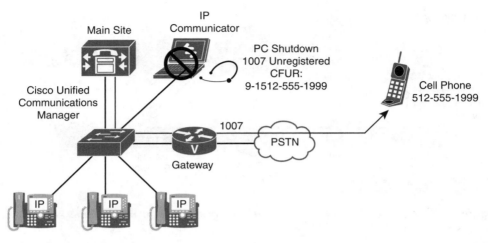

Figure 2-15 *AAR and CFNB*

Mobility Solutions

This section provides an overview of mobility solutions that solve issues that are the result of roaming users, roaming devices, and users with multiple telephones (office phone, cell phone, home phone, and so on).

When users or devices roam between sites, issues arise that can be solved by mobility solutions:

- **Device mobility:** Solves issues caused by roaming devices, including invalid device configuration settings such as regions, locations, SRST reference, AAR groups, calling search spaces (CSS), and so on. An example of device roaming is when an office worker gets a promotion or the dreaded demotion and moves to a different building or site. If that user moves her physical phone to a different location in CUCM, the phone retains the original site configuration. The device mobility feature of CUCM allows device settings that depend on the physical location of the device to be automatically overwritten if the device appears in a different physical location.

- **CUCM extension mobility:** Solves issues that are the result of roaming users using shared guest IP phones located in other offices. Typically, one would see this in a call center or "hot desk" environment where multiple users come into an office, sit down in an open cubicle or space, and log in to their PC and phone. If they cannot log in to their phone and their permanent extension and settings get applied, issues such as wrong DN, missing IP phone service subscriptions, CSS, and so on can occur. CUCM extension mobility allows users to log in to guest phones and replace the IP phone's configuration with the IP phone configuration of the logged-in user.

- **Cisco Unified Mobility:** Solves issues of having multiple phones and consequently multiple phone numbers, such as an office phone, cell phone, home (office) phone, and so on. Cisco Unified Mobility allows users to be reached by a single number, regardless of the phone that is actually used. This is sometimes referred to as single number reach (SNR). With Unified Mobility, when a user receives an incoming call to their direct inward dial (DID) extension, multiple simultaneous outbound calls can be made to various devices in an attempt to "find" the user. When multiple devices ring simultaneously, one typically has a higher success rate in connecting the call.

> **Note**   All of these mobility features are explained in detail in Chapter 8, "Implementing Cisco Device Mobility," Chapter 9, "Cisco Extension Mobility," and Chapter 10, "Implementing Cisco Unified Mobility." In fact, each technique has its own dedicated chapter.

Overview of Dial Plan Solutions

Dial plan issues in multisite deployments can be solved in the following ways:

- **Overlapping and nonconsecutive numbers:** You can implement access codes and site codes for intersite dialing. An example is a CUCM configuration that uses four-digit dialing internal to a site but then dials six digits (a two-digit site code) plus the four-digit extension to dial between sites. This allows call routing that is independent of DNs. Appropriate digit manipulation, the removing site codes in the Dialed Number Identification Service (DNIS) of outgoing calls, and prefixing site codes in Automatic Number Identification (ANI) of incoming calls are required. DNIS is a fancy term for the dialed number and ANI represents caller ID.

- **Variable-length numbering plans:** Dial string length is determined by timeout. If overlap sending and receiving is enabled, it allows dialed digits to be signaled one by one instead of being sent as one whole number. Overlap sending and receiving is a technique in which gateways can receive a call one digit at a time from the signaling device. Normally, gateways receive their digits en-bloc, meaning they receive all digits simultaneously for processing.

- **Direct inward dialing (DID) ranges and E.164 addressing:** Solutions for mapping of internal DNs to PSTN numbers include DID, use of attendants to transfer calls, and extensions added to PSTN numbers in variable-length numbering plans.

- **Different number presentation in ISDN (type of number [TON]):** Digit manipulation based on TON enables the standardization of numbers signaled using different TONs.

- **Least-cost routing, toll bypass, tail-end hop-off (TEHO), and PSTN backup:** Can be implemented by appropriate call routing and path selection based on priorities.

- **Globalized call routing:** In this dial plan implementation, all received calls are normalized toward a standardized format. The format that is used in call routing is globalized format, because all numbers are represented in +E.164 format. The process of normalizing the numbers as dialed by the end users (localized ingress) is therefore referred to as *globalization*. Once the localized input has been globalized during ingress, the call is routed based on the globalized number. After call routing and path selection, the call is localized during egress, depending on the selected egress device. The concept of globalization and +E.164 format of numbers is usually familiar to non-U.S. readers; the United States has traditionally been slow in adopting this standard. Chapter 3, "Overview of PSTN and Intersite Connectivity Options," focuses entirely on the concept of globalization and +E.164 addresses.

NAT and Security Solutions

When CUCM servers and IP phones need to connect via the Internet, CUBE can be used as an application proxy. When used in this way, CUBE splits off-net calls inside the CUCM cluster and outside the cluster in the PSTN into two separate call legs. CUBE also features signaling interworking from SIP to SIP, SIP to H.323, H.323 to SIP, and H.323 to H.323.

> **Note**   Cisco Unified Border Element (CUBE) in older manuscripts and documentation used to be called Cisco IP to IP Gateway.

The CUBE can function in two modes:

- **Flow-around:** In this mode, only signaling is intercepted by CUBE. Media exchange occurs directly between endpoints (and flows around CUBE). Only signaling devices (CUCM) are hidden from the outside.

- **Flow-through:** In this mode, signaling and media streams are both intercepted by CUBE (flowing through CUBE). Both CUCM and IP phones are hidden from the outside.

In flow-through mode, only CUBE needs to have a public IP address, so NAT and security issues for internal devices (CUCM servers and IP phones) are solved. Because CUBE is exposed to the outside, it should be hardened against attacks.

CUBE in Flow-Through Mode

In Figure 2-16, CUCM has a private IP address of 10.1.1.1, and the Cisco IP phone has a private IP address of 10.2.1.5 with a subnet mask of 255.0.0.0. A CUBE connects the CUCM cluster to the outside world (in this case, to an Internet telephony service provider [ITSP]). The CUBE is configured in flow-through mode and uses an internal private IP address of 10.3.1.1 and an external public IP address of A.

When CUCM wants to signal calls to the ITSP, it does not send the packets to the IP address of the ITSP (IP address B). Instead, it sends them to the internal IP address of the CUBE (10.3.1.1) via a SIP trunk configuration. CUBE then establishes a second call leg to the ITSP using its public IP address A as the source and IP address B (ITSP) as the destination. As soon as the call is set up, the CUBE terminates the RTP stream toward the ITSP using its public IP address and sends the received RTP packets to the internal IP phone using its internal IP address.

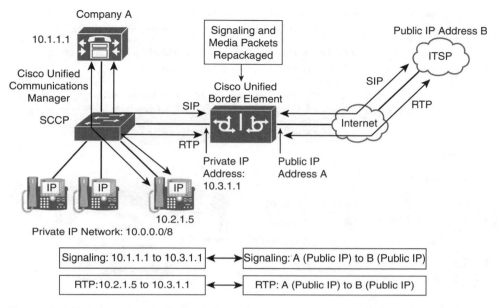

Figure 2-16 *CUBE in Flow-Through Mode*

This solution allows CUCM and IP phones to communicate only with the internal, private IP address of the CUBE. The only IP address visible to the ITSP or anyone sniffing traffic on the outside is the public IP address of CUBE.

Cisco Expressway C and Cisco Expressway E As a Solution to NAT and Security Issues in a Multisite Environment

Cisco Expressway is another solution to NAT and security issues in Cisco Unified Communications Manager deployments.

Cisco Expressway provides edge services in a CUCM deployment. These services include firewall traversal and remote access. Firewall traversal solves NAT issues that usually prevent a remote endpoint from directly exchanging media with another endpoint when PAT is used between the two endpoints.

Remote access provides secure access to remote endpoints without the need for a VPN. Similar to the Cisco Phone Proxy feature (a feature involving the Cisco ASA), Expressway acts as a proxy and connects to the remote endpoint via secure protocols (like TLS and SRTP) while the connection between Cisco Expressway and CUCM does not have to be encrypted.

A Cisco Expressway solution consists of two Cisco Expressway devices:

■ **Cisco Expressway E:** This server faces the untrusted outside network. It is located on a network segment that is close to the untrusted network. In most cases, this network segment is a demilitarized zone (DMZ), so there is one firewall (or a firewall ruleset) between Cisco Expressway E and the untrusted network.

- **Cisco Expressway C:** This server faces the inside network. It is located on an intermediate or internal network, typically a server network.

The set of two separate servers allows well-defined traffic patterns and tight access control (which device is allowed to send which kind of traffic to which other device).

Figure 2-17 illustrates a typical Cisco Expressway deployment.

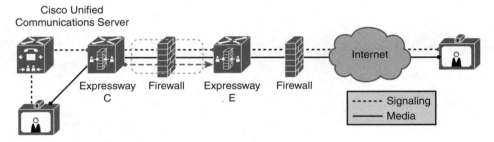

Figure 2-17   *Cisco Expressway Deployment Example*

The figure illustrates how Cisco Expressway E and Cisco Expressway C work together when providing remote access services to endpoints that are located in an untrusted network such as the Internet.

The remote endpoint connects to Cisco Expressway E, and Cisco Expressway E connects to Cisco Expressway C via a traversal zone. It is important to note that there is a firewall component and DMZ isolating the public and private networks, thus Expressway C and E. Cisco Expressway C connects to CUCM. An outside firewall is deployed between the Internet and Cisco Expressway E, and an internal firewall is deployed between Cisco Expressway E and Cisco Expressway C.

Summary

The following key points were discussed in this chapter:

- QoS solves voice and video quality issues in a multisite deployment by prioritizing voice and video above traditional data traffic. Packets are queued on a router or switch during times of congestion.

- Low-bandwidth codecs and RTP header compression can be utilized to reduce IP WAN bandwidth consumption for slow-speed links.

- Local conference bridges, transcoders, mixed conference bridges, call admission control, and multicast MOH can reduce IP WAN bandwidth consumption.

- In the event of an IP WAN failure, the PSTN can act as a backup path.

- Survivability for remote MGCP gateways.

- SRST can provide survivability to remote phones if the IP WAN has a failure. This can be used in conjunction with the default application and SRST on an MGCP gateway. CFUR can be used to forward calls to alternate numbers if the IP phones become unregistered.

- Cisco Expressway and Cisco Unified Border Element can provide NAT and security solutions for UC environments that support remote or mobile users.

References

For additional information, refer to the following:

Cisco Systems, Inc. Cisco Collaboration Systems 10.x Solution Reference Network Designs (SRND), May 2014. http://www.cisco.com/c/en/us/td/docs/voice_ip_comm/cucm/srnd/collab10/collab10.html

Review Questions

Use these questions to review what you have learned in this chapter. The answers appear in Appendix A, "Answers Appendix."

1. **Which of the following best describes QoS as a solution for multisite environments?**

   a. Ensures reliable PSTN calls

   b. Ensures that all forms of IP traffic receive excellent performance

   c. Ensures excellent data performance

   d. Ensures that selected traffic such as RTP audio traffic receives excellent performance at the expense of lower-priority traffic

2. **Which three of the following statements about bandwidth solutions in a multisite deployment are true? (Choose three.)**

   a. RTP header compression compresses the RTP header to 1 byte.

   b. WAN bandwidth can be conserved by using low-bandwidth codecs within a remote site.

   c. WAN bandwidth can be conserved by deploying local media resources.

   d. Voice payload compression is part of RTP header compression.

   e. Multicast MOH from branch router flash eliminates the need to send MOH over the WAN.

3. Which two of the following statements about availability are true? (Choose two.)

   a. CFNB is required to enable main site phones to call remote site phones during SRST fallback.

   b. SRST provides fallback for Cisco IP phones in the event of a WAN failure.

   c. MGCP fallback allows the gateway to use local dial peers when the call agent cannot be reached.

   d. AAR is required to enable phones to reroute calls to another CUCM cluster over the PSTN when the IP WAN is down.

   e. MGCP fallback and SRST cannot be implemented at the same device.

4. Which two of the following are not relevant dial plan solutions for multisite CUCM deployments? (Choose two.)

   a. Access and site codes

   b. TEHO

   c. PSTN backup

   d. Shared lines

   e. Overlap signaling

5. Which Cisco IOS feature provides a signaling and media proxy function that addresses the need for NAT?

   a. Cisco Unified Border Element

   b. Cisco PIX Firewall

   c. CUCM

   d. Cisco ASA

6. Which of the following most accurately describes the difference between flow-around and flow-through modes for CUBE?

   a. Signaling is intercepted in both modes, but media is intercepted only in flow through.

   b. Both modes intercept media and signaling.

   c. Both modes intercept only media.

   d. Both modes intercept only signaling.

7. Which of the following two protocol conversions are supported by CUBE? (Choose two.)

   a. SCCP to SIP

   b. H.323 to SIP

c. SIP to H.323

d. SIP to MGCP

e. MGCP to SCCP

8. **What is the best description of MGCP fallback?**

   a. If CUCM fails, MGCP fails over to SCCP dial peers for PSTN dialing.

   b. IF CUCM fails, MGCP fails over to MGCP dial peers for PSTN dialing.

   c. IF CUCM fails, MGCP fails over to H.323 dial peers for PSTN dialing.

   d. IF CUCM fails, all PSTN dialing fails.

9. **What is the best benefit of multicast for MOH in a branch router?**

   a. Multiple MOH files can be used in router flash in an SRST configuration.

   b. Router flash is not needed for MOH files, because the branch phones send multicast MOH to each other.

   c. MOH has only one stream of RTP to all listeners with multicast.

   d. MOH has several streams of RTP to all listeners to ensure optimal voice quality.

10. **What is the best benefit of standardizing on E.164 numbering?**

   a. E.164 only requires five digits.

   b. E.164 provides a standard of numbering that aligns with an international dial plan.

   c. E.164 numbering supports the + symbol, which can be dialed directly on Cisco IP phones.

   d. E.164 is mandatory for TEHO in CUCM v8.

Chapter 3

Overview of PSTN and Intersite Connectivity Options

Cisco Unified Communications Manager (CUCM) multisite deployments can use a vast range of connection options to the public switched telephone network (PSTN) and various wide-area network (WAN) technologies for connections between sites. At the turn of the century, the majority of U.S.-based companies were using primary rate interface (PRI ISDN) circuits and some flavor or WAN technologies such as Frame Relay, Asynchronous Transfer Mode (ATM), T1, and so on.

Fast forward just a few short years and Session Initiation Protocol (SIP) adoption has grown multifold, so have alternative WAN technologies that offer high-speed WAN connections between sites. With the increase in WAN speed, many organizations are able to take advantage of video and immersive video capabilities such as Cisco TelePresence, video endpoints, and Jabber.

Any time multiple sites are configured for a Unified Communications (UC) solution, several key issues must be addressed in the planning and design phase of the implementation. Multisite dial plans must address special issues such as overlapping and nonconsecutive directory numbers (DNs), PSTN connectivity and access, PSTN backup and failover techniques, and tail-end hop-off (TEHO).

This chapter describes multisite connection options and explains how to build multisite dial plans using Cisco Unified Communications Manager (CUCM) and Cisco IOS voice gateways. Although it is impossible to cover every technology in this chapter, the most common techniques, circuit types, and signaling protocols relating to an enterprise production environment are discussed.

Upon completing this chapter, you will be able to meet these objectives:

- Describe multisite connectivity options for the LAN and WAN
- Describe the main steps in the configuration of an H.323 voice gateway
- Describe the main steps in the configuration of an MGCP voice gateway

- Describe the characteristics of an SIP trunk

- Describe the characteristics of an H.323 trunk

- Describe the main steps in the configuration of an H.323 trunk

- Describe the main steps in the configuration of an SIP trunk

- Explain design considerations for multisite dial plans

- Describe how to use site codes for on-net calls

- Describe how to implement PSTN access

- Describe how to prioritize and implement selective PSTN trunks

- Describe how to implement TEHO

- Describe globalized call routing and +E.164 formatting

- Describe globalization of localized call ingress from gateways and trunks

- Describe globalization and localization of call egress to gateways, trunks, IP phones, and video endpoints

- Describe sample scenarios of globalized call routing

Overview of Multisite Connection Options

Multisite environments have several connection options. Figure 3-1 shows a CUCM cluster at the main site, with three connections to other sites.

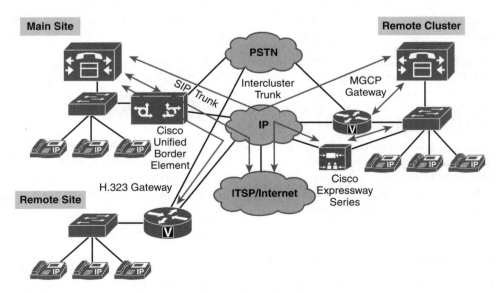

Figure 3-1 *Connection Options for Multisite Deployments*

The connections are as follows:

- Intercluster trunk (H.323) to another CUCM cluster located at a different site.

- An H.323 gateway located at a remote site.

- A Session Initiation Protocol (SIP) trunk connected to an Internet Telephony Service Provider (ITSP) via a Cisco Unified Border Element (CUBE), which also has an SIP trunk to CUCM.

- At the remote location, a Media Gateway Control Protocol (MGCP) gateway is configured to be controlled by the remote cluster CUCM.

- A Cisco Expressway Core and Edge solution located at a remote cluster for remote access for Cisco Jabber and select Video endpoints.

This connection does not require a virtual private network (VPN) solution.

CUCM Connection Options Overview

In CUCM, you can configure gateways and trunks for connections to the PSTN, an ITSP, or other Voice over Internet Protocol (VoIP) domains.

Gateways and phones are configured according to the VoIP protocol that they use. CUCM supports H.323 gateways, Media Gateway Control Protocol (MGCP) gateways, and Skinny Client Control Protocol (SCCP) gateways. Trunks can be configured as H.323 trunks or SIP trunks. The three types of H.323 trunks are as follows:

- H.225 trunk (gatekeeper controlled)

- Intercluster trunk (gatekeeper controlled)

- Intercluster trunk (non-gatekeeper controlled)

Trunks and gateways are configured when connecting to devices that allow access to multiple endpoints, such as an analog phone, Foreign exchange office (FXO) port, T1 channel associated signaling (CAS) line, or Integrated Services Digital Network (ISDN) primary rate interface (PRI). If the destination is a single endpoint, phones are configured. Phones can be configured as SCCP, SIP, or H.323 (although H.323 phones are not as common as SCCP and SIP phones). When CUCM routes calls to a device that uses MGCP, SCCP, or SIP, it is obvious which type of device to add because these protocols can be configured only with either a gateway or a trunk. In the case of H.323, however, an H.323 gateway and an H.323 trunk can be configured, and it is important to know whether to use the gateway or the trunk. Use only H.323 trunks when connecting to another CUCM server (either a cluster or a standalone CUCM server, in the case of CUCM Business Edition) or when using an H.323 gatekeeper. H.323 gateways are configured when connecting to any other H.323 device that is not an endpoint. Such devices can be Cisco IOS H.323 gateways or H.323 gateways of other vendors.

Cisco IOS Gateway Protocol Functions Review

Table 3-1 reviews Cisco IOS gateway (clients) protocol functions by protocol.

As shown in Table 3-1, the three main gateway signaling protocols (MGCP, H.323, and SIP) provide various features and functions when implemented with CUCM and Cisco IOS gateways.

Table 3-1   *Cisco IOS Gateway Protocol Functions Review*

Function	MGCP	H.323	SIP
Clients	Unintelligent	Intelligent	Intelligent
NFAS	Not supported	Supported	Supported
QSIG	Supported	Not supported	Not supported
Fractional T1/E1	More effort to implement	Easy to implement	Easy to implement
Signaling protocol	TCP and UDP	TCP	TCP or UDP
Code basis	ASCII	Binary (ASN.1)	ASCII
Call survivability	No	Yes	Yes
FXO caller ID	Yes*	Yes	Yes
Call applications usable	No	Yes	Yes

*Support introduced with CUCM Version 8.0 and later.

Table 3-2 reviews the advantages and disadvantages of H.323 gateways, MGCP-controlled gateways, and SIP gateways.

Table 3-2   *Cisco IOS Gateway Protocol Comparison Review*

	H.323	MGCP	SIP
Pros	Dial plan directly on the gateway	Centralized dial plan configuration	Dial plan directly on the gateway
	Translations defined per gateway	Centralized gateway configuration	Translations defined per gateway
	Regional requirements that can be met	Simple gateway configuration	Third-party telephony system support
	More specific call routing	Easy implementation	Third-party gateway interoperability
	Advanced fax support	Support of QSIG supplementary services	Third-party end device support
Cons	Complex configuration on the gateway	Extra call routing configuration for survivability	Less feature support

When compared with each other, each of the three gateway protocols has advantages and disadvantages. There is no generally "best" gateway protocol. Select the most appropriate protocol, depending on the individual needs and demands in a production-telephony environment with CUCM. Generally speaking, SIP and H.323 provide robust failover capabilities due to being peer-to-peer technologies, so they do not rely on the CUCM cluster for call processing. However, H.323 and SIP gateways are more complex to configure due to dial plan and dial peers that need to be implemented on the gateway or device for call processing to occur. MGCP is a client/server architecture. If the communications path between the gateway and CUCM fails, call processing will fail.

SIP Trunk Characteristics

SIP uses the distributed call-processing model, so an SIP gateway or proxy has its own local dial plan and performs call processing on its own. A CUCM SIP trunk can connect to Cisco IOS gateways, a CUBE, other CUCM clusters, or an SIP implementation with network servers (such as an SIP proxy).

Figure 3-2 shows some examples of SIP trunks with CUCM clusters, IOS gateways, and an ITSP.

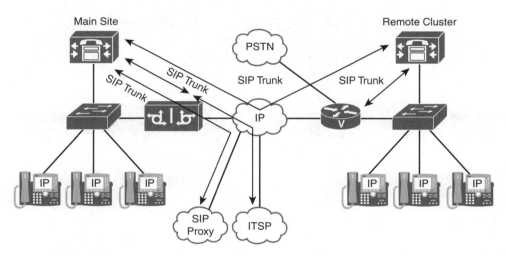

Figure 3-2 *SIP Trunk Examples*

SIP is a simple, customizable, and extensible signaling protocol with a rapidly evolving feature set. In recent years, the SIP protocol has become the preferred signaling protocol due to its rich feature set and multi-vendor interoperability.

Note When you use SIP trunks, media termination points (MTPs) might be required if the endpoints cannot agree on a common method of dual-tone multifrequency (DTMF) exchange or SIP protocol standards. MTPs can be software provisioned on CUCM or hardware provisioned from DSPs on gateways.

H.323 Trunk Overview

H.323 trunks come in three different and very distinct types:

- Non-gatekeeper-controlled ICT
- Gatekeeper-controlled ICT
- H.225 trunks

Figure 3-3 illustrates the various types of H.323 trunks.

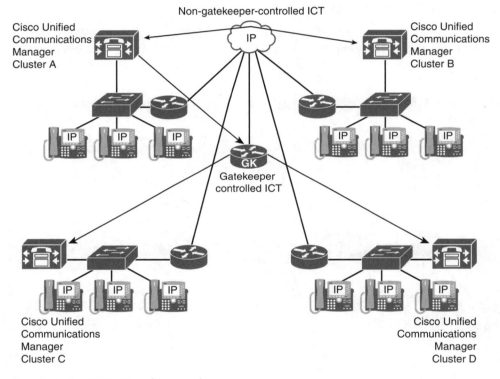

Figure 3-3 *H.323 Trunk Examples*

In Figure 3-3, the CUCM Cluster A uses a non-gatekeeper-controlled intercluster trunk (ICT) to CUCM Cluster B. In addition, CUCM Cluster A is configured with a gatekeeper-controlled ICT. The gatekeeper is a Cisco IOS router labeled GK in the figure. The gatekeeper-controlled ICT points to a gatekeeper, which is used for address resolution and potentially call admission control (CAC). In this example, the gatekeeper can route calls between CUCM Cluster A, C, and D.

Note   Hookflash transfer with H.323 gateways is possible via a Toolkit Command Language (Tcl) script, which you can download from Cisco.com.

Table 3-3 compares the characteristics of the three available H.323 trunk types in CUCM.

Table 3-3   *H.323 Trunk Comparison*

	Non-Gatekeeper-Controlled ICT	Gatekeeper-Controlled ICT	H.225 Trunk
IP address resolution	IP address specified in trunk configuration	IP address resolved by H.323 RAS (gatekeeper)	
Gatekeeper call admission	No	Yes, by H.323 RAS (gatekeeper)	
Scalability	Limited	Scalable	
Peer	CUCM	Before Cisco CallManager 3.2	Cisco CallManager Version 3.2 or later and all other H.323 devices

The non-gatekeeper-controlled ICT is the simplest because it does not use a gatekeeper. It requires the IP address of the remote CUCM server or servers to be specified because the dialed number is not resolved to an IP address by a gatekeeper. CAC can be implemented by locations but not by gatekeeper CAC. Scalability is limited because no address resolution is used and all IP addresses have to be configured manually. The non-gatekeeper-controlled ICT points to the CUCM server or servers of the other cluster.

You may define up to three remote CUCM servers in the same destination cluster. The trunk automatically load balances across all defined remote CUCM servers. In the remote cluster, it is important to configure a corresponding ICT (non-gatekeeper-controlled) that has a CUCM group containing the same servers that were defined as remote CUCM servers in the first cluster. A similar configuration is required in each CUCM cluster that is connected by the ICTs.

For a larger number of CUCM clusters, the gatekeeper-controlled ICT should be used instead of the non-gatekeeper-controlled trunk. The advantages of using the gatekeeper-controlled trunk are mainly the overall administration of the cluster and failover times. Non-gatekeeper-controlled trunks generally require that a full mesh of trunks be configured, which can become an administrative burden as the number of clusters increases. In addition, if a subscriber server in a cluster becomes unreachable, a 5-second (the default) timeout occurs while the call is attempted. If an entire cluster is unreachable, the number of attempts before either call failure or rerouting over the public switched telephone network (PSTN) depends on the number of remote servers defined for the trunk and on the number of trunks in the route list or route group. If many remote servers and many non-gatekeeper-controlled trunks exist, the call delay can become excessive.

With a gatekeeper-controlled ICT, you configure only one trunk that communicates via the gatekeeper with all other clusters that are registered to the gatekeeper. If a cluster or subscriber becomes unreachable, the gatekeeper automatically directs the call to another subscriber in the cluster or rejects the call if no other possibilities exist. This allows the call to be rerouted over the PSTN (if required) with little incurred delay. With a single Cisco gatekeeper, it is possible to have 100 clusters that each registers a single trunk to the gatekeeper, with all clusters being able to call each other through the gatekeeper. Of course, in an enormous enterprise environment with 100 clusters, multiple gatekeepers configured as a gatekeeper cluster eliminates the single point of failure. With non-gate-keeper-controlled ICTs, this same topology would require 99 trunks to be configured in each cluster. The formula for full-mesh connections is $N(N + 1)/2$. Therefore, without the gatekeeper, 100 clusters would require 4950 total trunks for complete intercluster connectivity. The gatekeeper-controlled ICT should be used to communicate only with other CUCMs because the use of this trunk with other H.323 devices might cause problems with supplementary services. In addition, a gatekeeper-controlled ICT must be used for backward compatibility with CUCM versions earlier than Release 3.2 (referred to as Cisco CallManager).

The H.225 trunk is essentially the same as the gatekeeper-controlled ICT, except that it can work with CUCM clusters (release 3.2 and later). It also can work with other H.323 devices, such as Cisco IOS gateways (including CUCM Express), conferencing systems, and clients. This capability is achieved through a discovery mechanism on a call-by-call basis. This type of trunk is the recommended H.323 trunk if all CUCM clusters are at least Release 3.2.

Trunk Implementation Overview

Figure 3-4 illustrates non-gatekeeper-controlled ICTs and SIP trunks.

Figure 3-4 illustrates the most common elements for implementing an SIP or non-gatekeeper-controlled ICT in CUCM. These elements are for the configuration of the trunk itself, in which you have to specify the following:

- IP address of the peer
- The route group
- Route list
- Route pattern configuration

This implementation is like that of a gateway.

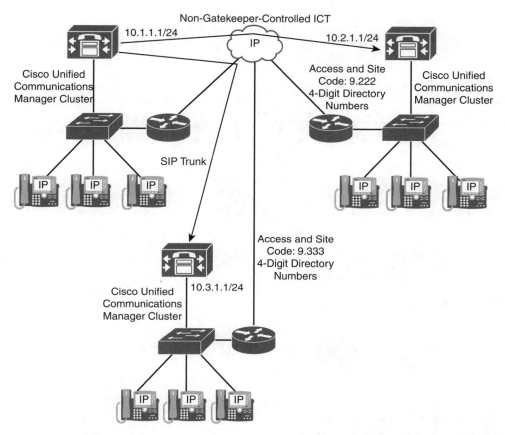

Figure 3-4 *Non-Gatekeeper-Controlled ICT and SIP Trunk Topology*

Gatekeeper-Controlled ICT and H.225 Trunk Configuration

Figure 3-5 illustrates the most common configuration elements when implementing a gatekeeper-controlled ICT or H.225 trunk in CUCM. These elements include the gatekeeper-controlled ICT as well as gatekeeper inside CUCM. Both of these components are logical connections inside CUCM that point to the underlying gatekeeper, which resides inside an ISR router.

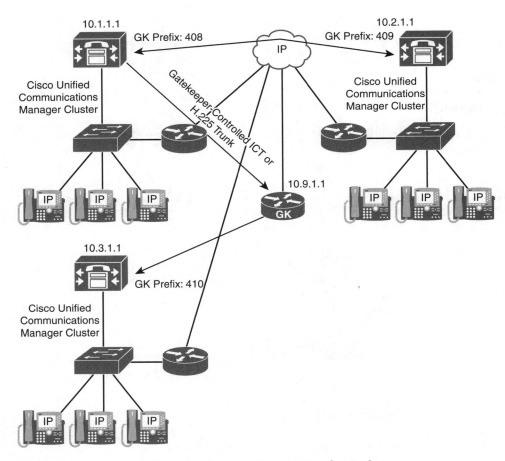

Figure 3-5   *Gatekeeper-Controlled ICT and H.225 Trunk Topology*

The required items are the configuration of the gatekeeper with its IP address and the gatekeeper-controlled ICT that points to the gatekeeper. CUCM also needs the route group connected to the gateway, route list, and route pattern configuration, similar to Figure 3-4.

Trunk Types Used by Special Applications

Some applications require special trunk types to be configured.

For example, when implementing Cisco extension mobility cross clusters (EMCC), a dedicated trunk has to be configured between the CUCM clusters that allow users of the remote cluster to log in locally using Cisco EMCC. These trunks, which are exclusively configured for Cisco EMCC, must use SIP; H.323 is not supported by Cisco EMCC.

Another application that requires special trunks to be configured is Call Control Discovery (CCD). When you use CCD, internal directory numbers (DNs) and the associated external PSTN numbers are advertised and learned from a Service Advertisement Framework (SAF)-enabled network. These trunks can be either SIP or H.323 and must be explicitly enabled for SAF.

Cisco Video Communications Server (VCS) is another application that may require a dedicated SIP trunk between the CUCM cluster and the VCS cluster with special SIP profile characteristics to allow SIP protocol interoperability between the disparate systems.

Inter-cluster Lookup Service (ILS) is another application that requires a special trunk to be configured between Cisco CUCM clusters.

> **Note**   Chapter 15, "Cisco Inter-Cluster Lookup Service (ILS) and Global Dial Plan Replication (GDPR)," and Chapter 16, "Cisco Service Advertisement Framework (SAF) and Call Control Discovery (CCD)," provide you with more information regarding Cisco ILS and SAF trunks.

Dial Plan Requirements for Multisite Deployments with Distributed Call Processing

Dial plan requirements for multisite deployments with distributed call processing are similar to those with centralized call processing. In multisite environments with distributed call processing, use the following dial plan solutions:

- Access and site codes
- Implementing PSTN access
- Implementing gateway and IP phone backup
- Scalability

Table 3-4 compares these topics in both centralized and distributed multisite deployments.

Table 3-4   *Dial Plan Requirements for Multisite Deployments*

Dial Plan Requirements	Multisite Deployments with Centralized Call Processing	Multisite Deployments with Distributed Call Processing
Access and site codes	Allows routing independent of directory number Solves overlapping and nonconsecutive DN ranges	
PSTN access	Gateways selected based on local route group	Single gateway or set of gateways per cluster
	For TEHO, gateways selected per destination with local gateway as a backup	For TEHO, ICT selected per destination with local gateways as backup
Gateway and IP phone backup	MGCP fallback for remote MGCP gateways	
	SRST for remote IP phones	
	CFUR for phones at remote sites	
Scalability		ILS, GDPR, SAF, and CCD

By adding an access code and a site code to DNs of remote locations, you can provide call routing based on the site code instead of the DNs. As a result, DNs do not have to be globally unique, although they must be unique within a site. Configuration elements include route patterns and translation patterns.

You implement PSTN access within a CUCM cluster by using route patterns, route lists, and route groups. The applicable gateway or gateways are selected by the local route group feature. When implementing TEHO within a CUCM cluster, you use the same dial plan configuration elements. However, you have to configure more route patterns (one per TEHO destination), which makes the configuration more complex. The primary route group refers to the TEHO gateway or gateways. The secondary route group is determined by the local route group feature.

In a multisite CUCM deployment with distributed call processing, each site has its own cluster. In such a deployment, implement PSTN access by configuring a single gateway (or a set of gateways) per cluster. When implementing TEHO in a multicluster deployment, configure ICTs between the clusters. Then you must add a route pattern per TEHO destination in each cluster. The route pattern refers to the corresponding TEHO trunk as the primary path and uses the local route group feature for the backup path.

Survivability features for remote MGCP gateways and remote phones are provided by MGCP fallback and Cisco Unified survivable remote site telephony (Cisco Unified SRST) or CUCM Express in SRST mode, and Call Forward Unregistered (CFUR).

In large distributed call-processing deployments, Global Dial Plan Replication (GDPR), an application that uses the ILS, and CCD, and application that uses Cisco SAF, can be used to simplify dial plan implementation.

Implementing Site Codes for On-Net Calls

When designing the dial plan for a multisite implementation, several key decisions about scalability must be made, including the following:

- How will users dial between sites?

- How will users handle overlapping directory numbers at more than one site?

- How will users handle non-consecutive number ranges or directory numbers at each site?

Each question presents a challenging design opportunity but can be easily solved by implementing site and access codes. Figure 3-6 demonstrates how site codes and access codes can be used.

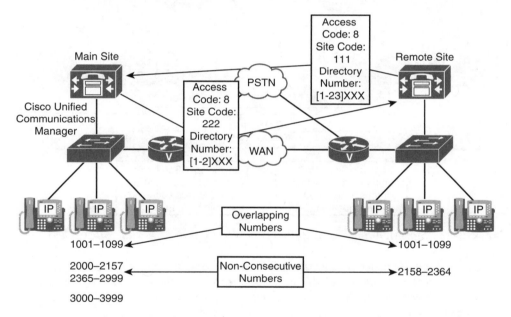

Figure 3-6 *Access and Site Codes Solve Issues with DNs Used at Different Sites*

In Figure 3-6, two sites have overlapping and nonconsecutive directory numbers. To accommodate unique addressing of all endpoints, site-code dialing is used. Users dial an access code (8, in this example), followed by a three-digit site code. A UC engineer can choose another site code access code as well. When calling the phone with DN 1001 at the remote site, a user located at the main site has to dial 82221001. For calls in the other direction, remote users dial 81111001. When distributed call processing is used, each CUCM cluster is only aware of its own DNs in detail. For all DNs located at the other site, the call is routed to a CUCM server at the other site based on the dialed site code.

Digit-Manipulation Requirements When Using Access and Site Codes

When you are using site codes in multisite environments with distributed call processing, as shown in Figure 3-7, call processors must add or strip access and site codes from the called and calling numbers as needed. The dial plan is designed so that both the called-party and calling-party numbers include access and site codes when they are sent over a trunk.

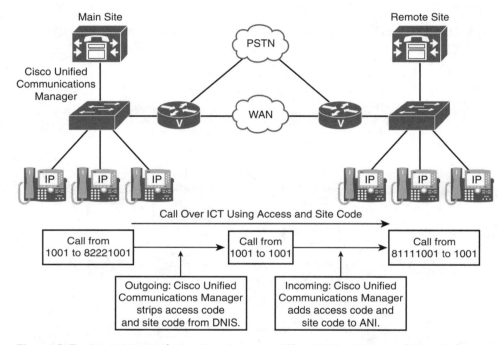

Figure 3-7 *Digit-Manipulation Requirements When Using Access and Site Codes*

If access and site codes are configured before the . (dot) in the route pattern, they can be easily stripped by using the discard digit instruction on the route pattern or route list. For incoming calls, add the access code and appropriate site code that are used to get to the caller's site.

Implement such a dial plan by adding the access and site codes to the calling-party number at the originating cluster on the outbound trunk. At the receiving cluster, on the inbound trunk, strip the access and site codes from the called-party number.

Regarding call routing, you can either configure a route pattern per destination site, or you can use GPDR or CCD to advertise the patterns.

Note   The automatic number identification (ANI) can also be properly manipulated on the outgoing CUCM servers at the main site.

Note   If the WAN link is not functional, additional digit manipulation is required to route the call through the PSTN because the PSTN does not understand the site code.

Access and Site Code Requirements for Centralized Call-Processing Deployments

When situations arise where two sites have overlapping directory numbers in separate partitions, it is possible to allow users to dial in between sites using site codes and access codes. A user would dial an access code and a site code, which would in turn match a translation pattern.

If overlapping DNs exist in a centralized call-processing deployment, access and site codes can be implemented to allow DN overlap as shown in Figure 3-8.

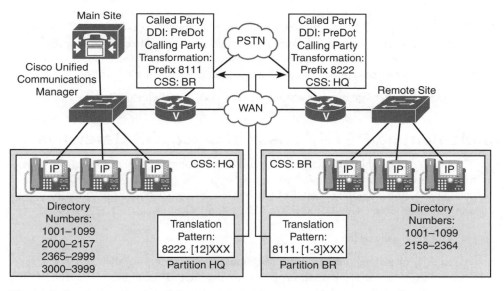

Figure 3-8   *Centralized Call-Processing Deployments: Access and Site Codes*

Figure 3-8 shows two sites with centralized call processing. DNs in the main site (headquarters, or HQ in the figure) and the remote site (branch, or BR in the figure) partially overlap. Access and site codes are used to solve the problem of overlapping DNs.

In Figure 3-8, partitions and CSSs can be deployed so that phones at the remote site do not see the DNs of phones that are located at the main site, and vice versa. Then a translation pattern is added per site.

The translation pattern of each site includes the access and site codes of the respective site. Phones at each site have a CSS assigned, which provides access to the DNs of the local site and the translation pattern for the other site or sites. The translation patterns are configured with a transformation mask that strips off the access code and site code. Further, each translation pattern must have a CSS, which provides access to only those DNs that are located at the target site of the respective translation pattern. This way, all phones can dial local DNs and site-code translation patterns for accessing other sites. After a user dials an intersite number (composed of the access code, site code, and DN), the corresponding translation pattern is matched. The translation pattern strips the site code and access code so that only the DN remains. This DN is matched again in the call routing table using a CSS that has access only to the DNs of the site, which were identified by the site code.

An alternative to a solution with translation patterns is the use of the enterprise alternative number that can be configured at the Directory Number Configuration page. As indicated in the name, the enterprise alternate number is an alternate number for the configured DN in the format that is used for enterprise-wide dialing. An example of Figure 3-8 is when intersite dialing is done by dialing 8, followed by a three-digit site code, followed by the four-digit DN, you need to configure an enterprise alternative number mask of 8111XXXX at all HQ DNs and an enterprise alternate number mask of 8222XXXX at all BR DNs. The resulting enterprise alternate numbers can be seen like an alias for the DN. Although you must add the DNs to a site-specific partition, make the enterprise alternate number reachable to all phones by adding it into an enterprise-wide partition.

The enterprise alternate number was introduced in CUCM Release 10.

Implementing PSTN Access in Cisco IOS Gateways

When you implement PSTN access in a multisite environment, you must perform digit manipulation, described in the following list, before the call is sent to the PSTN. Digit manipulation must be done in CUCM when you use an MGCP gateway. It can be performed either in CUCM or at the H.323 gateway when using an H.323 gateway:

- Outgoing calls to the PSTN:

  - **ANI or calling number transformation:** If no direct inward dialing (DID) range is used at the PSTN, transform all DNs to a single PSTN number in the ANI. If DID is used, extend the DNs to a full PSTN number.

  - **Dialed Number Identification Service (DNIS) or called number transformation:** Strip the access code.

- Incoming calls from the PSTN:

  - **ANI or calling number transformation:** Transform ANI into the full number (considering type of number [TON]), and add the access code so that users can easily redial the number.

  - **DNIS or called number transformation:** If DID is used, strip the office code, area code, and country code (if present) to get to the DN. If DID is not used, route the call to an attendant, such as a receptionist or an interactive voice response (IVR) application.

PSTN Access Example

When placing calls or receiving calls from the PSTN you must be concerned with digit manipulation performed both on the calling and called numbers. Figure 3-9 shows an example of digit manipulation performed for both incoming and outgoing PSTN calls.

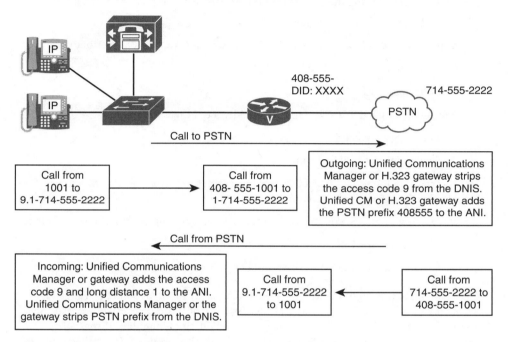

Figure 3-9 *PSTN Number Format and Access Requirements*

As shown in Figure 3-9, internal numbers have to be represented as valid PSTN numbers, and PSTN numbers should be shown with access code 9 internally. Recall from Chapter 1, "Cisco Collaboration Solution Multisite Deployment Considerations," that the ANI is the number calling from, and the DNIS is the number calling to.

Note Adding the access code (and changing 10-digit PSTN numbers to 11-digit PSTN numbers, including the long-distance 1 digit) to the ANI of incoming calls is not required. Adding it, however, allows users to call back the number from call lists (such as received calls or missed calls) without having to edit the number by adding the required access code.

Transformation of Incoming Calls Using ISDN TON

On ISDN circuits a special element can be included in the signaling and call setup messages called type of number (TON). The ISDN TON can specify the format of a number, such as how ANI or DNIS is represented for calls to and from the PSTN. TONs can also be used by CUCM for digit-manipulation purposes. TONs are especially useful in the concept of globalized dial plan and +E.164 dial plans. To achieve a globalized dial plan, TONs can be used to read an incoming number and append or strip digits based on the dial plan's needs. To have a unique, standardized way to represent PSTN numbers in CUCM, the numbers have to be transformed based on the TON.

U.S. TONs in ISDN setup messages provides information about called and calling number in the following formats:

- **Subscriber:** Subscriber TONs are represented by a seven-digit subscriber number:
  - Three-digit exchange code
  - Four-digit station code
- **National:** National TONs are represented by a ten-digit number:
  - Three-digit area code
  - Seven-digit subscriber number
- **International:** International TONs are represented by a variable length number (11 digits or more for U.S. numbers):
  - One digit for U.S. country code; one, two, or three digits for all other countries
  - Three digits for U.S. area code
  - Seven digits for U.S. subscriber number

For example, in the United States, if the calling number of an incoming PSTN call is received with a TON subscriber, a PSTN access code of 9 can be prefixed so that the user can place a callback on his IP phone without editing the number. If the calling number is in national format, a PSTN access code of 91 is prefixed. If a calling number is received with an international TON, the PSTN access code 9011 is prefixed.

In countries with fixed-length numbering plans, such as the United States and Canada, transforming the numbers is not required because users can identify the type of calling number that is based on the length. In this case, users can manually prefix the necessary access codes. The United States and Canada use the North American Number Plan (NANP), which provides a fixed-length number of digits for certain types of calls.

In countries with variable-length numbering plans, such as Great Britain, it can be impossible to identify whether the call was received from the local area code, from another area code of the same country, or from another country by just looking at the number itself. In such cases, the calling numbers of incoming PSTN calls have to be transformed based on the TON.

Note The NANP applies to the United States, Canada, and several Caribbean nations, as described at http://www.nanpa.com.

ISDN TON Example: Calling Number Transformation of Incoming Call

On ISDN-based circuits such as T1 E1 PRI and BRI, PSTN providers can deliver information or metadata based on a called or calling number in the form of a type of number (TON).

Figure 3-10 shows an example of performing TON-based digit manipulation based on the incoming call's ANI.

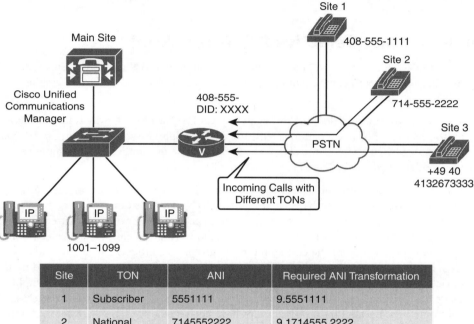

Site	TON	ANI	Required ANI Transformation
1	Subscriber	5551111	9.5551111
2	National	7145552222	9.1714555 2222
3	International	49404132673333	9.01149404132673333

Figure 3-10 *ISDN TON: ANI Transformation of an Incoming Call*

In Figure 3-10, three different calls are received at the main site gateway:

- The first call is received from the same area code (site 1) with a subscriber TON and a seven-digit number. This number only needs to be prefixed with access code 9.

- The second call, received with national TON and ten digits (site 2), is modified by adding access code 9 and the long-distance 1, both of which are required for placing calls back to the source of the call.

- The third call is received from Germany (site 3), with an international TON. For this call, the access codes 9 and 011 have to be added to the received number, which begins with Germany's E.164 country code of 49. Note that 011 is the NANP international access code, which is different for calls originating outside the NANP.

The end result benefits an internal user who receives but misses any calls from these sites and wants to easily call back any of these numbers without editing it from the missed call list.

> **Note**   Figure 3-10 demonstrates the commonly used access code 9 to dial out to the PSTN. It is perfectly acceptable for an organization to choose another access code, such as 8, or no access code. The required ANI transformation digits in this example would be changed accordingly.

Implementing Selective PSTN Breakout

A centralized multisite deployment typically has multiple PSTN gateways, usually one per site. Selective PSTN breakout ensures that local gateways are used to access the PSTN.

There are two ways to select the local gateway for PSTN calls. One way is to configure a site-specific set of route patterns, partitions, CSSs, route lists, and route groups. If you apply a site-specific CSS at the end, a site-specific route group is used. This implementation model was the only one available before CUCM Version 7.

With CUCM Version 7 and later, the local route group feature was introduced. With local route groups, all sites that share the same PSTN dial rules can use one and the same route pattern (or set of route patterns). The route pattern (or set of route patterns) is put into a system-wide route list, which includes the local route group. At the device pool of the calling device, one of the configured route groups is configured to be the standard local route group for this caller. In this model, the route group used is determined by the device pool of the calling device, not by its CSS. The local route group feature simplifies dial plans because it eliminates the need for duplicate CSSs, partitions, route patterns, and route lists. Because local route groups have been introduced, they are the preferred method for local gateway selection because they can significantly reduce the total number of route patterns in a multisite dial plan.

Beginning with CUCM Release 10, more than one local route group is allowed. Assume that you add three local route groups (LRG1 to LRG3) to the standard local route group (which exists by default). At each route list, you can now refer to one of the four local route groups. At each device pool, you can set the local route group for each of the four local route groups. The ability to configure multiple local route groups increases flexibility because you can use the local route group feature in scenarios where some matched patterns should use a different local route group than others.

Configuring IP Phones to Use Local PSTN Gateway

In a multisite deployment, there are typically multiple PSTN gateways, usually one per remote site. Figure 3-11 demonstrates that selective PSTN breakout ensures that local gateways are used to access the PSTN.

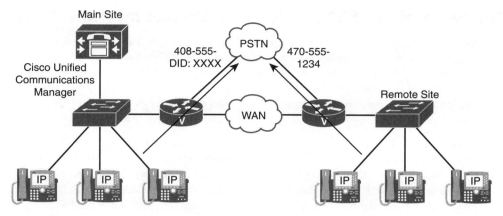

Figure 3-11 *Configure IP Phones to Use a Local PSTN Gateway*

From a dial plan perspective, you can create one 9.@ route pattern (assuming that the NANP is used). This route pattern is in a partition that is part of a global CSS used by all phones. The route pattern refers to a system-wide route list that is configured to use the local route group. At the site-specific device pools, the standard local route group is set to the route group that includes the site-specific gateway.

In Figure 3-11, there is a device pool for the main site and one for the remote site. There is a main site route group, including the main site gateway, and a remote site route group, including the remote site gateway. IP phones at the main site and remote site can now be configured with the same CSS. They will all match the same route pattern and, hence, use the same route list. Based on the local route group feature, however, they will always use their local PSTN gateway for PSTN breakout because the local route group uses the device pool for each location.

> **Note** The local route group is configured with NANP PreDot digit stripping, by default. If the H.323 gateway expects calls that are received from CUCM and that would be routed to the PSTN to include the PSTN prefix 9, appropriate digit manipulation has to be configured in CUCM. In this case, the best solution is to configure the called-party transformation patterns and apply gateway-specific called-party transformation CSS at the gateways in CUCM.

> **Note** If greater control over restricting outbound dialing is required by implementing CSSs and partitions, more specific route patterns should be created instead of those using the generic @ wildcard.

Implementing PSTN Backup for On-Net Intersite Calls

Many organizations have multiple paths that IP telephony calls can traverse, these include both the WAN where the call travels across a private network such as Multiprotocol Label Switching (MPLS) or T1 and even the PSTN. It is important to note that toll charges may exist for call paths across the PSTN. In such cases the ideal first choice is to route the calls via the WAN, then use the PSTN as a backup call routing path.

Figure 3-12 shows a multisite distributed deployment with two sites. Each site has its own CUCM cluster. Intersite calls use the intercluster trunk (ICT) or an SIP trunk over the IP WAN to the other cluster. If the IP WAN link fails for any reason, because both sites have access to the PSTN, the PSTN is used as a backup for intersite calls. Note that proper digit manipulation must take place if the calls are routed through the PSTN.

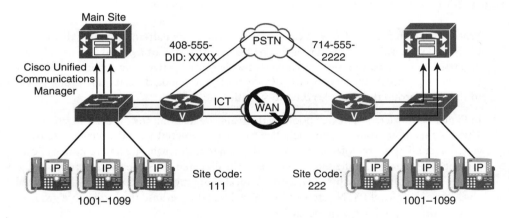

Figure 3-12   *Implementing PSTN Backup for On-Net Intersite Calls*

To ensure that phones at different sites always use their local gateway for PSTN backup, a route list is configured that includes the ICT as the first option and the local route group as the second option. With this approach, there is no need to have multiple, site-specific route lists with a different, site-specific route group as the second entry. An example of this scenario is when the primary and the backup path require the called numbers to be formatted differently, use internal site-code dialing over the ICT and PSTN numbering for the PSTN backup path, the called number must be modified depending on the chosen path.

Digit-Manipulation Requirements for PSTN Backup of On-Net Intersite Calls

PSTN backup for on-net calls can be easily provided by route lists and route groups giving priority to the ICT over the PSTN gateway, as shown in Figure 3-13.

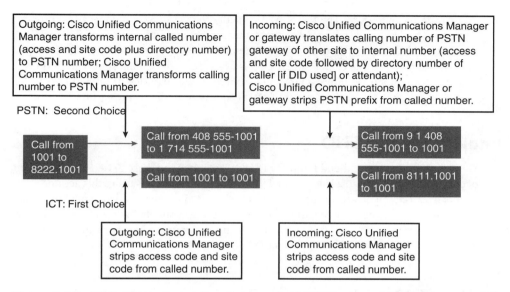

Figure 3-13   *Digit-Manipulation Requirements for PSTN Backup of On-Net Intersite Calls*

When using a PSTN backup for on-net calls, you must address internal versus external dialing. Although on-net calls usually use site codes and DNs, calls sent through the PSTN have to use E.164 numbers. Digit-manipulation requirements vary, depending on the path that is taken for the call:

- Digit-manipulation requirements when using the ICT, which is the first choice in the route list and route group:

    - **At the calling site:** The access and site code are removed from DNIS.

    - **At the receiving site:** The access and site code are added to the ANI. (This can also be done on the calling site.)

- Digit-manipulation requirements when using the PSTN (secondary choice in the route list and route group):

    - **At the calling site:** The internal DNIS comprising an access code, site code, and DN is transformed into the PSTN number of the called phone. The ANI is transformed into the PSTN number of the calling phone.

Note   If DID is not supported, the site's PSTN number is used in DNIS and ANI instead of the IP phone's PSTN number.

When different digit-manipulation configuration is required, depending on the selected path, the digit-manipulation settings are either configured at a path-specific route group or by using global transformations.

■ **At the receiving site:** The PSTN ANI is recognized as a PSTN number of an on-net connected site and is transformed into the internal number: access and site code, followed by the DN of the calling phone (if DID is used at the calling site) or of the attendant of the calling site (if DID is not used at the calling site). The DNIS is transformed into an internal DN and is routed to the IP phone (if DID is used at the receiving site) or to an attendant (if DID is not used at the receiving site).

Implementing TEHO

When you implement tail-end hop-off (TEHO), as shown in Figure 3-14, PSTN breakout occurs at the gateway closest to the dialed PSTN destination. Basically, this action occurs because you create a route pattern for each destination area that can be reached at different costs. These route patterns refer to route lists that include a route group for the TEHO gateway first and the local route group as the second entry so that the local gateway can be used as a backup when the IP WAN cannot be used.

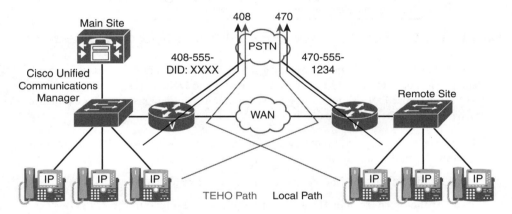

Figure 3-14   *Implementing TEHO*

Using TEHO might not be permitted in your country or by your provider. There can also be issues with emergency calls. Therefore, ensure that your planned deployment complies with the legal requirements of each location.

When using backup TEHO, consider the following potential issues regarding the number you want to use for the calling number of the outgoing call. Basically, there are two options:

■ **Use the PSTN number of the originating site at the TEHO gateway:** When using the PSTN number of the originating device for the caller ID of a TEHO call, the called party is not aware that TEHO has been used. Standard numbering is maintained for all PSTN calls, regardless of the egress gateway; callbacks to the calling number are possible. However, sending calls to the PSTN with PSTN caller IDs of other sites might not be permitted, or the receiving PSTN provider might remove caller IDs from the signaling messages.

Note Sending calls out of a gateway with the calling number of another site might not be permitted in your country or by your provider. There can also be issues with emergency calls. Therefore, ensure that your planned deployment complies with legal requirements.

- **Replace the PSTN number of the originating site by the PSTN number of the TEHO site:** When using the calling number of the backup gateway, called parties may get confused about the number that should be used when calling back. For example, they might update their address books with the different number and inadvertently end up sending calls to the TEHO site every time they call. Furthermore, DID ranges must include remote phones or IVR scripts (automated attendants) to be able to route calls to phones located in any site, regardless of where the PSTN call was received.

Note Using a remote gateway for PSTN access might not be permitted in your country or by your provider. There can also be issues with emergency calls. Therefore, ensure that your planned deployment complies with legal requirements.

In general, it is highly recommended that you use the local route group feature when implementing TEHO. To provide a local backup for TEHO calls, call processing must route all calls differently, based on the source (physical location) and on the dialed number, when the TEHO path cannot be used. When you are not using local route groups, this approach can require a huge number of route patterns, partitions, CSSs, and route lists, resulting in complex dial plans. Such dial plans are difficult to maintain and troubleshoot.

Note You also must consider call admission control (CAC) when implementing TEHO. When the primary (TEHO) path is not admitted as a result of reaching the CAC call limit, calls should be routed through the local gateway. More information about CAC is provided in Chapter 7, "Call Admission Control (CAC) Implementation."

TEHO Example Without Local Route Groups

Traditional dial plans in CUCM can include multiple sites. With this addition the complexity of the dial plan increases due to the number of route patterns, route lists, and route groups required. Nearly all the sites have similar dial plans, which include an emergency, long-distance, and international route pattern, which can lead to duplication of patterns and dial plan bloat.

In Figure 3-15, there are five sites in a centralized call-processing deployment. Each site uses identical call routing policies and numbering plans, but the site-specific details of those policies prevent customers from provisioning a single set of route pattern and route list that works for all sites. This principle applies when no local route groups are used (as it was the case before CUCM Version 7). Although the primary path for a given TEHO PSTN destination is always the same (the appropriate TEHO gateway), the backup path is different for each site (the local gateway of the site where the call has been placed). Without a backup path, TEHO requires only one route pattern per TEHO destination number and refers only to the corresponding TEHO gateway from its route list and route group. However, as the IP WAN is used for TEHO calls, it is not recommended that you only configure a single path. Therefore, TEHO configurations easily end up in huge dial plans where each site requires a different route pattern and route list for each of the other sites. In addition, each site has one generic route pattern for non-TEHO PSTN destinations (using the local gateway).

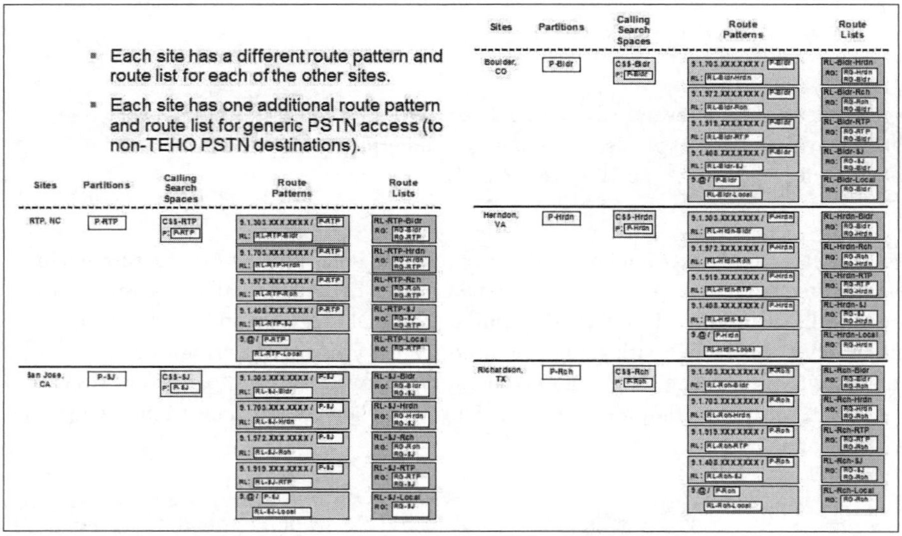

Figure 3-15   *TEHO Without Local Route Groups*

Some route patterns in Figure 3-15 include the . character multiple times (for example, 9.1.703.XXX.XXXX). In this case, the . character illustrates the different components of the number patterns in order to make it easier to interpret the patterns. In reality, the . in route patterns is used only once when being referenced by a corresponding DDI (for example, the PreDot DDI).

Figure 3-16 illustrates the configuration for one site (Boulder). There is a TEHO route pattern for area code 703 (Herndon) that refers to the route list RL-Bldr-Hrdn. This route list uses the Herndon gateway first and the (local) Boulder gateway as a backup. There is also a route pattern for area code 972 (Richardson), again using a dedicated route list for calls from Boulder to Richardson (with the Richardson gateway preferred over the local Boulder gateway). There are two more such constructs for the other two sites. Finally, there is a generic PSTN route pattern (9.@) for all other PSTN (non-TEHO) calls. The generic PSTN route pattern refers to a route list that contains only the local gateway. All five route patterns are in the Boulder partition (P-Bldr) so that they can be accessed only by Boulder phones (using the Boulder CSS CSS-Bldr).

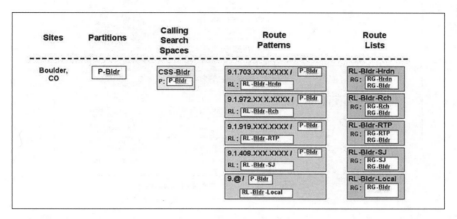

Figure 3-16   *TEHO Without Local Route Groups, Continued*

In summary, when using CUCM v6.x or earlier, for each TEHO destination, a route pattern per originating site refers to a dedicated route list using the appropriate TEHO gateway before the local gateway. For n sites, there are $n * (n + 1)$ of these patterns. In addition, each site has a generic route pattern referring to a dedicated route list containing the local gateway only. This generic route pattern increases the total number of route patterns and route lists to $n * n$. In large TEHO deployments, this approach does not scale.

Some route patterns in the figure include the . character multiple times (for example, 9.1.703.XXX.XXXX). In this case, the . character illustrates the different components of the number patterns in order to make it easier to interpret the patterns. In reality, the . in route patterns is used only once when being referenced by a corresponding DDI (for example, the PreDot DDI).

TEHO Example with Local Route Groups

When using TEHO in conjunction with local route groups, the dial plan can be dramatically simplified. This technique also permits greater scalability and flexibility for selective PSTN breakout. Using local route group and TEHO allows for a reduction in the number of route patterns and route lists.

As shown in Figure 3-17, when implementing TEHO with local route groups, you can reduce the number of route patterns and route lists from $n * n$ to $n + 1$.

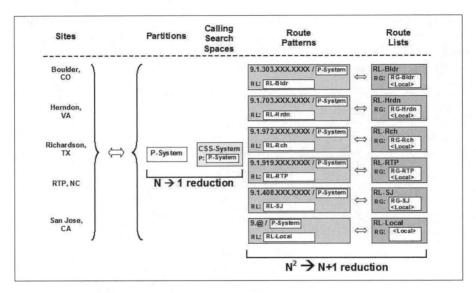

Figure 3-17 *TEHO with Local Route Groups*

> **Note** Some route patterns in the figure include the . character multiple times (for
> example, 9.1.703.XXX.XXXX). In this case, the . character illustrates the different
> components of the number patterns to make it easier to interpret the patterns. In reality,
> the . in route patterns is used only once when being referenced by a corresponding DDI
> (for example, the PreDot DDI).

In Figure 3-17, with five remote sites and one local site, using local route groups
simplifies the dial plan as follows:

- The number of route patterns and route lists for TEHO destinations is reduced from
  $n * (n + 1)$ to n. In this example, the reduction is from 20 to 5.

- The number of route patterns and route lists for non-TEHO destinations is reduced
  from n to 1 (5 to 1, in this example).

- Thus, the total number of route patterns and route lists is reduced from n * n to
  $n + 1$ (25 to 6).

- The number of partitions and CSS is reduced from n to 1 (5 to 1).

- The number of gateways, route groups, and device pools remains the same: n.

Implementing Globalized Call Routing

Many UC engineers tend to overthink the concept of globalized dial plans and +E.164
dialing. This topic is perhaps one of the more complex in all of UC. But if you break it
down into the core building blocks and functions, it can be looked at in smaller, more

manageable individual tasks. At the core, globalized dial plans involved one concept: globalizing ingress or inbound, and localizing egress or outbound. The trick is to always assume you are the CUCM cluster. (That is, view or pretend you are the CUCM when determining what transformations, translations, and digit manipulations are required.)

For example, when a call arrives ingress or inbound from a telco or ITSP, the DNIS and ANI need to be observed and the dialed number and caller ID need to be adjusted or "globalized" ingress or inbound into the CUCM environment. This often involves reading the TON of an ISDN call and appending digits to globalize the call in the +E.164 format. The same concept applies when a user with an IP phone dials a local or long-distance number. CUCM needs to read the digits dialed by the user and globalize that call into the +E.164 format. The globalization of a call can occur using calling-party and called-party transformations, translation patterns, and digit appending on a gateway.

After a call has been globalized in the CUCM cluster, the call will likely route to a gateway, ICT, or SIP trunk based on the dialed digits performed by the user. The key item to remember is that this call is in the +E.164 format, the call is globalized! Many telcos and ITSPs will not accept calls in a globalized format. Therefore, you must localize the call back to a standards-based number format before the call leaves our gateway or trunk exiting to the telco. This concept refers to localization of a call or "localize" egress or outbound. The same concept can be applied to a call that arrived via gateway or trunk and was globalized and then routed to an IP phone or video endpoint. Do you think an average end user wants to see the call display on a phone or video endpoint show the call in a +E.164 format? Most end users are used to how they normally dial traditional numbers from perhaps a cell phone or plain old telephone system (POTS) telephone line. In this case, you need to localize the display or localize egress to display the number in a normal format. You can invoke calling- and called-party transformations, translations, voice translation profiles, and digit appending on a gateway to achieve these techniques.

Globalized call routing and +E.164 format can dramatically simplify UC multisite and multinational dial plan design and implementation. Using the +E.164 format for your dial plan approach allows for easier dialing as staff travels around the globe, results in a smaller number of dial plan elements, and makes adding new sites more of a cookie-cutter approach.

In summary, with globalized call routing, all calls that involve external parties are based on one format with the purpose of simplifying international multisite dial plans. All numbers are normalized as follows:

- **Normalized called-party numbers:** E.164 format with the + prefix is used for external destinations. Therefore, called-number normalization is the result of globalization. Internal DNs are used for internal destinations. Normalization is achieved by stripping or translating the called number to internally used DNs.

- **Normalized calling-party numbers:** E.164 global format is used for all calling-party numbers, except calls from an internal number to another internal number. Such purely internal calls use the internal DN for the calling party number.

> **Note**   Except for the internal calls that were mentioned (where the destination is a DN and, in the case of an internal source, the source is a DN), all numbers are normalized to the E.164 global format. Therefore, call routing based on the normalized numbers is referred to as globalized call routing.

If sources of calls (users at phones, incoming PSTN calls at gateways, calls received through trunks, and so on) do not use the normalized format, the localized call ingress must be normalized before being routed. This requirement applies to all received calls (coming from gateways, trunks, and phones), and it applies to both the calling- and called-party numbers.

After the call is routed and path selection (if applicable) is performed, the egress device typically must change the normalized numbers to the local format. This situation is referred to as *localized call egress*.

Localized call egress applies to these situations:

- **Calling- and called-party numbers for calls that are routed to gateways and trunks:** If the PSTN or the telephony system on the other side of a trunk does not support globalized call routing, the called- and calling-party numbers must be localized from the global format. For example, the called-party number +494012345 is changed to 011494012345 before the call is sent out to the PSTN in the United States.

- **Calling-party numbers for calls that are routed from gateways or trunks to phones:** This situation applies to the phone user who does not want to see caller IDs in a global format. For example, if a user at a U.S. phone wants to see the numbers of PSTN callers who are in the same area code, that user may want to see each number as a seven-digit number and not in the +1XXXXXXXXXX format.

Localized call egress is not needed for the called-party number of calls that are routed to phones, because internal DNs are the standard (normalized) format for internal destinations (regardless of the source of the call). These numbers might have been initially dialed differently, However, in that case, the localized call ingress was normalized before call routing. Localized call egress is also not required for the calling-party number of internal calls (internal to internal) because, again, the standard for the calling-party number of such calls is to use internal DNs.

Globalized call routing simplifies international dial plans because the core call-routing decision is always based on the same format, regardless of how the number was initially dialed and how the number looks at the egress device.

Globalized Call Routing: Number Formats

Globalized call routing involved the globalization and localization of calls as they arrive into and flow out of the environment. It is sometimes referred to as *globalize ingress and localize egress*. This can be broken down into the different digit manipulations that occur for inbound from an IP phone or gateways as well as the digit manipulations for outbound calls to a gateway or IP phone. Figure 3-18 indicates several digit manipulation techniques.

Figure 3-18 describes the number formats that are used by globalized call routing and explains some commonly used expressions.

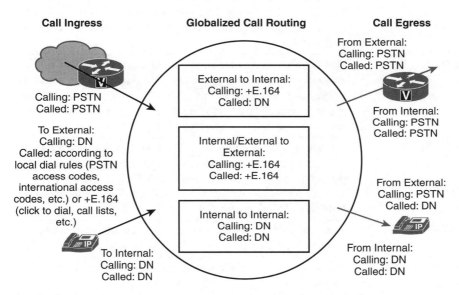

Figure 3-18 *Ingress and Egress Globalized Call Routing Number Formats*

Table 3-5 explains some commonly used expressions to describe globalized call routing, as illustrated in Figure 3-18.

Table 3-5 *Globalized Call Routing Terminology*

Term	Description
Number normalization	The process of changing numbers to a well-defined, standardized (normalized) format. In this case, all external phone numbers are changed to global +E.164 format. The +E.164 indicates an E.164 number with a plus [+] prefix.
Number globalization	The process of changing numbers to global +E.164 format. Example: Because the normalized format is +E.164, you normalize a called number (for example, 4085551234) by globalizing the number; that is, by changing the number to global format (for example, +14085551234).

Table 3-5 *continued*

Term	Description
Number localization	The process of changing from normalized format (in this case, global format) to local format. Usually, the local format is the shortest possible format that does not conceal relevant information. An example of local format is 555-1234 instead of +1 408 555-1234, or 972 333-4444 instead of +1 972 333-4444 (assuming that the device where localization occurs is located in the +1408 area).
Incoming PSTN call	Call from PSTN to internal phone. Like all calls, such a call consists of two call legs (incoming and outgoing). See also call ingress and call egress in this table. On an incoming PSTN call, the incoming call leg (call ingress) is PSTN gateway to CUCM; the outgoing call leg (call egress) is CUCM to internal phone.
Outgoing PSTN call	Call from internal phone to PSTN. Like all calls, such a call consists of two call legs (incoming and outgoing). On an outgoing PSTN call, the incoming call leg (call ingress) is internal phone to CUCM; the outgoing call leg (call egress) is CUCM to PSTN gateway.
Call ingress	This term refers to the incoming call leg, which is the call received by CUCM
Call egress	This term refers to the outgoing call leg, which is the call routed to destination by CUCM
Localized E.164 (number)	PSTN number in partial (subscriber, national, international) E.164 format.
E.164 (number)	PSTN number in complete E.164 format with + prefix.
PSTN Format	This format is the PSTN number in full (international) or partial (Subscriber, National) E.164 format. Depending on the PSTN provider, it may or may not include a + prefix (in the case of full E.164 format). It may also include additional information elements inside the TON.

Note In Cisco IOS gateways, each call has an incoming and an outgoing dial peer, one for each call leg. In CUCM, these call legs are referred to as call ingress and call egress.

On the left side of Figure 3-18, call ingress is illustrated by these two types of call sources:

- **External callers:** Their calls are received by CUCM through a gateway or trunk. In the case of a PSTN gateway, calling- and called-party numbers are usually provided in localized E.164 format.

■ **Internal callers:** Their calls are received from internal phones. In the case of calls to internal destinations (for example, phone to phone), calling- and called-party numbers are typically provided as internal DNs. In the case of calls to external destinations (for example, phone to PSTN), the calling number is the DN (at call ingress time), and the called number depends on the local dial rules for PSTN access. These dial rules can differ significantly for each location.

Note   As mentioned earlier, it is also possible to configure DNs in +E.164 format. In this case, at call ingress the calling-party number of internal calls would be in +E.164 format and the called-party number would be in the format that is used for internal dialing (example, 4-digit dialing).

The center of Figure 3-18 illustrates the standards defined for normalized call routing. As previously mentioned, because most calls use global E.164 format, this type of call routing is also referred to as globalized call routing. Here are the defined standards:

■ **External to internal:**

  ■ **Calling-party number:** E.164

  ■ **Called-party number:** Directory number

Note   If DNs or one of their aliases (enterprise alternate number or +E.164 alternate number) are specified in +E.164 format, then the +E.164 format is also used for the called party.

■ **External to external (if applicable):**

  ■ **Calling-party number:** E.164

  ■ **Called-party number:** E.164

■ **Internal to internal:**

  ■ **Calling-party number:** Directory number

  ■ **Called-party number:** Directory number

Note   Similar to calls from external to internal endpoints, +E.164 format is used if DNs are configured in +E.164 format.

- **Internal to external:**

  - **Calling-party number:** E.164

  - **Called-party number:** E.164

At the right side of Figure 3-18, call egress is illustrated by two types of call targets:

- **Gateways:** When sending calls to the PSTN, localized E.164 format is used for both the calling- and called-party numbers. The format of these numbers (especially of the called-party number) can significantly differ based on the location of the gateway. For example, the international access code in the United States is 011, and in most European countries, it is 00.

- **Phones:** When a call from an internal phone is sent to another internal phone, the call should be received at the phone with both the calling and called number using internal DNs. Because this format is the same format that is used by globalized call routing, there is no need for localized call egress in this case. When a call from an external caller is sent to an internal phone, most users (especially users in the United States) prefer to see the calling number in localized format. (For example, national and local calls should be displayed with ten digits.) The called number is the directory number.

It should be evident from Figure 3-18 that there are several situations where the numbers provided at call ingress do not conform to the normalized format to be used for call routing. These situations also apply to call egress, where the normalized format is not always used when the call is delivered. Therefore, localized call ingress has to be normalized (that is, globalized), and the globalized format has to be localized at call egress.

Normalization of Localized Call Ingress on Gateways

When calls arrive from the PSTN ingress into the UC environment, they are often represented in the local dialing format. The calls arrive with local, long-distance, or international caller ID and are delivered to the gateway with a defined number of dialed digits. Once the call arrives, it must be normalized, meaning that the DNIS or dialed number as well as the caller ID must be normalized into the +E.164 format and then placed into a globalized partition for further processing by CUCM.

Figure 3-19 illustrates how localized call ingress on gateways gets normalized.

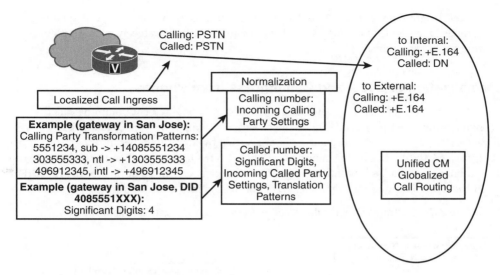

Figure 3-19 *Normalization of Localized Call Ingress on Gateways*

The requirements for normalizing localized call ingress on gateways are as follows:

- Changing the calling number from PSTN format to global E.164 format

- Changing the called number from PSTN format to DNs for calls to internal destinations

- Changing the called number from PSTN format to global E.164 format for calls to external destinations (if applicable)

As shown in Figure 3-19, the calling number can be normalized by incoming calling-party settings. They are configured at the gateway or at the device pool, or they can be configured as CUCM service parameters. Figure 3-19 provides an example for a gateway in San Jose. Assuming that the calling-party number is provided in the shortest-possible format (subscriber, national, international) and with the corresponding type of number information, the relevant calling-party transformation patterns are changed as follows:

- **Prefix for incoming called-party numbers with number type subscriber:** +1408

  - 5551234 marked as type Subscriber TON becomes +14085551234.

- **Prefix for incoming called-party numbers with number type national:** +1

  - 303555333 marked as type International TON becomes +1303555333.

- **Prefix for incoming called-party numbers with number type International:** +

  - 496912345 marked as type International TON becomes +496912345.

The called number can be normalized by significant digits that are configured at the gateway (applicable only if no calls to other external destinations are permitted and a fixed-length number plan is used), translation patterns, or by incoming called-party settings (if available at the ingress device). In Figure 3-19, the gateway is configured with four significant digits.

Normalization of Localized Call Ingress from Phones

When an end user dials a number from their IP phone or video endpoint, the called number must be normalized. This is the process of taking caller input including any access or site codes and normalizing or converting the dialed number into a +E.164 format. After the call has been normalized, it is placed into a global partition that contains all globalized routing patterns. The same can be done to normalize the caller ID at the same time, and thus the caller ID is normalized into a +E.164 format.

Figure 3-20 illustrates how localized call ingress on phones gets normalized.

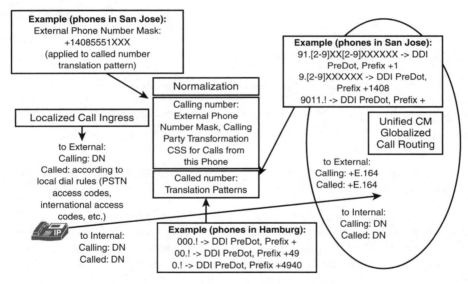

Figure 3-20 *Normalization of Localized Call Ingress from Phones*

The requirements for normalizing localized call ingress on phones are as follows:

- **For calls to external destinations:** Changing the calling number from an internal DN to +E.164 format. Changing the called number to +E.164 format if any other format was used (according to local dial rules).

- **For calls to internal destinations:** No normalization is required unless the destination DN and none of its aliases (enterprise alternate number, +E.164 alternate number) match the number as dialed at the source. An example would be that users must dial access and site codes, but the DNs do not include access and site codes and they do not have aliases configured.

As shown in Figure 3-20, the normalization of the called-party number can be achieved by translation patterns. The normalization of the calling-party number for calls to external destinations can be reached by configuring an external phone number mask (in the +E.164 format) at the phone and applying the external phone number mask to the relevant translation patterns that are used to normalize the called-party number. Another way of normalizing the calling number is to use the calling-party transformation CSS for calls from this phone. Because this transformation CSS is applicable to all calls, it is only useful if the +E.164 format is also used for internal calls. In Figure 3-20, examples for phones that are located in Hamburg (Germany) and San Jose (United States) are given.

Localized Call Egress at Gateways

Globalized +E.164 dial route calls to gateways, trunks, and off network destinations in a globalized format by default. Many times the far-end system such as the telco or PSTN cannot accommodate the DNIS or dialed number in the globalized format, thus requiring you to localize the call as it routes to the gateway. This section discusses the concept of localizing a call egress or outbound as it routes through the gateway destined for the PSTN. Figure 3-21 shows several called- and calling-party transformations as well as DDIs that are in place to achieve this functionality.

Figure 3-21 illustrates how to implement localized call egress at gateways.

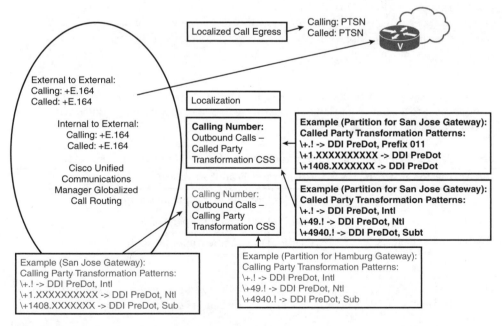

Figure 3-21 *Localized Call Egress at Gateways*

The only requirement is to change the calling and called number from global +E.164 format to the localized PSTN format or the format the PSTN provider expects. You can change the format by configuring called- and calling-party transformation patterns, putting them into partitions, and assigning the appropriate called- and calling-party transformation CSS to gateways. You can configure called- and calling-party transformation CSS at the device (gateway or trunk) and at the device pool.

Table 3-6 shows the configuration of the called-party transformation patterns that are applicable to the San Jose gateway (based on partition and called-party transformation CSS).

Table 3-6   *Called-Party Transformation Patterns for the San Jose Gateway*

Transformation Pattern	Performed Transformation
\+.!	DDI PreDot, Prefix 011
\+.1XXXXXXXXXX	DDI PreDot
\+1408.XXXXXXX	DDI PreDot

Note   In Figure 3-21, the San Jose gateway does not support number types, but it requires the use of national and international access codes. Therefore, 011 must be prefixed on international calls. For national calls, the plus (+) sign and country code (1) must be replaced by the national access code. Even though the country code and the national access code are identical, only by coincidence, the transformation can be simplified by removing only the + sign.

Table 3-7 shows how to configure the called-party transformation patterns that are applicable to a gateway in Hamburg, Germany (based on partition and called-party transformation CSS).

Table 3-7   *Called-Party Transformation Patterns for the Hamburg, Germany Gateway*

Transformation Pattern	Performed Transformation
\+.!	DDI PreDot; number type: international
\+49.!	DDI PreDot; number type: national
\+4940.!	DDI PreDot; number type: subscriber

Note   In Table 3-7, the Hamburg gateway uses number types instead of international (00) or national (0) access codes (in contrast to the San Jose gateway, which does not use number types).

Table 3-8 shows how to configure calling-party transformation patterns that are applicable to the San Jose gateway (based on partition and calling-party transformation CSS).

Table 3-8   *Calling-Party Transformation Patterns for the San Jose Gateway*

Transformation Pattern	Performed Transformation
\+.!	DDI PreDot; number type: international
\+1.XXXXXXXXX	DDI PreDot; number type: national
\+1408.XXXXXXX	DDI PreDot; number type: subscriber

> **Note**   In Table 3-8, subscriber, national, and international number types are used at the San Jose gateway for the calling-party number. If no number types were used, because of the fixed-length numbering plan, the number type could also be determined by its length (seven-digit numbers when the source of the call is local, ten-digit numbers when the source of the call is national, or more than ten digits when the source of the call is international). In reality, however, countries that use the NANP typically use ten-digit caller IDs for both national and local callers.

> **Note**   Having nonlocal calling-party numbers implies the use of TEHO or PSTN backup over the IP WAN. This scenario is not permitted in some countries or by some PSTN providers. Some providers verify that the calling-party number on PSTN calls they receive matches the locally configured PSTN number. If a different PSTN number is set for the caller ID, either the call is rejected or the calling-party number is removed or replaced by the locally assigned PSTN number.

Table 3-9 shows how to configure calling-party transformation patterns that are applicable to a gateway in Hamburg, Germany (based on partition and calling-party transformation CSS).

Table 3-9   *Calling-Party Transformation Patterns for the Hamburg, Germany Gateway*

Transformation Pattern	Performed Transformation
\+.!	DDI PreDot; number type: international
\+49.!	DDI PreDot; number type: national
\+4940.!	DDI PreDot; number type: subscriber

Localized Call Egress at Phones

Another component of +E.164 globalized dial plan deals with the localization of calls as they arrive and are displayed on IP phones and video endpoints. When a call arrives on an endpoint, the call can be displayed in the globalized format, but often end users like seeing calls presented in their locally accepted formats. For example, in the United States, end users prefer to see the caller ID displayed as ten digits for a local call

version in the +E.164 globalized format. Figure 3-22 further illustrates this point and illustrates how to implement localized call egress at phones.

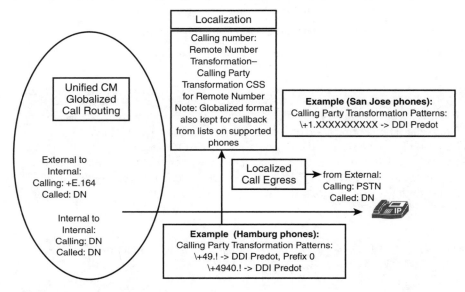

Figure 3-22   *Localized Call Egress at IP phones*

The only requirement is that you change the calling number from global +E.164 format to the preferred format of the end users. You can change the format by configuring calling-party transformation patterns, putting them into partitions, and assigning the appropriate calling-party transformation CSS to IP phones. As previously mentioned, you can configure calling-party transformation CSS at the phone and at the device pool.

Table 3-10 shows how to configure the calling-party transformation patterns that are applicable to a phone that is located in San Jose (based on partition and calling-party transformation CSS), as illustrated in Figure 3-22.

Table 3-10   *Calling-Party Transformation Patterns for San Jose Phones*

Transformation Pattern	Performed Transformation
\+1.XXXXXXXXXX	DDI PreDot

In Table 3-10, international calls are shown in standard normalized format (E.164 format with + prefix), because there is no \+! calling-party transformation pattern. National calls are shown with ten-digit caller IDs, and local calls are shown with seven-digit caller IDs.

Table 3-11 shows how to configure the calling-party transformation patterns that are applicable to a phone located in Hamburg, Germany (based on partition and calling-party transformation CSS), as illustrated in Figure 3-22.

Table 3-11 *Calling-Party Transformation Patterns for Hamburg, Germany Phones*

Transformation Pattern	Performed Transformation
\+49.!	DDI PreDot, Prefix 0
\+4940.!	DDI PreDot

Note Because there is no \+! calling-party transformation pattern, international calls are preserved in normalized format (E.164 with + prefix). As opposed to the San Jose example, phones located in Hamburg do prefix the national access code (using 0, which is equivalent to the long-distance 1 in the NANP). The reason is that, in Germany, variable-length PSTN numbering plans are used. Therefore, national and local numbers cannot be distinguished based on their length (like in the United States, with seven- and ten-digit numbers). When the national access code 0 is prefixed to numbers that are used by national callers, a user can identify national calls by their leading 0.

Note When users call back PSTN callers, the globalized number is used for the outgoing call. Therefore, there is no need to edit the localized number from a call list and add PSTN access codes and national or international access codes.

Globalized Call Routing Example: Emergency Dialing

In a multisite deployment with centralized call processing, it might be desirable to simplify emergency dialing by introducing a globalized emergency number (or one globalized emergency number for each emergency service).

Having a globalized emergency number allows roaming users who might not be aware of the local emergency dial rules to use a corporate emergency number that is accessible from all sites.

In addition, localized emergency dialing should still be supported, so a user can dial either the locally relevant emergency number or the corporate emergency number.

Here is how to implement such a solution:

- Introduce one or more corporate emergency numbers.

- Allow localized emergency dialing. It can be limited to local emergency dialing rules per site (for example, an Austrian emergency number can be dialed only from phones that are located in Austria), or you can globally enable all possible local emergency numbers.

Having all possible local emergency numbers that are globally enabled allows a roaming user to employ the emergency number that is local to the site where the user is located or the emergency number that the user knows from the home location of the user (for example, a UK user dials 999 while roaming in Austria), or the corporate emergency number.

Note Making all possible local emergency numbers available to all users may not be possible because of overlapping dial plans. For example, the emergency number of one country may be identical with the first digits of a valid local PSTN number.

- If a user dials a localized emergency number, that number is first normalized (that is, translated) to the corporate emergency number. A route pattern exists only for this corporate emergency number, and you configure the corresponding route list to use the local route group.

- At the gateway that processes the call, localize the corporate emergency number (the globalized emergency number) by using called-party number transformations at the gateway. This localization ensures that, regardless of which emergency number was dialed, the gateway that sends out the emergency call uses the correct number as expected at this site.

Note In deployments with more complex emergency calls, like in the United States with E911, such a solution is not applicable because there are other requirements for emergency calls. In such a scenario, the emergency call is routed via a dedicated appliance (Cisco Emergency Responder) that is reached via a computer telephony integration (CTI) route point. Cisco CER provides location awareness for 911 calls.

In Figure 3-23, a corporate emergency number of 888 has been established. In addition, Australian, European Union, and UK emergency numbers are supported at all sites of the enterprise. The appropriate numbers (000, 112, and 999) are translated (normalized) to the corporate (global) emergency number 888. A route pattern 888 exists, which refers to a route list that has been configured to use the local route group. Consider two sites in this example: one in the EU and one in the United Kingdom. Each site has its own PSTN gateway (GW in the figure); phones at each site are configured with a site-specific device pool. The device pool of each site has its local route group that is set to a site-specific route group.

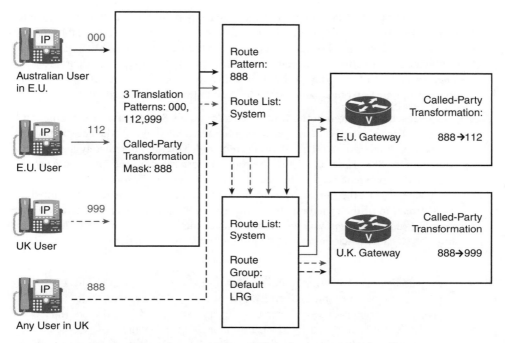

Figure 3-23 *Globalized +E.164 Emergency Dialing Call-Routing Example*

The following scenarios examine four emergency calls:

■ **An Australian user, currently located at an EU site, dials 000 (Australian emergency number):** The dialed Australian emergency number 000 is translated to the corporate emergency number 888. After translation, the 888 route pattern is matched. The route list of the route pattern refers to the local route group. Because the emergency call was placed from an EU phone, the local route group in the device pool of the phone refers to the EU gateway. At that gateway, a global transformation of the called number (from 888 to 112) is configured. Therefore, the call exits the EU gateway with a destination number of 112, which is the emergency number in the European Union.

■ **An EU user dials 112 (EU emergency number):** The dialed EU emergency number 112 is translated to the corporate emergency number 888. After translation, the 888 route pattern is matched. The route list of the route pattern refers to the local route group. Because the emergency call was placed from an EU phone, the local route group in the device pool of the phone refers to the EU gateway. At that gateway, a global transformation of the called number (from 888 to 112) is configured. Therefore, the call exits the EU gateway with a destination number of 112, which is the appropriate emergency number to be used in the EU.

■ **A UK user dials 999 (UK emergency number):** The dialed UK emergency number 999 is translated to the corporate emergency number 888. After translation, the 888 route pattern is matched. The route list of the route pattern refers to the local route group. Because the emergency call was placed from a UK phone, the local route group in the device pool of the phone refers to the UK gateway. At that gateway, a global transformation of the called number (from 888 to 999) is configured. Therefore, the call exits the UK gateway with a destination number of 999, which is the appropriate emergency number to be used in the United Kingdom.

- **Any user who is located in the United Kingdom dials 888 (corporate emergency number):** Because no local emergency number was dialed except the corporate emergency number 888, no translation is required. The call immediately matches route pattern 888. The route list of the route pattern refers to the local route group. Because the emergency call was placed from a UK phone, the local route group in the device pool of the phone refers to the UK gateway. At that gateway, a global transformation of the called number (from 888 to 999) is configured. Therefore, the call exits the UK gateway with a destination number of 999, which is the appropriate emergency number to be used in the United Kingdom.

Note   The Australian user with an EU phone (with an EU extension) can use his own device with device mobility enabled, or an EU phone with his own extension (by using Cisco extension mobility). In all three scenarios, the emergency call would work fine, as described earlier. The reason is that the device pool of the phone will be the EU device pool in all three scenarios (with device mobility enabled, the home device pool would be replaced by the roaming device pool), and hence the local route group is always the EU-GW.

Note   The only problem would be if the Australian user was using his own device with device mobility disabled. In this case, the local route group would refer to the Australian gateway, and the call would be sent through the Australian gateway instead of the local EU gateway. The localized egress number would be appropriate for an Australian gateway (transformed to 000), so the user would get connected to an Australian emergency service.

Considering Globalized Call Routing Interdependencies

Globalized call routing simplifies the implementation of several dial plan features in an international deployment. The affected dial plan features include:

- TEHO
- Automated alternate routing (AAR)
- Cisco Unified SRST and CFUR

- Cisco device mobility

- Cisco extension mobility

If TEHO is configured, the appropriate TEHO gateway is used for the PSTN call. The TEHO route list can include the Default Local Route Group setting as a backup path. In this case, if the primary (TEHO) path is not available, the gateway that is referenced by the local route group of the applicable device pool will be used for the backup path. If the device pool selection is not static, but Cisco Unified device mobility is used, the gateway of the roaming site will be used as a backup for the TEHO path.

The same situation applies to Cisco extension mobility. When a user roams to another site and logs in to a local phone, PSTN calls use the local gateway (if TEHO is not configured) or the local gateway is used as a backup (if TEHO is configured). The local gateway selection is not based on the Cisco extension mobility user profile, but on the device pool of the phone where the user logs in. The line CSS, however, is associated with the user profile, and therefore the user can dial PSTN numbers the same way that they do at home. The localized input is then globalized. After call routing and path selection occur, the globalized number is localized again based on the requirements of the selected egress device. The localized input format that the user used can be completely different from the localized format that is used at call egress.

Globalized Call Routing and TEHO Advantages

As previously discussed, when you use local route groups, there is no need to have duplicated TEHO route patterns for each originating site. Instead, the local PSTN gateway is selected by the local route group feature when the TEHO path cannot be used.

When combining globalized call routing with local route groups, you do not have to consider the various possible input formats for the TEHO call routing decision. No matter how the user dialed the number, it is changed to globalized format before it is routed. Because the called number is then localized after call routing and path selection, you can localize the called- and calling-party number differently at the primary gateway (TEHO gateway) and the backup gateway (local gateway). However, the global transformations that you configure for each egress gateway all refer to a single globalized format regardless of how the user dialed the destination. This globalized format that is combined with local route groups for local backup gateway selection makes implementing TEHO much simpler. Without globalized call routing, you need to perform localization at the egress gateway differently for each originating site.

Globalized Call Routing TEHO Example

Globalization of calls and the use of TEHO can dramatically simplify multinational dial plan deployments. The benefit of Globalization and +E.164 dial plan is paired with TEHO for selective PSTN breakout of calls. Route patterns play a crucial role in selective PSTN breakout with globalization, as shown in Figure 3-24. Figure 3-24 shows an example of TEHO when globalized call routing is used.

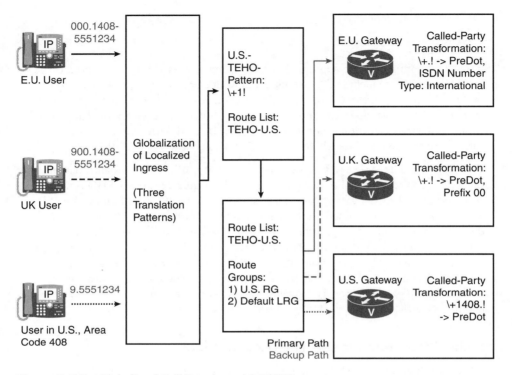

Figure 3-24 *Globalized Call Routing with TEHO*

At the call ingress side, there are three PSTN dial rules: European Union, United Kingdom, and United States The same rules apply to the egress gateways: The EU, UK, and U.S. gateways all require different digit manipulation when you are sending calls to the PSTN.

As long as users are allowed to roam between sites and TEHO with local backup in place, users can dial each PSTN destination differently at each site. In addition, if the TEHO path is not available, the local gateway (which again can be any of the three) is used for backup. With globalized call routing, you do not have to consider all the possible combinations of ingress and egress, but you must consider call ingress and call egress independent of each other.

All that you need to configure is translation patterns for each of the PSTN dial rules (European Union, United Kingdom, and United States). Then, create TEHO route patterns that refer to the TEHO gateway as the first choice, and to the local gateway as the backup, using the local route group feature. At the egress gateways, configure the called- and calling-party transformations, where you do not match on all possible input formats again, but on a globalized format only.

As an example, suppose that a user travels to the San Jose location and makes a call to 555-1234 by dialing the 9 access code first. The primary path will be the U.S. gateway, which sends the user's call to the San Jose local exchange carrier (LEC). The 9 access code is stripped off, 1408 is added, and the number 14085551234 is presented to the LEC.

Summary

The following key points were discussed in this chapter:

- Connectivity options for multisite deployments include gateways and trunks such as MGCP, SCCP, H.323, and SIP.

- When implementing MGCP gateways, you do most of the configuration in CUCM.

- When implementing H.323 gateways, you must configure both CUCM and the IOS gateway with a dial plan in the form of dial peers.

- Trunk support in CUCM includes SIP and three types of H.323 trunks.

- When configuring non-gatekeeper-controlled ICTs, you must specify the IP address of the peer. Gatekeeper-controlled ICTs and H.225 trunks require that you configure an H.323 gatekeeper and route calls to the IP address of the gatekeeper.

- Multisite dial plans should support selective PSTN breakout with backup gateways, PSTN backup for on-net calls, TEHO, and intersite calls using access codes and site codes.

- With the addition of an access code and site code to DNs at each site, DNs no longer have to be globally unique.

- When calls are routed to the PSTN, calling DNs must be transformed to PSTN numbers, and access codes that are used on dialed patterns must be removed to ensure that calling number and called number are in accordance with PSTN numbering schemes.

- Selective PSTN breakout means that different gateways are used for PSTN access, depending on the physical location of the caller.

- When the PSTN is used as a backup for intersite calls, internal DNs and internally dialed patterns must be transformed to ensure that calling numbers and called numbers are in accordance with PSTN number schemes.

- When you implement TEHO, calls to the PSTN are routed differently, based on the physical location of the caller and the PSTN number that was dialed. This difference ensures that the call uses the IP WAN as much as possible and breaks out to the PSTN at the gateway that is closest to the dialed PSTN destination.

- Globalized call routing is a dial plan concept in which the call routing is based on numbers in +E.164 format.

- Globalized call routing reduces the complexity of international dial plans substantially and makes it easier to implement features such as device mobility, extension mobility, AAR and CFUR, or TEHO in international deployments.

References

For additional information, refer to the following:

Cisco Systems, Inc. Cisco Collaboration Systems 10.x Solution Reference Network Designs (SRND), May 2014. http://www.cisco.com/c/en/us/td/docs/voice_ip_comm/cucm/srnd/collab10/collab10.html.

Cisco Systems, Inc. Cisco Collaboration Systems 11.x Solution Reference Network Designs (SRND), July 2015. http://www.cisco.com/c/en/us/td/docs/voice_ip_comm/cucm/srnd/collab11/collab11.html

Cisco Systems, Inc. Features and Services Guide for Cisco Unified Communications Manager, Release 10.0(1), May 2014. http://www.cisco.com/c/en/us/td/docs/voice_ip_comm/cucm/admin/10_0_1/ccmfeat/CUCM_BK_F3AC1C0F_00_cucm-features-services-guide-100/CUCM_BK_F3AC1C0F_00_cucm-features-services-guide-100_chapter_0110011.html

Cisco Systems, Inc. Cisco Unified Communications Manager Administration Guide Release 11.0(1), July 2015. http://www.cisco.com/c/en/us/td/docs/voice_ip_comm/cucm/admin/11_0_1/administration/CUCM_BK_A0A10476_00_administration-guide-for-cisco-unified.html

Cisco Systems, Inc. *Cisco IOS Voice Configuration Library* (with Cisco IOS Release 15.0 updates), July 2007. http://www.cisco.com/en/US/docs/ios/12_3/vvf_c/cisco_ios_voice_configuration_library_glossary/vcl.htm

Review Questions

Use these questions to review what you have learned in this chapter. The answers appear in Appendix A, "Answers Appendix."

1. **What is not configured in IOS when you configure an H.323 PSTN gateway?**

   a. The IP address used for the H.323 interface

   b. The VoIP dial peer pointing to CUCM

   c. The IP address of the call agent for centralized call control of H.323

   d. The POTS dial peers pointing to the PSTN

2. **Which two types of trunks are configured directly with the IP address of CUCM in another CUCM cluster? (Choose two.)**

   a. SIP gateway

   b. Non-gatekeeper-controlled intercluster trunk

   c. SIP trunk

   d. Gatekeeper-controlled intercluster trunk

   e. MGCP trunk

3. **What do you need to specify at the gatekeeper configuration page when adding a gatekeeper to CUCM?**

   a. Hostname of the gatekeeper

   b. H.323 ID of the gatekeeper

   c. IP address of the gatekeeper

   d. Zone name

   e. Technology prefix

4. **What is the best reason to implement a gatekeeper in a multisite implementation?**

   a. To allow IP phones to register to the gatekeeper to implement centralized call routing

   b. To prevent IP phones from registering to the incorrect CUCM server

   c. To implement centralized call routing with CAC

   d. To implement centralized call routing, ensuring that CAC is not implemented

5. **Which of the following statements about implementing PSTN backup for the IP WAN is true?**

   a. In distributed deployments, PSTN backup for intersite calls requires CFUR.

   b. Route groups including the on-net and off-net path are required for PSTN backup in a centralized deployment.

   c. PSTN backup requires a PSTN gateway at each site.

   d. CFUR allows remote-site phones to use the PSTN for calls to the main site.

6. **When implementing TEHO for national calls and using the local PSTN gateway as a backup, how many route patterns are required for a cluster with three sites located in different area codes?**

   a. Three, when not using the local route group feature

   b. Six, when using the local route group feature

   c. Nine, when not using the local route group feature

   d. Four, when using the local route group feature

7. **Which two are not valid type of number (TON) codes for incoming ISDN PSTN calls?**

   a. International

   b. National

   c. Subscriber

   d. Directory number

   e. Operator number

8. **Which two statements about PSTN gateway selection are true? (Choose two.)**

   a. Using TEHO minimizes PSTN costs.

   b. Using local PSTN gateways minimizes PSTN costs.

   c. When you use TEHO, you must pay special attention to the configuration of the calling number.

   d. Using remote PSTN gateways for backup is never recommended.

   e. It is recommended that you use remote PSTN gateways as a backup when the IP WAN is overloaded.

9. **You should perform digit manipulation in CUCM at the _____ when digit-manipulation requirements vary based on the gateway used for the call.**

   a. Gateway

   b. Route list as it relates to the route group

   c. Route pattern

   d. Trunk

   e. Translation pattern

10. **The PSTN egress gateway can be selected in which two of these ways? (Choose two.)**

    a. By the partition of the calling device

    b. Based on the CSS of the gateway

    c. By the local route group feature

    d. Based on the matched route pattern when route patterns exist once per site

    e. By the standard local route group that is configured at the gateway device pool

11. **Where can digit manipulation be performed when digit manipulation requirements vary for the on- and off-net paths?**

    a. Per route group of the route list

    b. Route pattern

    c. Directory number

    d. Translation pattern

Chapter 4

URI-Based Dial Plan for Multisite Deployments

The way in which people communicate has evolved at an exponential rate over the past few years. Fifteen years ago, had anyone heard of Facebook? Immersive video abilities were a dream, and many organizations were using primary rate interface (PRI) as their primary means to interface with the public switched telephone network (PSTN) and their telco provider. Today, immersive video capabilities are prevalent, and instant messaging is often preferred over traditional calls. Cisco WebEx is used as a replacement to the traditional conference room, and organizations are embracing alternative ways of keeping users connected.

Today, users can be reached on multiple devices and via different technologies (voice calls, video calls, instant messages, WebEx, e-mails, and so on). From an addressing point of view, these various devices and technologies can result in lots of different user addresses and address formats. For example, many users have a direct inward dial (DID) extension, corporate e-mail address, and instant messaging usernames. Session Initiation Protocol (SIP) makes it possible to use the Uniform Resource Identifier (URI) format for addressing an end user. The URI format consists of a user and a host portion, such as an e-mail address, and provides a simple, uniform way of identifying an end user. An example of a URI is username@company.com. SIP URI dialing has been used for Internet-based video calling for years and is gaining universal popularity. Part of the reason of the rise in popularity is that SIP URI dialing is globally routable via Domain Name Service (DNS) and friendly to end users. The traditional dial plan and +E.164 dialing are not going away anytime soon. However, it gives users another method of communication.

Starting with Cisco Unified Communications Manager (CUCM) Release 8.6 and later, SIP-based URI dialing is now a mainstream feature that is easy to deploy within the enterprise. URI dialing is often easier for end users than having to remember long, complex telephone numbers. If you think about it, everyone remembers e-mail addresses, right? URI dialing enables elegant business reachability for voice and video calls within the enterprise and for extending beyond it. Coupling SIP-based URI dialing with a Cisco Unified Border Element (CUBE), Cisco Expressway series, or Video Communication Server (VCS) can extend URI dialing out to the Internet and perform business-to-business transactions all via URI dialing.

When implementing SIP URI dialing in a multisite or multicluster environment, route calls to other SIP-based call control systems such as a private branch exchange (PBX) or another CUCM cluster by using SIP route patterns and SIP trunks. Certain SIP parameters may have to be tuned to ensure proper operation and interoperability between different SIP call control systems.

Upon completing this chapter, you will be able to meet these objectives:

- Provide an overview of URI dialing

- Describe how directory URIs are assigned to an IP phone

- Describe how partitions and CSSs relate to directory URIs

- Describe the possible sources of URI calls and related configuration elements

- Describe how to support blended addressing of directory URIs and directory numbers

- Describe how to present remote directory URIs and directory numbers

- Describe how URI calls are routed

URI Dialing Overview

Figure 4-1 depicts multiple CUCM clusters that have URI dialing enabled, along with various IP phones that are dialing via a URI-enabled dial plan.

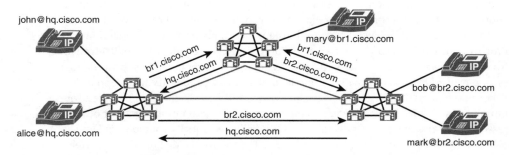

Figure 4-1 *Example of URI Dialing in a Unified Communications Environment*

CUCM supports dialing by using directory URIs for call addressing. An example is if a user dials john.doe@example.com. Rather than having to remember a lengthy on premise directory number (DN) or direct inward dial (DID) number, certain model phones and video endpoints support dialing via URI. A directory URI is a string of characters that can identify a DN. Directory URIs look like e-mail addresses and follow the username@ host format, where the host portion is an IP address or a fully qualified domain name

(FQDN). It is important to remember that CUCM treats URIs as aliases for DNs. Some examples of URIs are john@example.com and john.doe@eg.example.com.

If that DN is assigned to a phone, CUCM can route calls to that phone by using the directory URI. URI dialing is available for specific Session Initiated Protocol (SIP) and Skinny Client Control Protocol (SCCP) endpoints that support directory URIs. Not all model IP phones support URI dialing. Therefore, it is recommended to check the specific phone model's release notes before attempting to configure URIs to ensure the feature is supported. A potential workaround solution for lack of support on certain IP phone models is to use speed dials that specify URI-based destinations.

A call to a URI behaves as if the call was made directly to a DN. The endpoint does not treat the call any differently. There is no "special ringtone" or feature enhancement for URI dialing. However, the receiving endpoint does need to account for differences in which caller ID may be displayed. A call from an endpoint includes the URI in the caller ID when a URI is associated with a DN. Due to the lack of support for the display of directory URIs across all product lines, the DN is always included as part of the caller ID so that it can be presented when a device does not support URIs. Delivering both the URI and DN is required for callback purposes.

When designing a URI dial plan for a CUCM cluster, it is important to note that URIs are functionally just "overlays" to existing numeric-based dial plans. It is still required to implement DNs on lines of phones and build out proper dial plans for traditional call routing.

There are some basic rules for SIP URIs inside CUCM. Once an IP phone or video endpoint is configured with a DN, up to five SIP URI aliases can be configured. One of the SIP URIs is classified as the primary SIP URI for caller ID purposes. Figure 4-2 shows the concepts previously discussed for fictitious user john.doe@example.com.

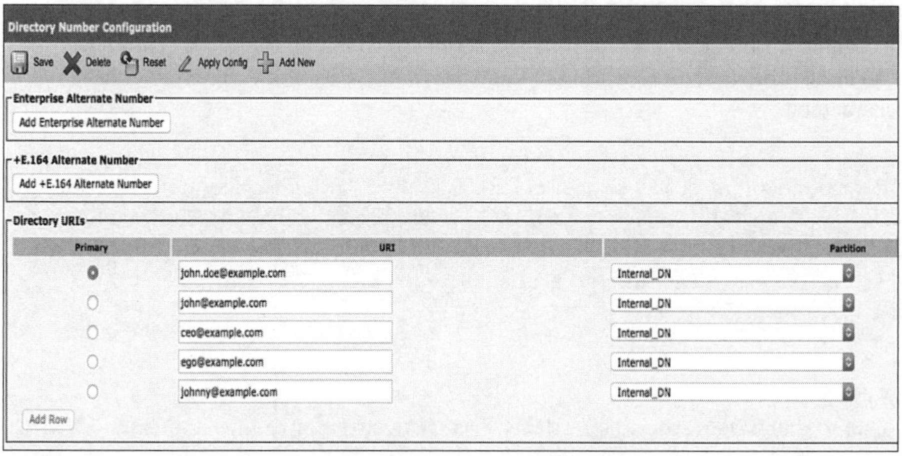

Figure 4-2 *Directory URI Aliases Configured for a DN*

CUCM supports the following formats for directory URIs:

■ user@domain (for example, bob@example.com)

■ user@ip_address (for example, bob@10.1.1.1)

> **Note**   Due to database restrictions in the Informix database, the directory URI field has a maximum length of 254 characters. In addition, for compatibility with third-party call control system, it is recommended to use lowercase letters for directory URIs. URI designation using the user@domain nomenclature is the most common.

The format of the user and the host portions of a URI must follow certain rules as well. CUCM supports all the following formats in the user portion of a directory URI (the portion before the @ symbol):

■ Accepted characters are a–z, A–Z, 0–9, !, $, $, %, &, *, _, +, ~, -, =, \, ?, \, ', comma (,), period (.), and /.

■ The user portion has a maximum length of 47 characters.

■ The user portion accepts percent encoding from %2[0–9A–F] through %7[0–9A–F].

Percent encoding refers to how special characters such as spaces or perhaps a pound symbol are stored in the IBM Informix database. The special characters are often "converted." For example, a pound symbol (#) in a directory URI is stored in the Informix database as a %23. For some accepted characters, CUCM automatically applies percent encoding. These characters are as follows:

#, %, ^, `, {,}, |, \, :, ", <, >, [,], \, ', and spaces. When percent encoding is applied, the digit length of the directory URI increases.

For example, if you input joe smith#@cisco.com (20 characters) as a directory URI, CUCM stores the directory URI in the database as joe%20smith%23@cisco.com (24 characters).

> **Note**   Percent encoding is also known as URL encoding. URL encoding translates special characters called reserved characters into ASCII codes preceded by a percent symbol, so that they are not interpreted as URL syntax in the URI string. For example, a forward slash (/) refers to a page in a URL, but may be included in a URI if translated into URL encoding as %2F.

> **Note**   Due to database restrictions, CUCM rejects any attempt to save a directory URI greater than 254 characters in total length. The column in the Informix database can only hold 254 characters.

By default, the user portion of the directory URI is case sensitive. This can be changed as a service parameter inside CUCM. The URI Lookup Policy Enterprise parameter is set to case sensitive by default. Case sensitivity can often be a challenge when deploying a URI overlay to an existing CUCM dial plan. Figure 4-3 demonstrates the enterprise parameters that can be modified to determine the case sensitivity when dialing a URI.

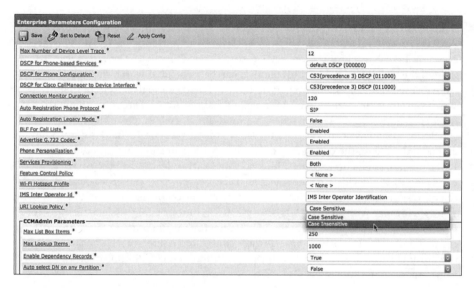

Figure 4-3 *URI Lookup Policy Case-Sensitivity Options*

CUCM supports the following formats in the domain or host portion of a directory URI (the portion after the @ symbol):

■ The host portion supports IP addresses or FQDNs.

■ Accepted characters are a–z, A–Z, 0–9, hyphens, and periods.

■ The host portion cannot start or end with a hyphen.

■ The host portion cannot have two periods in a row. The host portion has a minimum of two characters. The host portion is not case sensitive.

URI Endpoint Addressing Overview

Each phone that registers with CUCM is associated with all of its DNs. If an IP phone has a phone button template containing multiple lines, all lines are associated with that device. CUCM can contain virtual DNs that are not tied to any specific phone. However, they are not associated with any phones or video endpoints. Once the IP phone has associated DNs, the URI aliases can be configured. Directory URIs are associated with IP phones and video endpoints in the following three primary ways:

■ **Manually configured at the End User Configuration page:** In CUCM Administration, you can associate or configure a single directory URI at the end-user level. In CUCM Administration, if you browse to **User Management > End User** and drill into an end-user account, you can associate only a single URI. To achieve this, the end user must be associated with a device, and the primary extension of the device must be selected. As a result, the configured directory URI is associated with the primary extension.

■ **Synchronized via LDAP:** When a user is provisioned via Lightweight Directory Access Protocol (LDAP), one directory URI can be imported from LDAP. The most common LDAP synchronization is Microsoft Active Directory. The directory URI field in the CUCM end-user configuration can be populated with either the Microsoft Active Directory msRTCSIP-primaryuseraddress field or the mail field during LDAP synchronization. It is important to remember that the Active Directory administrator must configure one of these fields with a valid directory URI. Once the LDAP synchronization has been enabled on CUCM, the directory URI field is grayed out and noneditable on the end-user account. Like with manually configured directory URIs, the imported end user must be associated with a device, and the primary extension of the device must be selected. The directory URI can be associated with the corresponding DN only when the primary extension of the end user is set. When you synchronize with an LDAP directory, CUCM automatically assigns the directory URI value that you choose from the LDAP directory as the primary directory URI for that end user. Even if you have already configured a directory URI as the primary directory URI for that end user's primary extension, the LDAP value overrides the value that is configured in CUCM Administration.

■ **Manually associated with a DN:** At the Directory Number Configuration page, up to five directory URIs can be associated with the DN. One directory URI per DN functions as the primary directory URI. A directory URI that is configured on an end-user page always becomes the primary directory URI of the primary extension of the end user. If no end-user-based directory URI exists, you can select which of the directory URIs that are configured at the Directory Number Configuration page should be the primary directory URI. By default, the first directory URI that is configured at the Directory Number Configuration page is the primary URI. The primary URI is used as the calling URI on outbound calls.

■ **Bulk administration:** In CUCM, it is possible to bulk administer and update the directory URIs at both the end-user and DN levels. One consideration when performing bulk administration for directory URIs deals with the encoding of double quotes. Within CUCM Administration, you can enter directory URIs with embedded double quotes or commas. When you use bulk administration to import a comma-separated values (CSV) file that contains directory URIs with embedded double quotes and commas, you must perform special encoding. When you use the Bulk Administration Tool (BAT) to import a CSV file that contains directory URIs with embedded double quotes and commas, you must enclose the entire directory URI in double quotes. To further complicate matters, each embedded double quote in the URI must be escaped with a double quote. For example, if you want to use BAT to insert the directory URI Jared, "Jerry", Smith@example.com, the directory URI must be entered into the BAT CSV file as "Jared," "Jerry," "Smith@test.com".

URI Partitions and Calling Search Spaces

Configuring a directory URI is similar to that of a regular DN. Once you enter the pattern, you have an option of assigning different partitions. The analogy that many engineers learn about calling search spaces (CSSs) and partitions is that of a lock and key. A partition is a lock that hides numbers, patterns, or URIs. A CSS is the key that grants endpoints the ability to dial the URI. Call routing decisions are enforced in the same manner as a regular DN.

When a directory URI is provisioned via a local end user or an LDAP end user, a default partition is applied to the URI. The default partition is created during the CUCM installation and is named directory URI. This partition cannot be deleted, but it can be renamed. In addition to renaming the directory URI partition, CUCM allows you to define an alias partition for the directory URI partition. By setting the enterprise parameter Directory URI Alias Partition to an existing partition that all CSSs have access to, the selected alias partition automatically has access to the directory URI partition and its associated DNs. The directory URI alias partition can be found under End User Parameters in the Enterprise Parameters on a CUCM server, as shown in Figure 4-4.

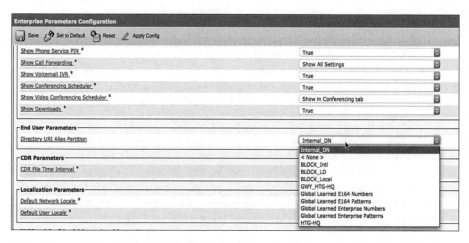

Figure 4-4 *The Directory URI Alias Partition Enterprise Parameter in CUCM*

To allow calls to end user-based directory URIs, a CSS that includes either the directory URI partition or the directory URI alias partition must be applied to the calling device. The calling device can be any number of sources in CUCM such as gateways, trunks, phones, phone lines, and so on.

> **Note**   This concept differs from the implementation of Presence-enabled call lists. In Presence-enabled call lists, a DN has only one partition, which is used for call routing requests and for requests to watch the Presence status. The CCS that is used for Presence-enabled call lists is configured separately from the CSS for call routing. In addition to the standard CSS, which is used for call routing, there is a subscribe CSS that is configured at the watching phone, which is used to watch the presence status of a DN.

The call routing logic for directory URIs is also identical to the logic for DNs. The best or closest matching pattern is used. If there are multiple equally qualified matching patterns, the order of partitions in the calling device's CSS is used as a tie-breaker (with priority given to partitions that are listed first). One other item of interest is there is no concept of URI manipulation inside CUCM. With traditional DNs, an administrator could invoke a translation pattern, transformation pattern, or various digit manipulations at the call routing level. With directory URIs, there is no concept of a translation pattern.

URI Call Sources Overview

In CUCM, there can be several different sources of URI calls. These sources can be IP phones, Jabber, video endpoints, SIP trunks, CUBE, and VCS, just to name a few. If you take a look at the entire Cisco IP phone portfolio, there are some model IP phones that do not support SIP URI dialing natively, such as IP Communicator, and Generation 1 7900 series such as 7940 and 7960. As a possible workaround for these endpoints, calls to directory URIs can be initiated from any Cisco IP phone by using a speed-dial button that is configured with a destination in URI format.

In addition to using speed dials, many IP phone models allow you to enter URIs manually. As mentioned previously, it is a best practice to check the release notes and features guides to see whether the IP phone or video endpoint model supports URI dialing. Cisco Jabber clients and modern IP phones such as Cisco Unified IP Phones 8961, 9951, and 9971 each support URIs and URI dialing.

You can also place calls to call list entries in URI format at supported phones, including Cisco Unified IP Phones 7942, 7963, 7945, 7965, and 7975, and all Cisco Unified IP Phone 6900, 8900, and 9900 series. When a call list entry is used to place an outbound call and the list entry is in URI format, the call is placed to the corresponding directory URI. An example of a call list entry is redialing a missed call using a URI. A user could check his missed call log and one-touch redial a missed call via URI format.

CUCM also routes calls to directory URIs when the calls have been received or placed through a SIP trunk. In the latest XE IOS releases, CUBE supports both traditional dial peers and now URI-based dial peers. With emerging industry trends, many ITSPs and telco providers are routing calls via SIP and URIs. CUBE allows you to configure URI-based destinations inside dial peers.

With proper planning, it is possible to enable users to dial without entering the domain or hostname in a URI. This can be particularly convenient with long domain names.

When a dialed URI does not include the @ symbol followed by a host portion (that is, only the user portion is dialed), the call fails unless you have configured the enterprise parameter organization top-level domain (OTLD). The OTLD is a setting located in the Enterprise Parameters on the CUCM Administration web page. By default, the OTLD is unset. If the OTLD is set, the @ symbol followed by the configured OTLD is appended to the dialed URI. Setting the OTLD allows users to place calls to URIs by entering only the user portion. One additional setting is the Cluster Fully Qualified Domain Name (CFQDN). These settings are illustrated in Figure 4-5. The CFQDN is directly involved with routing of numeric URIs, as discussed later in this chapter.

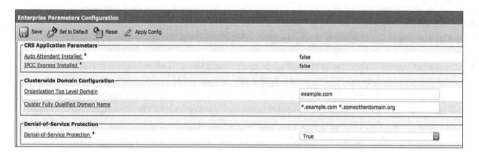

Figure 4-5   *OTLD and CFQDN Enterprise Parameters Configuration Settings*

The OTLD and the CFQDN are important concepts inside Cisco UC. These two settings are not only used for URI and blended addressing, but play key roles in Global Dial Plan Replication (GDPR) and Inter-cluster Lookup Service (ILS). Both of these topics are discussed in depth in Chapter 15, "Cisco Inter-Cluster Lookup Service (ILS) and Global Dial Plan Replication (GDPR)."

Blended Addressing

When a call is placed in CUCM via SIP, a SIP endpoint, or from a SIP trunk, SIP provides a SIP INVITE message that contains a To and From field. This To and From field is extremely important for call routing purposes. Blended addressing refers to the manner in which CUCM delivers or receives SIP INVITE messages when URIs are present in a call. When blended addressing is enabled across the network, CUCM inserts both the directory URI and the DN of the sending party in outgoing SIP INVITE messages or in responses to SIP INVITE messages. In essence, the signaling message for a call contains both the DN dialed and the primary directory URI.

CUCM uses the x-cisco-number tag in the SIP identity headers to communicate a blended address. The x-cisco-number tag identifies the DN that is associated with the directory URI. An easy way to see the SIP INVITE field of a call is to run a Wireshark trace or perhaps view the CUCM trace files using Real Time Monitoring Tool (RTMT), which is downloadable from the **CUCM Tools > Plugins** page. Another excellent resource to start viewing SIP traces is using a tool called TranslatorX, which enables you to parse CUCM log files easily and filter out for specific protocol stacks such as SIP. For more details on logs, debugs, and troubleshooting, refer to CTCOLLAB publication from Cisco Press.

When both a directory URI and DN are available for the sending phone and blended addressing is enabled, CUCM uses the directory URI in the From fields of the SIP message and adds the x-cisco-number tag with the accompanying DN to the SIP identity headers. When the call is received on the destination phone, a user can display the blended address information by pressing the Show Detail softkey on an IP phone. Keep in mind that this feature is dependent on the firmware and the model of the phone. It is recommended to check the release notes for your specific IP phone model to see whether directory URIs and blended addressing are supported. Earlier in this chapter, we mentioned various IP phone models that cannot support directory URI capabilities.

There are a few rules about the display of blended addressing at an endpoint. For an endpoint to display a blended address, the following conditions must be met:

- At all SIP trunks along a given call path, the Calling and Connected Party Info Format parameter must be set to Deliver URI and DN in Connected Party, if available. This parameter can be set in a SIP trunk under the Outbound Calls section. An example is a CUBE connected to an SBC or ITSP. The SIP trunk from CUCM to the CUBE must have this value set to deliver the blended address to the far end. Another example is a SIP trunk connecting to CUCM clusters for ILS. Each end of the SIP trunk must also have this parameter set in the Outbound Calls section of the configuration.

- The phone that should be identified with blended addressing must be configured with a directory URI that is associated with the DN of the calling line. The relationship between directory URI and DN can be configured on the line appearance on an IP phone. As stated earlier, up to five directory URIs can be configured per DN. Only one directory URI can be deemed the primary directory URI. With blended addressing, only the primary directory URI is placed into the x-cisco-number tag field of the SIP identity header.

- The receiving phone must support blended addressing. (That is, the x-cisco-number tag needs to be supported.) Many Cisco IP phone models do, in fact, support the delivery of blended addressing. As mentioned previously, there are several older Generation 1 model phones, such as the 7940 and 7960 series, that cannot display the x-cisco-number field.

FQDNs in Directory URIs

By default, CUCM replaces the host or domain portion of a directory URI with the IP address of the local CUCM that routes the call when the identity of the remote endpoint is delivered as URI only. CUCM modifies the SIP INVITE to inject the IP address in place of the hostname or domain name. For intracluster SIP URI calls, the host or domain portion is not changed or modified in any way.

As Figure 4-6 shows, if you want to deliver the actual hostname of a directory URI of the remote endpoint to a local phone, you must set the Use Fully Qualified Domain Name parameter to Enabled in the SIP profile applied to a SIP trunk or a SIP endpoint. The parameter must be applied to all outbound devices (SIP trunks and the terminating phone) along the path. If you want to use the FQDN for the presentation on all devices or endpoints (bidirectionally), you must apply the parameter at all outbound devices or trunks along the paths of both directions.

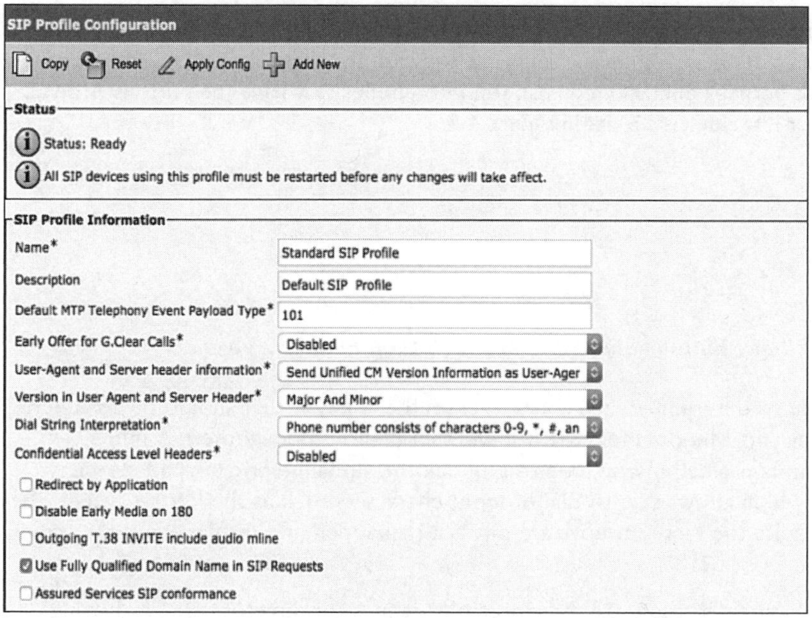

Figure 4-6 *SIP Profiles and FQDNs in Directory URIs*

If you use blended addressing in one direction only, you do not need to apply the Use Fully Qualified Domain Name parameter to the outbound devices in the direction to the endpoint that receives the blended addresses. CUCM only replaces the host portion with a local server IP address if it receives a URI only. An example of this is an outbound call from a SIP-based IP phone to an ITSP via a CUBE. CUCM replaces the domain name outbound to the SIP trunk with the IP address of the local CUCM that originated the routing request.

URI Call Routing

CUCM routes calls to URIs differently depending on the type of URI dialed. In UC, URIs can be considered numeric (and include numbers) as well as non-numeric (and include strings and characters).

The URI call routing process is rather complex and depends on several conditions. The first thing that is checked is whether the user portion of the called URI is numeric. An example of a numeric URI might be 1000@domain.com. CUCM uses the SIP profile's Dial String Interpretation parameter applied to the SIP endpoint to determine whether the called number is a numeric URI. By default, a called URI is considered to be numeric if the user portion of the dialed string consists of the characters: 0–9, *, #, and +. If the user portion includes any other characters, the URI is considered to be non-numeric.

One important design consideration for directory URIs is how dial strings are interpreted in CUCM. On certain analog phones (a prime location for these devices is old financial institutions or perhaps military), you can run into phones that have the buttons A,B,C,D on the right or left side, as shown in Figure 4-7.

Figure 4-7 *Auxiliary Buttons Such As A Through D on Phone Keypads*

It is important to determine when a user presses the A key if that should be considered a string and as part of a directory URI or whether it should be ignored. A more modern example is a cell phone; you can unlock the alphanumeric keypad on many cell phones, which allows you to dial or input characters such as the letter C versus the number 2 despite the fact that both are physically mapped and displayed on the same phone button.

As shown in Figure 4-8, there are three possible settings for directory URI dial string interpretations, as follows:

- **Always treat all dial strings as URI addresses:** This option will treat any URI as a "string" and will count letters A,B,C,D as strings, not numbers. As mentioned previously, in older military bases or older financial institutions, there are analog phones that have buttons A–D on them; these buttons could be programmed as numbers or characters in CUCM.

- **Phone number consists of characters 0–9, *, #, and + (others treated as URI addresses):** This option treats any URI as a "string" and does not observe buttons A–D on an older keypad.

- **Phone number consists of characters 0–9, A–D, *, #, and + (others treated as URI addresses):** This option will treat any URI as a "string" and will count letters A,B,C,D as numbers.

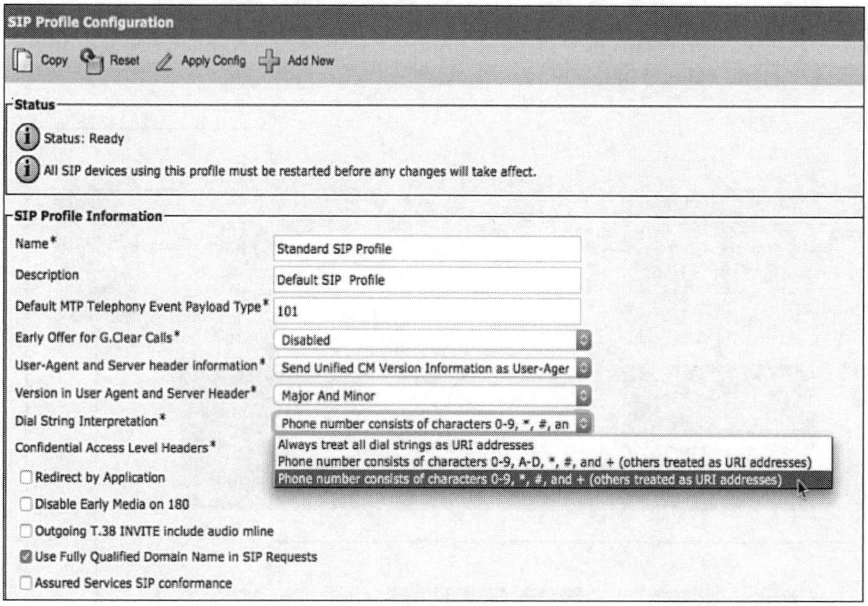

Figure 4-8 *SIP Profile Dial String Interpretations*

> **Note** Setting the Directory URI Dial String Interpretation rule to Always treats all dial strings as URI addresses and effectively disables numeric URI-based call routing.

In the call routing flow, if a URI is determined to be numeric, further call processing differs for local versus remote URIs. A pattern is considered to be local when one of the following conditions apply:

- The host or domain portion of the URI matches the IP address of a cluster member or server. CUCM clusters are aware of each of the server's IP addresses. This value is stored in a table in the IBM Informix database called the process_node table. This table is replicated between all cluster members.

- The host or domain portion of the URI matches the CFQDN. The CFQDN must be properly planned out in a multilevel domain environment. Many large enterprises have regional FQDNs, such as usa.cisco.com and emea.cisco.com. Setting the FQDN of a cluster to usa.cisco.com versus simply cisco.com must be well planned out, especially in terms of IM and Presence, Jabber, ILS, and GDPR.

- The host or domain portion of the URI matches the OTLD.

> **Note** The CFQDN and the OTLD parameters are both set on the enterprise parameters inside the CUCM Administration web page.

The complete call routing process for a directory URI is shown in Figure 4-9. Pay close attention to the numeric versus non-numeric logic in this flow chart. Each of the options is explained in greater detail in the next section.

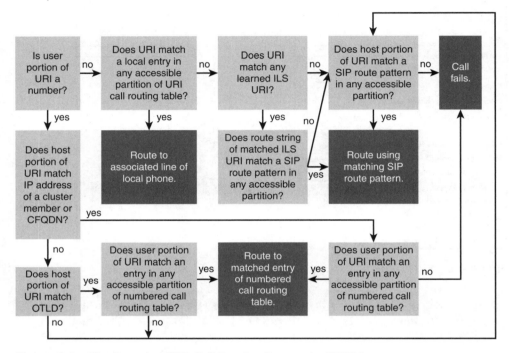

Figure 4-9   *The Complete URI Call Routing Process in CUCM*

Non-Numeric URI Call Routing Process

The first determination of CUCM call routing logic is if the dialed URI is a numeric or non-numeric URI. Based on the dial string interpretations discussed earlier in this chapter, assume a non-number URI was dialed, such as user@domain.com. When a SIP call is placed to a URI, and the user portion of the called URI is not a number, CUCM routes the call as follows:

Step 1.   The full URI is looked up in the URI routing table. The URI routing table is a database table in CUCM that contains all locally configured directory URIs. A directory URI is considered "local" when it is applied to a DN at the Directory Number Configuration page or via an end user in CUCM Administration. If a local match is found in the URI routing table, the call is sent to the corresponding phone line.

Note   It is important to note that this type of local lookup enforces the concept of CSS and partitions. Like DNs, a partition can be assigned to a directory URI. The CSS that is used for numeric call routing is also applicable or evaluated for the URI routing. In effect, the IP phone or endpoint making the call will use the line-over-device CSS methodology, where the line and device CSSs are evaluated for a partition that contains a closest-matching URI. In an event where CUCM finds multiple URI patterns that are equal, the tie-breaker is the order of the partition in the CSS with the highest partition in the list serving as the tie-breaker.

Step 2.   If no match is found and the dialed directory URI does *not* match an entry in the URI routing table, the full URI is looked up against the ILS-learned URIs. ILS is a feature where multiple CUCM clusters are joined via SIP trunk, and each cluster can "advertise" or "learn" the directory URIs and DNs of the other clusters.

Note   The lookup against the ILS-learned URIs always considers *all* ILS-learned URIs regardless of partitions assigned. In fact, ILS-learned URIs are *not* assigned a partition, because they are learned via other CUCM clusters. In fact, there is no way for CUCM to assign a partition to a URI belonging to a foreign CUCM cluster. The CSS and partition logic described in Step 1 does not apply to lookups of ILS-learned URIs.

Step 3.   If the full URI matches an ILS-learned URI, the dialed directory URI string is matched against all accessible SIP route patterns. A SIP route pattern is similar to a traditional route pattern in CUCM, only a SIP route pattern allows the routing of directory URIs. If no explicitly matching SIP route pattern (that is, a pattern that matches both the user@domain.com) is matched, the host or domain portion of the URI is looked up against all accessible SIP route patterns. In the case of the latter, the domain name @domain.com is evaluated for a possible matching SIP route pattern.

Step 4.   If the full dialed URI does not match an ILS-learned URI, the host or domain portion of the dialed URI is looked up against all accessible SIP route patterns.

Note   A partition can be assigned to a SIP route pattern. When a route string or the host part of a called URI is matched against the SIP route pattern, CSS and partition logic applies. SIP route patterns in partitions that are not included in the CSS of the calling device are ignored.

Step 5.   If a matching SIP route pattern is found, the call is routed using the matching SIP route pattern. If no matching SIP route pattern is found, the call fails, and the caller hears the annunciator or a fast busy signal.

In summary, the non-numeric URI call routing process depends on multiple factors and conditions. The calling party must have a CSS with the proper partitions involved to access a SIP route pattern. The call routing algorithm for URIs also depends on the type of URI lookup being performed. The full URI is first searched in the local URI routing table. ILS-learned URI patterns are examined second if no local match is found. Finally, regular SIP route patterns are searched and given the lowest priority.

Numeric URI Call Routing Process

The second determination CUCM call routing logic makes is if the dialed URI is indeed a numeric URI. Based on the dial string interpretations discussed earlier in this chapter, assume a numeric URI was dialed, such as 2000@domain.com. When a SIP call is placed to a URI, and the user portion of the called URI is in fact a number, CUCM routes the call as follows:

Step 1.   If the host or domain portion of the called URI is the IP address of one of the CUCM cluster servers or matches the CFQDN, the user portion of the URI is looked up in the accessible partitions of the standard numerical call routing table in the CUCM Informix database. If a match is found, the call is routed to the matched entry. If no match is found, the call fails, and the user hears the annunciator or a fast busy signal.

Step 2.   If the host or domain portion of the called URI matches the OTLD, the user portion of the URI is looked up in the accessible partitions of the standard numeric call routing table in the CUCM Informix database. If a match is found, the call is routed to the matched entry.

Step 3.   The host or domain portion of the called URI is looked up against the accessible SIP route patterns. If a match is found, the call is routed using the matched SIP route pattern. If no matching SIP route pattern is found, the call fails.

Note   All three steps described perform lookups that enforce the CSS and partition logic. If a match is found, the calling device must have a CSS with access to a partition containing the DN or SIP route pattern.

Routing URI Calls over SIP Trunks

In UC, separate CUCM clusters can be joined together via SIP trunks. SIP trunks can be used to route both numeric and URI-based calls between clusters.

There are several ways in which CUCM could potentially route a URI over a SIP trunk. Keep in mind that this scenario matches a URI routing to another CUCM cluster or perhaps to a CUBE and ultimately to an ITSP or telco. An alternate scenario is when

CUCM matches the host or domain portion of a URI that is learned via ILS. When the URI call routing process finds a match of the called host or domain portion of a URI and it matches an ILS route string or a SIP route pattern, the call is sent to a remote cluster or to a remote SIP call control system.

The SIP route pattern can be configured as an IP address or, more likely, as an FQDN. When you configure a SIP route pattern as an IP address, you must set the Pattern Usage to IP Address Routing. When you configure a SIP route pattern as an FQDN, you must choose Pattern Usage Domain Routing. Figure 4-10 provides a SIP route pattern configuration example inside CUCM.

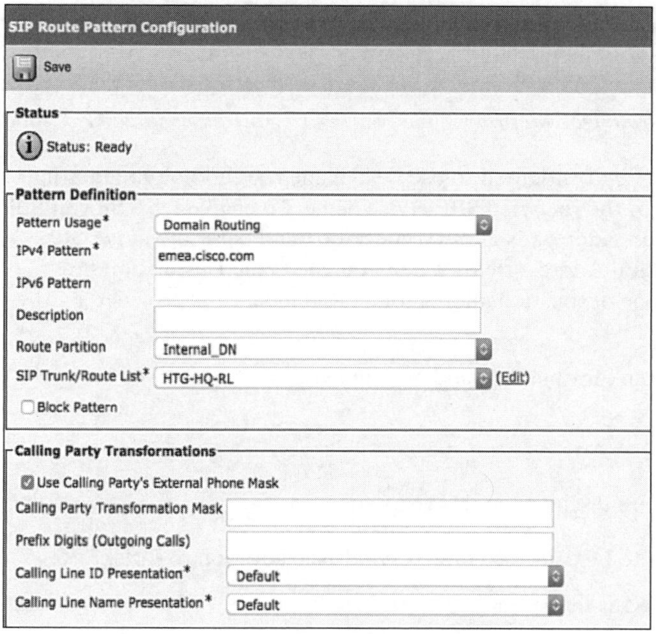

Figure 4-10 *SIP Route Pattern in CUCM Example*

You can configure a SIP route pattern to refer directly to a SIP trunk, or use the route list and route group construct. You must use the route list and route group construct when multiple SIP trunks are used for load-sharing or backup purposes.

Each SIP trunk refers to one or more destinations by IP address or FQDN. In the case of an FQDN, the FQDN is not only a domain (like in the case of a SIP route pattern) but also the FQDN of a host, specifically the host of the remote call control server. You can configure more than one destination per SIP trunk. If the remote call control system is a CUCM cluster, you must configure all members of the remote cluster that can control the interconnecting SIP trunk. If you omit a remote server, your cluster will not accept incoming calls from that unknown host. Figure 4-11 demonstrates multiple destination addresses as part of a SIP trunk inside CUCM.

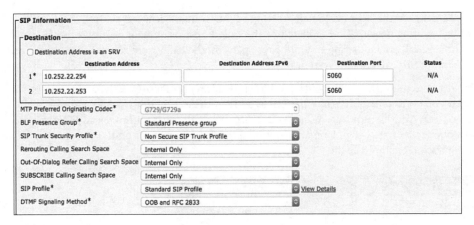

Figure 4-11   *Example of Multiple-Destination CUCM Server IP Addresses in a SIP Trunk*

Alternatively, a SIP trunk can be configured to use DNS name resolution by requesting the SIP server SRV records of the specified SIP service name. To enable DNS SRV at SIP trunks, check the **Destination Address Is an SRV** check box and enter the target SIP service name in the Destination Address field. When you check the Destination Is an SRV box, you can specify one destination address only. Redundancy is provided by the DNS server in this case.

You can use the same SIP trunk for numeric and non-numeric calls.

Summary

The following key points were discussed in this chapter:

- URI dialing allows calls to be placed to directory URIs in addition to DNs.

- Only one URI is assigned as the primary URI per DN.

- One or more directory URIs (up to five total) can be assigned to a DN via the Directory Number Configuration page. Only one directory URI can be assigned via the End User Configuration page.

- End user-based directory URIs are always put into a default partition called directory URI. Directory URIs that are configured at the DN configuration page can be placed into any partition.

- Calls to directory URIs can be placed by using speed dials or call list entries and they can be manually entered.

- When blended addressing is enabled, the remote phone is shown by its DN and its directory URI.

- By default, CUCM replaces the host portion of the remote URI with its own IP address. This behavior can be deactivated by enabling the Use Fully Qualified Domain Name SIP profile parameter.

- The URI call routing process is different for numeric and non-numeric URIs. SIP route patterns match on URI host portions by FQDN or IP address. They refer to a route list or directly to a SIP trunk. A SIP trunk refers to the remote SIP call control system by IP address or FQDN or to the destination SIP service name to be resolved by DNS SRV records.

References

For additional information, refer to these resources:

Cisco Systems, Inc. Cisco Collaboration Systems 10.x Solution Reference Network Designs (SRND), May 2014. http://www.cisco.com/c/en/us/td/docs/voice_ip_comm/cucm/srnd/collab10/collab10.html

Cisco Systems, Inc. Cisco Collaboration Systems 11.x Solution Reference Network Designs (SRND), July 2015. http://www.cisco.com/c/en/us/td/docs/voice_ip_comm/cucm/srnd/collab11/collab11.html

Cisco Systems, Inc. Features and Services Guide for Cisco Unified Communications Manager, Release 10.0(1), May 2014. http://www.cisco.com/c/en/us/td/docs/voice_ip_comm/cucm/admin/10_0_1/ccmfeat/CUCM_BK_F3AC1C0F_00_cucm-features-services-guide-100/CUCM_BK_F3AC1C0F_00_cucm-features-services-guide-100_chapter_0110011.html

Review Questions

Use these questions to review what you have learned in this chapter. The answers appear in Appendix A, "Answers Appendix."

1. True or false: Directory URIs that are assigned at the end-user level are automatically placed into the default directory URI partition, where directory URIs configured at the Directory Number Configuration page can be placed into any partition.

   a. True

   b. False

2. How many directory URIs can be assigned or associated to a directory number?

   a. 1

   b. 2

   c. 3

   d. 4

   e. 5

   f. 6

   g. 7

   h. 8

3. True or false: All Cisco IP phone models support directory URI dialing capabilities.

   a. True

   b. False

4. Which two of the following are not acceptable characters in the user portion of a directory URI (the portion before the @ symbol)? (Choose two.)

   a. Y

   b. B

   c. ^

   d. !

   e. (

   f. %

5. What is the maximum allowed length of a directory URI?

   a. 128 characters

   b. 64 characters

   c. 23 characters

   d. 254 characters

   e. 256 characters

6. Which two of the following are valid LDAP attributes that can be synchronized into CUCM as a directory URI? (Choose two.)

   a. First name

   b. Telephone number

   c. Mail

   d. msRTCSIP-primaryuseraddress

   e. Last name

   f. Title

7. Which of the following SIP header identity attributes identifies the associated directory number in blended address dialing?

   a. User-Agent

   b. x-cisco-number

   c. m=audio RTP/AVP

    d. PRACK

    e. NOTIFY

8. True or false: ILS-learned URIs can be assigned to a partition.

    a. True

    b. False

9. Which of the following SIP profile attributes determines if CUCM will evaluate the dialed URI as a numeric versus non-numeric SIP URI?

    a. Use fully qualified domain name in SIP requests

    b. User agent and server header information

    c. Calling line identification presentation

    d. Send send-receive SDP in mid-call INVITE

    e. Dial string interpretation

10. True or false: By default, URI lookup policy is set to case insensitive.

    a. True

    b. False

Chapter 5

Remote Site Telephony and Branch Redundancy Options

Sometimes it is not feasible or cost-effective to have a full-blown Cisco Unified Communications Manager (CUCM) solution for remote sites or smaller office locations. To integrate these remote sites, small to medium-size businesses (SMBs), or small office/home office (SOHO) with enterprise networks, Cisco IOS Unified Communications (UC) gateways offers a wide variety of functionality, ranging from call control features, survivability features, and redundancy options for remote site infrastructure.

This chapter covers the fundamentals of Cisco remote site telephony options and survivable remote site telephony (SRST) and its variations.

Upon completing this chapter, you will be able to meet these objectives:

- Describe remote site telephony options
- Describe Cisco SRST
- Describe Cisco E-SRST
- Describe Cisco SIP SRST
- Describe MGCP fallback
- Describe CFUR

Cisco Unified Communications Manager Express

Cisco Unified Communications Manager Express (Cisco Unified CME) is the express version of Cisco's enterprise call control solution. A Cisco Unified CME-based solution is ideal for small businesses or remote sites in an enterprise environment. Figure 5-1 illustrates a Cisco Unified CME-based telephony solution connecting two SOHO/SMB sites.

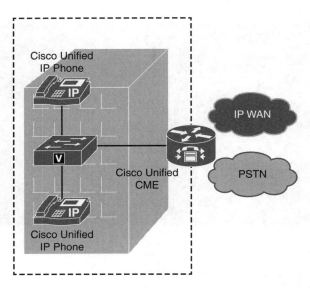

Figure 5-1   *Cisco Unified CME-Based Telephony Solution*

A Cisco Unified CME is capable of the following:

- Supporting Skinny Client Control Protocol (SCCP) and Session Initiation Protocol (SIP) phones

- Terminating public switched telephone network (PSTN) (traditional E1/T1) or SIP trunks

- Tight integration with CUCM and Unity Connection

- Integrating with Cisco Unity Express

- Offering features like Call-Park, Ad-Hoc, or Meet-Me conferences, and Transcoding

- Offering an interactive self-service solution: Basic Automatic Call Distribution (B-ACD)

Hence, Cisco Unified CME can successfully service a site ranging from 25 to 1500 users (or even more depending on the model of router and licenses chosen). As discussed later in this chapter, Cisco Unified CME supports both SIP and SCCP endpoints and enterprise-grade features during regular and survivability mode, including video endpoints, conferencing, music on hold (MOH), paging, call-park, and so on.

Cisco Business Edition

Cisco offers another solution for SMB and remote offices in the form of Cisco Business Edition. Cisco Business Edition (5000/6000/7000) is a feature-rich, enterprise-class collaboration solution aimed at midsize businesses, where the full-blown capability of CUCM and relevant enterprise applications would be much more than required. Cisco Business Edition can scale from a few hundred to about 1000 users and also offers contact center ability for the agents. Cisco Business Edition comes with preloaded images of core applications, such as the following:

- CUCM

- Instant messaging/presence with Jabber

- Cisco Unity Connection

- Unified Contact Center Express

- Unified Attendant Console

The applications and licenses depend on the model of Cisco Business Edition, and the size of users/agents it can support varies as per the underlying platform (for example, Cisco UCS server [B-Series/C-Series]). Figure 5-2 depicts the Cisco Business Edition Collaboration setup on a Cisco B-Series UCS server.

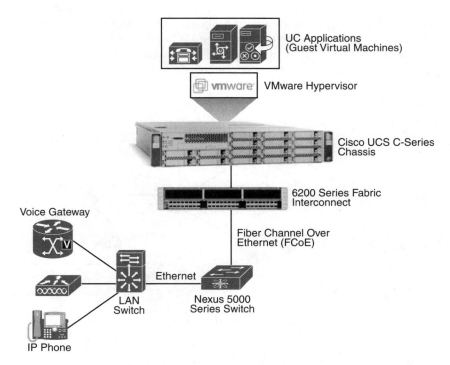

Figure 5-2 *Cisco Business Edition: Collaboration Setup for a Remote Site*

Survivable Remote Site Telephony

All Cisco endpoints register with CUCM by default unless there is another call control that can support the same or if an endpoint has specific call control requirements (for example, video endpoints that register with VCS). In the unlikely event of a WAN failure, the remote site endpoints need a backup call control. In such case, Cisco Unified survivable remote site telephony (SRST) can provide necessary call control features based on the Cisco IOS router, such as a Cisco Integrated Services Router (ISR) or Aggregation Services Router (ASR). Cisco Unified SRST (hereafter referred to as SRST) is an IOS gateway-based (license) feature that enables administrators to configure redundant call control for sites that do not have a local CUCM server. Cisco Unified SRST function can be co-resident with many other IOS features, including Cisco Unified Border Element (CUBE).

Cisco IOS gateways support two types of SRST:

- SRST (call manager fallback, traditional)

- Cisco Unified CME-based SRST or enhanced SRST (E-SRST)

The major difference between these is that traditional SRST offers basic telephony features and E-SRST provides an enhanced feature set for the endpoints. E-SRST builds on top of SRST by automatically synchronizing user, phone, hunt group, partitions, and so on from CUCM to Cisco Unified CME. Any change on CUCM is propagated to the branch sites without admin intervention. So when the WAN is down, the phone resembles the last changes performed on CUCM. When in E-SRST mode, Cisco Unified CME syncs automatically with CUCM to get updates as and when they happen on CUCM. Figure 5-3 depicts Cisco Unified SRST solution.

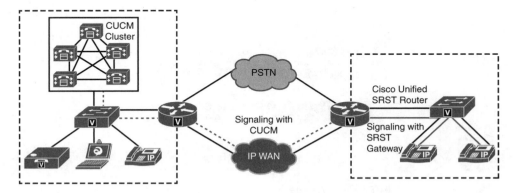

Figure 5-3   *Cisco Unified SRST solution*

The Figure 5-4 illustrates Cisco Unified CME-based SRST.

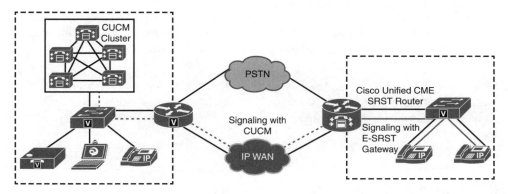

Figure 5-4 *Cisco Unified CME SRST Solution*

When the endpoints lose connectivity with CUCM (WAN failure or remote server is unavailable), SRST kicks in and allows the endpoints to register with the configured SRST reference—the IOS voice gateway (as defined in CUCM). In the case of traditional SRST, the router builds on the configured application service. The SRST gateway detects newly registered IP phones and queries these IP phones for their configuration, and then autoconfigures itself using the Simple Network Auto Provisioning (SNAP) technology. Only one instance of SRST is allowed per device pool in CUCM.

Cisco Unified SRST is mostly leveraged with Media Gateway Control Protocol (MGCP) fallback, as explained later in this chapter. Cisco Unified SRST supports the following features:

- Simple one-time configuration of basic SRST functions

- Customizable programmable line keys and button layout control

- Support for Forced Authorization Code (FAC)

- Normalized +E.164 support

In case the SRST reference is Cisco Unified CME at the remote site/branch office, the IOS router first searches for an existing configured ephone with the MAC address of the phone that tries to register. If an ephone is found, the stored configuration is used. The phone configuration settings provided by SNAP are not applied any ephone template. If no ephone is found for the MAC address of the registration phone, the router adds the ephone and applies the ephone template (similar to configuration using SNAP). Following are some E-SRST features that distinguish it from traditional SRST:

- Automatic provisioning of remote branch sites.

- E-SRST router is in sync with CUCM that pushes the updates to the branch routers. Automatic sync for moves, adds, and deletions from CUCM to router.

- GUI interface for provisioning, monitoring, reporting, and troubleshooting.

- On-demand information sync with CUCM.

Cisco Unified CME supports two modes in which it can be configured using E-SRST: Skinny Client Control Protocol (SCCP) and Session Initiation Protocol (SIP). This allows support for analog, SIP, and SCCP phones while in SRST mode.

> **Note** Cisco Unified CME can be deployed as an IOS feature on a Cisco ISR or ASR router, or as a Cisco Services Ready Engine (SRE) module. See http://www.cisco.com/c/en/us/products/interfaces-modules/services-ready-engine-sre-modules/index.html for more information.

SRST and E-SRST Configuration

As explained in the previous section, SRST and E-SRST are the two possible ways to provide call control redundancy for the remote sites. This section details the configuration for the these two scenarios.

Example 5-1 outlines SRST configuration.

Example 5-1 *SRST Configuration*

```
SRST-Router(config)# call-manager-fallback
SRST-Router(config-cm-fallback)# ip source-address 10.76.108.78 port 2000
SRST-Router(config-cm-fallback)# max-ephones 10
SRST-Router(config-cm-fallback)# max-dn 100
SRST-Router(config-cm-fallback)# max-conferences 4 gain -6
SRST-Router(config-cm-fallback)# transfer-system full-consult
SRST-Router(config-cm-fallback)# secondary-dialtone 9
SRST-Router(config-cm-fallback)# moh music-on-hold.au
SRST-Router(config-cm-fallback)# time-format 24
SRST-Router(config-cm-fallback)# date-format dd-mm-yy
SRST-Router(config-cm-fallback)# system message SRST Mode
```

Example 5-1 defines IOS commands that help configure SRST on Cisco IOS router. The key commands are as follows:

- **ip source-address** defines the address of a physical or virtual (loopback) interface on the router with which the phones should register while using the SRST process. This is defined along with a port (2000 for SCCP).

- **max-ephones** defines the total number of ephones, and **max-dn** defines the maximum number of directory numbers under **call-manager-fallback** for the phones at the remote site.

- **secondary-dialtone** gives the same dialing experience for plain switched telephone network (PSTN) calls to end users (as they would experience when phones are registered with CUCM). In Example 5-1, pressing 9 gives an outside dialtone (it can be set to 0 for Europe and other regions as required).

- **moh** (music on hold) command defines the filename that is used for music on hold. This in turn allows playing MOH when remote site is in SRST mode.

- **time** and **date** format set the right format for the phones.

- **system message** helps display a message (configurable by administrator) that can help make users aware that the endpoints are running in SRST mode.

Example 5-2 explains Cisco Unified CME based SCCP based E-SRST.

Example 5-2 *SCCP E-SRST Configuration*

```
CUCME-Router(config)# telephony-service
CUCME-Router(config-telephony)# srst mode auto-provision none
CUCME-Router(config-telephony)# srst dn line-mode dual
CUCME-Router(config-telephony)# srst ephone template 1
CUCME-Router(config-telephony)# srst dn template 1
CUCME-Router(config-telephony)# max-ephone 20
CUCME-Router(config-telephony)# max-dn 40
CUCME-Router(config-telephony)# ip source-address 10.76.108.76 port 2000
CUCME-Router(config-telephony)# moh music-on-hold.au
CUCME-Router(config-telephony)# max-conferences 4 gain -6
CUCME-Router(config-telephony)# secondary-dialtone 9
CUCME-Router(config-telephony)# system message SRST Mode
!
CUCME-Router(config)# ephone-template 1
CUCME-Router(config-ephone-template)# keep-conference local-only
!
CUCME-Router(config)# ephone-dn-template 1
CUCME-Router(config-ephone-template)# hold-alert 25 idle
```

Example 5-2 has a number of IOS commands that define the following tasks:

- **telephony-service** initiates Cisco Unified CME configuration.

- **srst** command before usual Cisco Unified CME commands defines Cisco Unified CME mode for E-SRST provisioning.

- The mode for ephone-dns is set to dual-line.

- **ephone-template** applies common ephone configuration for any newly configured phone on Cisco Unified CME in SRST mode.

- **dn template** refers to the template used for defining common properties for a DN for a new ephone configured.

Other options are similar to traditional SRST configuration, as discussed earlier in Example 5-1.

Example 5-3 describes the configuration for Cisco Unified CME based SIP SRST.

Example 5-3 *SIP E-SRST Configuration*

```
CUCME-Router(config)# voice register global
CUCME-Router(config-register-global)# mode srst
CUCME-Router(config-register-global)# source-address 10.76.108.76 port 5060
CUCME-Router(config-register-global)# max-dn 40
CUCME-Router(config-register-global)# max-pool 10
CUCME-Router(config- register-global)# system message SRST Mode
```

As shown in Example 5-3, the following commands help define various tasks:

- SIP configuration starts with change of mode from **cme** to **srst**.

- **max-dn** defines the maximum directory numbers (DNs) to be supported during SRST.

- **max-pool** defines the maximum number of SIP voice register pools.

SRST IOS Dial Plan

When the endpoints enter SRST/E-SRST, a dial plan is required such that in absence of CUCM the router can route calls to/from PSTN/ITSP (SIP provider). Example 5-4 shows a basic dial North America Numbering Plan (NANP) for the SRST gateway to accept incoming calls and enable users to dial emergency, local, national, and international numbers. The configuration also includes routing for incoming calls.

> **Note** The configuration shown is a basic IOS dial plan. An expanded IOS dial plan consists of voice translations and rules, number expansions, class of service, and so on. Refer to "Implementing Cisco Unified Communications Manager, Part 1 (CIPTv1)" for details of IOS dial plan.

Example 5-4 *SRST Router Dial Plan Configuration*

```
SRST-Router(config)# dial-peer voice 100 pots
SRST-Router(config-dial-peer)# description all incoming calls
SRST-Router(config-dial-peer)# incoming called-number .
SRST-Router(config-dial-peer)# direct-inward-dial
SRST-Router(config-dial-peer)# port 0/1:23
SRST-Router(config-dial-peer)# forward-digits all
!
SRST-Router(config)# dial-peer voice 10 pots
SRST-Router(config-dial-peer)# description Local outgoing calls
SRST-Router(config-dial-peer)# destination-pattern 9[2-9]......
SRST-Router(config-dial-peer)# port 0/1:23
SRST-Router(config-dial-peer)# forward-digits all
```

```
!
SRST-Router(config)# dial-peer voice 20 pots
SRST-Router(config-dial-peer)# description Long Distance outgoing calls
SRST-Router(config-dial-peer)# destination-pattern 91[2-9]..[2-9]......
SRST-Router(config-dial-peer)# port 0/1:23
SRST-Router(config-dial-peer)# forward-digits all
!
SRST-Router(config)# dial-peer voice 30 pots
SRST-Router(config-dial-peer)# description International outgoing calls
SRST-Router(config-dial-peer)# destination-pattern 9011T
SRST-Router(config-dial-peer)# port 0/1:23
SRST-Router(config-dial-peer)# forward-digits all
!
SRST-Router(config)# dial-peer voice 911 pots
SRST-Router(config-dial-peer)# description Emergency calls to 911
SRST-Router(config-dial-peer)# destination-pattern 911
SRST-Router(config-dial-peer)# port 0/1:23
SRST-Router(config-dial-peer)# forward-digits all
```

Note   A similar dial plan is required on a Cisco Unified CME router for routing calls to remote site phones. When you are using Cisco Unified CME router for E-SRST, the existing dial plan can be leveraged for SRST/E-SRST as well. In the case of an MGCP gateway, dial plan configuration is explicitly required.

The next section describes the CUCM side of configuration to support SRST.

CUCM SRST Configuration

CUCM SRST configuration is similar for both Cisco Unified SRST or Cisco Unified CME-based SRST. The IOS router must be defined as an SRST reference in CUCM. Then this SRST reference consecutively needs to be added to the appropriate device pool to enable the phones at remote sites to leverage SRST. To configure an SRST reference and add it to a device pool, follow these steps:

Step 1.   Go to **CUCM Administration GUI > System > SRST**, click **Add New**, and enter the required details about the remote site gateway, as shown in Figure 5-5. Ensure that the SRST gateway's hostname/IP address is defined and the correct protocol (SCCP/SIP) port is chosen. If the gateway is referenced by hostname, be sure that CUCM can resolve the name and has DNS servers configured.

Note   SRST reference should be defined for every gateway that is expected to run SRST or E-SRST service.

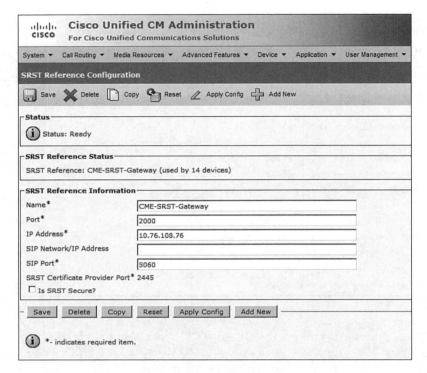

Figure 5-5   *Cisco SRST Reference Configuration*

Step 2.   Browse to **System > Device Pool** and select a device pool for remote site phones. Let's assume that you have created a device pool per remote site. Assign the appropriate SRST reference per device pool. Configure the SRST reference in the device pool and save the configuration.

At this time, the CUCM part of SRST reference configuration is complete and the remote site gateway or Cisco Unified CME needs to be configured (described in earlier sections of this chapter) for SRST or E-SRST, respectively.

Multicast Music on Hold in SRST

While in SRST, the IOS router can play an audio file from its flash as a multicast MOH to users in the remote sites (similar to multicast MOH explained in Chapter 6, "Cisco Collaboration Solution Bandwidth Management"). This feature has the following limitations:

- A single MOH source should be used for all phones at remote sites. Multicast MOH can only support one file.

- Cisco Unified SRST multicast MOH supports only G.711, and MOH from an SRST router cannot be unicast. Multicast MOH can work in one of the following two ways:

- **Nonfallback mode:** This occurs when the WAN link is up and the phones are controlled by CUCM. It allows the phones to consult a local MOH file instead of reaching out to CUCM across the WAN. This option is covered in detail in Chapter 6.

- **Fallback mode:** This occurs when SRST is active; i.e., the remote site has lost connectivity to the central-site CUCM. In this case, the branch router can continue to provide multicast MOH.

For configuring multicast MOH, CUCM and voice gateway at remote sites must be configured to support multicast. See multicast configuration for supporting MOH in Chapter 6. Assuming that multicast routing in the router is enabled and PIM dense mode is configured for identified interfaces, both CUCM and the IOS router need to be configured to support multicast MOH. To configure CUCM for supporting multicast MOH, follow these steps:

Step 1.   Go to **CUCM Administration GUI > Media Resources > Music on Hold Audio Source** and select the audio source you are enabling for multicast. Ensure that the **Allow Multi-casting** check box is checked, as shown in Figure 5-6.

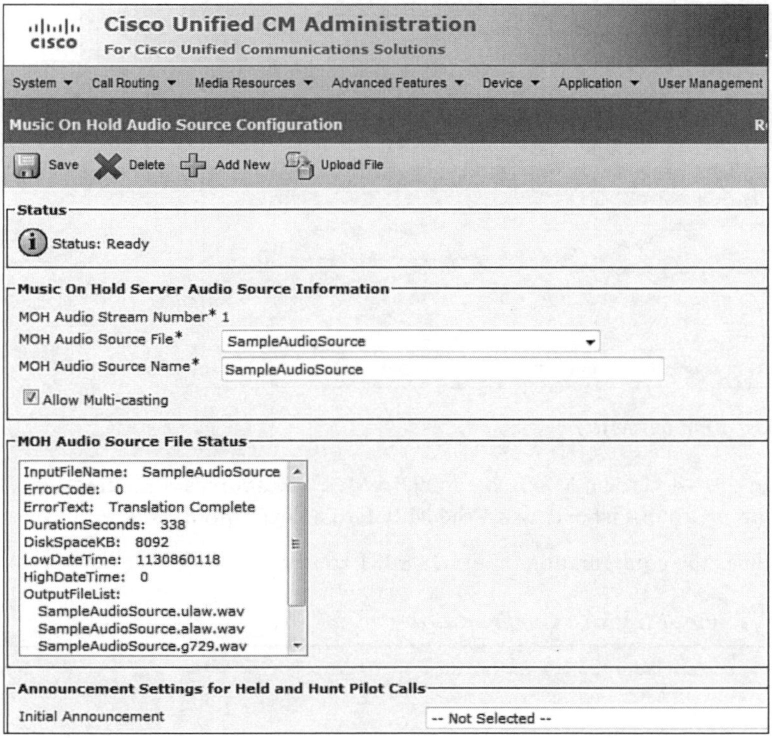

Figure 5-6   *Multicast MOH Audio Source*

Step 2.   Go to **Media Resources > Music on Hold Server Configuration**. Enable multicast support and select the option between port number or IP address. Ensure that the MOH server has a G.711 only enabled device pool assigned to it.

Step 3.   Go to **Media Resources > Media Resource Group (MRG)** and click **Add New**. Add a new MRG and ensure that this MRG has the multicast-enabled MOH server assigned to it and is multicast enabled, as shown in Figure 5-7.

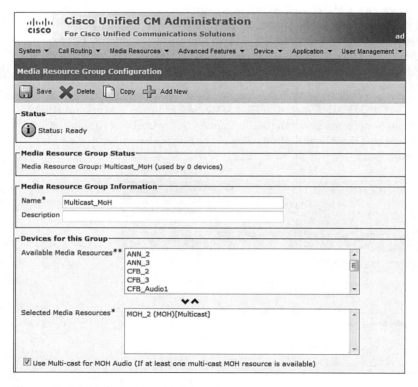

Figure 5-7   *Multicast-Enabled MRG*

Step 4.   Assign the MRG to an MRGL by going to **Media Resources > Media Resource Group List** and assign the MRGL to a device pool.

Example 5-5 outlines the configuration of an IOS SRST router.

Example 5-5   *SRST Router Multicast Configuration*

```
SRST-Router(config)# ccm-manager music-on-hold
!
SRST-Router(config)# interface loopback 1
SRST-Router(config-if)# ip address 10.86.108.82 255.255.255.255
!
```

```
SRST-Router(config)# call-manager-fallback
SRST-Router(config-cm-fallback)# ip source-address 10.76.108.78 port 2000
SRST-Router(config-cm-fallback)# moh music-on-hold.au
SRST-Router(config-cm-fallback)# multicast moh 239.1.1.1 port 16384 route
   10.86.108.82 10.76.108.78
```

MGCP Fallback

MGCP fallback provides call control functionality using application alternate H.323 operation when a gateway is configured for MGCP in CUCM. An MGCP gateway registers with CUCM like SCCP/SIP phones. MGCP fallback enables a gateway to act as local call control when the CUCM server to which the remote site phones and gateway register is offline or WAN connectivity is lost (in which case, Cisco Unified SRST kicks in and offers call control functionality). This is specifically useful when a centralized call control using MGCP is desired for all voice gateways.

Note   MGCP is a master-slave protocol, where CUCM (call control) acts as a master and the voice gateway (IOS router) acts as a slave. All dial plan intelligence is with CUCM, and the gateway receives commands from CUCM on the actions to be performed.

Figure 5-8 shows MGCP fallback (that is, MGCP application to default application H.323).

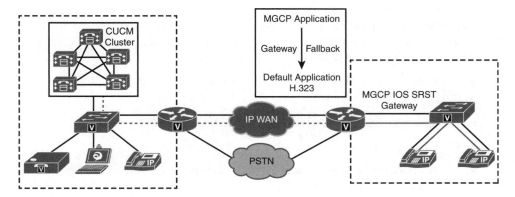

Figure 5-8   *MGCP Fallback Operation*

To configure MGCP fallback and enable default H.323 application-based SRST, perform the following steps:

Step 1.   Configure SRST reference in CUCM and assign it to the appropriate CUCM device pool (refer to the CUCM SRST Configuration section).

Step 2.   Configure the Call Forward Unregistered (CFUR) internal and external destinations on line of the remote site phones to an E.164 number (for example, a shared line on the main site or voicemail number). This is accomplished by going to **Device > Phone**, selecting a phone, selecting a line, and setting CFUR.

Step 3.   Configure the MGCP fallback and Cisco Unified SRST on the remote site gateways and implement the SRST dial plan on the remote site gateways. MGCP fallback configuration is shown in Example 5-6.

Example 5-6   *MGCP Fallback Configuration*

```
MGCP-Router(config)# ccm-manager fallback-mgcp
MGCP-Router(config)# application
MGCP-Router(config-app)# global
MGCP-Router(config-app-global)# service alternate default
```

At this time, the IOS router is ready for fallback to H.323 application (defined as service alternate default in Example 5-6). Upon loss of connectivity (server/CallManager service is unavailable or WAN connection is down) to the CUCM server, the phones register with the gateway and continue to operate as normal (in SRST mode) until the link or CUCM service is restored.

Cisco Call Forward Unregistered

One of the problems with SRST is that the CUCM is unable to identify if a whole site has gone offline and that it may possibly be in SRST mode. In such a case, a user trying to call a phone at remote site gets either a busy signal or rolls to voicemail (or the call handling behavior depending on the configuration). How do you reach the remote site from the HQ or main site if the remote site is in SRST mode? The problem is solved by the Call Forwarding Unregistered (CFUR) feature of CUCM.

On each device line directory number (DN), CFUR is a call forwarding option that is engaged when a device is not registered. This is true in the following instances:

■ The phone is unregistered, which implies the phone was registered earlier

■ The phone is unknown, which implies the phone was never registered

■ A user device profile is not logged in, which implies the user profile configured with CFUR for a physical phone or softphone that is not logged into extension mobility (and this is a segue to a later discussion of another potential use of CFUR).

Note   CFUR is configured per DN, but it gets activated when the physical phone or softphone associated with the DN is affected by the scenarios described above.

When an active DN is unregistered, a call to this line is redirected based on the CFUR configuration. Figure 5-9 explains the CFUR process in conjunction with SRST scenario.

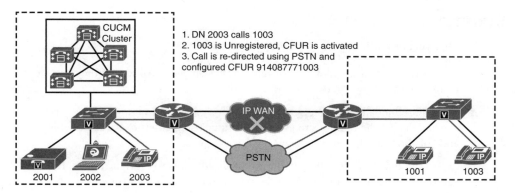

Figure 5-9 *CFUR-Based Call Routing*

The following steps explain the CFUR operation in conjunction with SRST:

Step 1. The phone with DN 2003 goes off-hook and dials 1003 at the remote site. The CUCM dial plan comes to action associated with the digits dialed (digit analysis).

Step 2. Because the remote site is in SRST mode, all phones at that site have the status of unregistered, which is one of the preliminary conditions for CFUR to begin.

Step 3. The CUCM call processing node handling DN 2003's call setup request passes the call decision process to the ForwardManager subprocess for a call routing decision. In turn, the ForwardManager subprocess determines that DN 1003 has a CFUR setting of 914087771003 (a valid E.164 number routable on the PSTN).

Step 4. The call is routed over PSTN to the remote site, where the call is handled as a direct inward dialing (DID) call (assuming that the calling search space [CSS] configured for CFUR has valid public switched telephone network [PSTN] access). After digit manipulation by the remote site's SRST/E-SRST router, a call setup is sent to DN 1003 that is registered with the SRST/E-SRST router as ephone, and the audio channel is established thereafter (following the normal call setup process).

Another potential use case for CFUR apart from use of CFUR in conjunction with SRST for availability is that CFUR can also be implemented when a user is leveraging a soft client. For example, if a mobile user has a laptop with a softphone (for instance, Cisco IP Communicator or Cisco Jabber) and the user shuts down the laptop or places it into standby/sleep mode, CFUR can forward calls destined for the unregistered softphone to the user's cell phone. The user does not have to set up Call Forward All (CFA) manually before closing the softphone application. If the softphone is not registered, calls are forwarded to the user's cell phone. This is another application of the CFUR feature that improves availability in CUCM deployments.

Summary

The following key points were discussed in this chapter:

- Cisco offers a host of solutions for remote sites/smaller sites so that they can connect to and leverage the services from the enterprise network.

- Keeping redundancy in mind, Cisco IOS gateways host the endpoints at the remote site and enable the remote site users to continue using enterprise-grade telephony and collaboration features.

- SRST (and E-SRST) allow the SCCP (and SIP) endpoints to register to local IOS gateways and enable calling within the remote site as well as outside using PSTN or SIP trunks.

References

For additional information, refer to these resources:

Cisco Systems, Inc. Cisco Collaboration Systems 10.x Solution Reference Network Designs (SRND), May 2014. http://www.cisco.com/c/en/us/td/docs/voice_ip_comm/cucm/srnd/collab10/collab10.html

Cisco Systems, Inc. Cisco Collaboration Systems 11.x Solution Reference Network Designs (SRND), July 2015. http://www.cisco.com/c/en/us/td/docs/voice_ip_comm/cucm/srnd/collab11/collab11.html

Review Questions

Use these questions to review what you have learned in this chapter. The answers appear in Appendix A, "Answers Appendix."

1. **Which of the following best describes Cisco Collaboration Solution's extension for remote site telephony?**

   a. Cisco Unified Communications Manager Express (Cisco Unified CME)-based call control

   b. Cisco Business Edition

   c. Cisco Unified Attendant Console

   d. a and b

   e. b and c

2. **Which of the following are two most commonly used remote site survivability solutions? (Choose two.)**

   a. Cisco Unified SRST

   b. Cisco Unified CME-based SRST

   c. Cisco Business Edition

   d. CUBE

3. **What is the default fallback application in Cisco IOS routers?**

   a. SIP

   b. H.323

   c. ISDN

   d. SCCP

4. **True or false: Multiple multicast MOH sources can be used in SRST mode.**

   a. True

   b. False

5. **Cisco Unified CME can support which three type of endpoints? (Choose three.)**

   a. Analog

   b. Digital

   c. SCCP

   d. SIP

   e. MGCP

6. **How many instances of SRST can be defined in a device pool on CUCM?**

   a. 1

   b. 2

   c. 3

   d. As many as configured in CUCM

7. **Which two commands are used for SRST configuration at the Cisco IOS router?**

   a. telephony-server

   b. ccm-manager fallback

   c. service alternate default

   d. call-manager-fallback

8. Which of the following five commands are configured in telephony service E-SRST configuration mode? (Choose five.)

a. create cnf-files

b. ephone

c. srst ephone-dn

d. max-ephones

e. max-dn

f. ip source-address

g. moh

h. number

Cisco Collaboration Solution Bandwidth Management

A Cisco Collaboration solution, like any IP-based communications network, relies heavily on transport of information (signaling and media) over IP-enabled links. The links can be from Multiprotocol Label Switching (MPLS) to Frame Relay to older ATM links. Every organization has a certain investment in the bandwidth over these physical links and bandwidth is an expensive asset. To optimize bandwidth, consumption over WAN links is a priority of the IT work force in every organization. However, a LAN can be seen as a pool of "virtually" unlimited bandwidth, with local traffic traversing over switches that can handle traffic in order of a few megabits per second to many gigabits per second. Quality of service (QoS), however, can only be guaranteed end to end if it is enabled from source to destination (that is device-LAN-WAN-LAN-device). In this chapter, you will learn about various bandwidth management options and their implementation in a Cisco Collaboration network.

Upon completing this chapter, you will be able to meet these objectives:

- Define bandwidth management options

- Define Cisco Unified Communications Manager codecs and regions

- Define local conference bridge implementation

- Define transcoder implementation

- Define multicast MOH from branch router flash

Bandwidth Management Options

There are multiple ways to tune the network for optimal and efficient bandwidth consumption in an enterprise Collaboration setup. The most commonly used mechanism is setting coders-decoders (codecs) to optimally match the local or remote site requirements. For example, using a high compression codec such as G.729r8 over a WAN is a better choice than using a low-compression codec such as G.711u-law.

The reason is that even though the latter offers higher quality, it consumes almost eight times the bandwidth consumed by G.729r8.

Another method that is used for bandwidth management and at the same time to conserve the quality of ongoing calls is call admission control (CAC). The following are a few common flavors of Cisco Unified Communications Manager (CUCM) CAC:

- **Location-based CAC (LCAC):** Static, intracluster, and topology-unaware CAC mechanism used in CUCM.

- **Enhanced LCAC:** As the name suggests, this is the advanced form of location-based CAC. It is topology aware, but it is dynamic in nature and can be used for intercluster bandwidth management.

- **Resource Reservation Protocol (RSVP):** Topology-aware CAC used in conjunction with IOS routers.

For details on these CAC mechanisms, see Chapter 7, "Call Admission Control (CAC) Implementation."

Quality of service (QoS) is another way to optimize bandwidth utilization and forms an important part of bandwidth management. By marking/re-marking traffic to different classes and assigning them to different priority queues, voice and video traffic can be optimized for timely transfer over the WAN links. Moreover, QoS offers policing to drop unwanted traffic. This reduces scavenger (unwanted or spam traffic [for example, P2P]) traffic. For details on QoS, refer to the *Implementing Cisco IP Telephony and Video, Part 1 (CIPTV1) Foundation Learning Guide*. There are many ways to reduce WAN bandwidth consumption and to ensure that successful Cisco Collaboration networks can be deployed without compromising quality of communication and keeping the budget under control.

Another bandwidth management solution is to have remote sites play music on hold (MOH) from a local IOS router's flash instead of streaming it (unicast) from the main site. This saves on *n* number of MOH streams (where *n* = number of IP phones), thereby protecting precious WAN bandwidth, bypassing the usual way of CUCM-based MOH server serving each IP phone. When using a local MOH source, it is multicast MOH from the branch router flash.

Another innovative solution is to leverage the local IOS router's digital signal processor (DSP) resources, which is also known as Packet Voice DSP Module (PVDM) to deploy a conference bridge, thereby allowing the remote/branch site's phones to join a conference call that is locally hosted on the branch router. For example, if three or more phones, all located at a remote site, try to establish an ad hoc conference, and if the conference resource is located at the main site, it has to be accessed over the IP WAN, and as a result the WAN bandwidth is used ineffectually. The conference bridge across the WAN mixes the received audio from the phones and then streams it back to all conference members in three separate unicast streams. Now, add to the fact if all phones are in a region that allows G.711 calls, at least 3x G.711 sessions have to be maintained by the main site conference bridge, and subsequently the WAN bandwidth takes a hit. In contrast to a local conference bridge on site, there is no bandwidth used over the WAN for conferencing (unless the local conference bridge is down or full and the phones are allowed to leverage the main site conference bridge, purely for redundancy purposes).

Another efficient way to reduce WAN bandwidth consumption is to use transcoders. Transcoders are based on Cisco IOS router DSP resources (PVDM)-like conference bridges. A transcoder can transcode IP voice streams (that is, they can change the way audio payload is encoded while it traverses the WAN). For example, using a transcoder, a G.711 audio stream can be changed into a G.729 audio stream such that lesser bandwidth is consumed (per call basis over the WAN).

The next sections describe the various bandwidth optimization mechanisms.

Voice and Video Codecs

Codec is an abbreviated form of coder-decoder. Codecs perform encoding and decoding on a digital data stream or signal and translate Voice over IP (VoIP) media streams into another format such as analog to digital, digital to analog, and digital to digital. A digital signal processor (DSP) is required to process voice signals from one format to another. For example, a DSP in an IP phone converts analog to digital (IP) voice and vice versa. However, a DSP on a PVDM module can be used for converting from one coding format to another. The most commonly used codecs in a Cisco Collaboration network include the following:

- **G.711:** G.711 is the default pulse code modulation (PCM) standard for IP-based private branch exchanges (PBXs) and for the public switched telephone network (PSTN). G.711 digitizes analog voice signals, producing output at 64 kbps. G.711 has two formats: G.711ulaw is used in North America and Japan; G.711alaw codec is more common in the rest of the world. G.722 is supported for Skinny Client Control Protocol (SCCP) and Session Initiation Protocol (SIP) endpoints.

- **G.729:** G.729 is also one of the most commonly used and popular codecs in IP PBXs. G.729 digitizes analog voice signals, producing output at 8 kbps using conjugate-structure algebraic code-excited linear prediction (CS-ACELP). G.729 comes with various extensions; the most common are G.729A and G.729AB. G.722 is supported for SCCP and SIP endpoints.

- **G.722:** G.722 is an ITU-T standard wideband audio codec. Compared to G.711, which has a sampling rate of 8 kHz and maximum frequency range of 3.44 kHz, G.722 has a sampling rate of 16 kHz and frequency range of approximately 7 kHz. G.722 codec can operate at variable bit rates of 48, 56, or 64 kbps. G.722 is supported for SCCP and SIP endpoints.

- **H.264:** H.264 is also known as MPEG-4 Part 10 and Advanced Video Coding (AVC). It is one of the most commonly used formats for the recording, compression, and distribution of video content.

- **Internet Low Bitrate Codec (iLBC):** iLBC is a narrowband, high-complexity speech codec that was developed by Global IP Solutions. iLBC has built-in error correction functionality. iLBC is supported by SIP, SCCP, MGCP, and H.323 endpoints.

- **Internet Speech Audio Codec (iSAC):** iSAC is a robust, bandwidth-adaptive, wideband, and super-wideband voice codec also developed by Global IP Solutions. iSAC is used in many VoIP and streaming audio applications. iSAC is supported for SCCP and SIP endpoints. iSAC has a sampling frequency of 16 kHz and supports an adaptive bit rate from 10 to 32 kbps.

- **Low Overhead Audio Transport Multiplex (LATM):** LATM is an Advanced Audio Coding-Low Delay (AAC-LD) MPEG-4 Part 3 media type. LATM is a super-wideband audio codec that provides superior sound quality for voice and music. LATM codec provides equal or improved sound quality, even at lower bit rates. LATM codec can operate at variable bit rates of 48, 56, 64, or 128 kbps. LATM is supported for SIP endpoints as well as Tandberg endpoints.

There are a number of other codecs that are used by Cisco Unified IP phones, Cisco gateways, and other call-processing/control entities.

With an overview of the various codecs, the next section addresses the codec selection mechanism to help ensure that the traffic traversing WAN is assigned to the right codecs and therefore conserving WAN bandwidth.

Codec Selection

CUCM offers an array of codecs to help manage bandwidth within an enterprise collaboration solution. When no specific codec is chosen, CUCM leverages a system default codec. An administrator, however, can deterministically specify codec order, so a range of codecs are available to negotiate before a voice or a video call can be established. The codec selection can be based at a global level or gateway, trunk, or endpoint level.

Codec selection and preference can be applied to the following type of endpoints (phones, gateways, and trunks):

- SIP

- SCCP

- MGCP

- H.323

Moreover, codecs can be defined for extension mobility cross cluster (EMCC)-capable endpoints. Codec preference can be set based on built-in Factory Default lossy or Factory Default low loss options, or the administrator can create a new category to define the codec selection criteria. This is accomplished by browsing to **CUCM Administration GUI > System > Region Information > Audio Codec Preference List.** Figure 6-1 gives an overview of Factory Default Lossy and Factory Default low loss options.

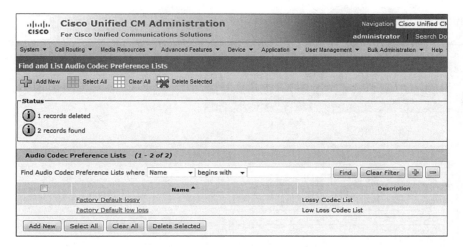

Figure 6-1   *Factory Default Low Loss and Lossy Options*

Figure 6-2 illustrates the factory default lossy codecs.

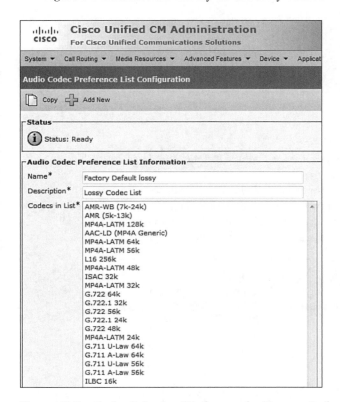

Figure 6-2   *Codec Selection/Preference for Factory Default Lossy Option*

Figure 6-3 depicts the factory default low loss codecs.

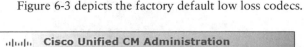

Figure 6-3   *Codec Selection/Preference for Factory Default Low Loss option*

Figure 6-4 illustrates the setup for lossy (G.729 and G.729ab) codecs for remote site phones. As observed, the two G.729 codecs have been moved to the top of the list, making them the most preferred choice for negotiation of codecs when a call is attempted from a main site (campus site) to a remote site.

Note   It is important to understand that codec preference is still determined by the region that a device/endpoint is assigned. An audio codec preference list can be applied to a region, and in turn the region is applied to endpoints/devices via device pools.

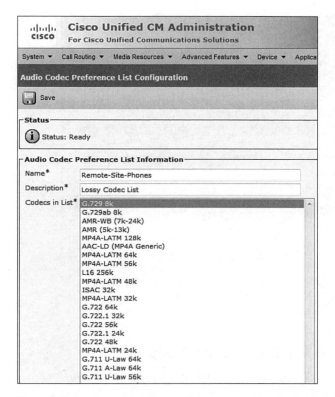

Figure 6-4   *Codec Preference for Remote Site Endpoints*

As shown in Figure 6-5, codecs can be assigned on a per-region basis. Go to **System > Region Information > Region**. Create new regions and assign codecs between the regions. This allows narrowing down the choice of codecs for a specific region (referred by device pool and endpoints), and thus allowing only certain bandwidth codecs over the WAN link. In Figure 6-5, the region Main-Site has been set to communicate with region Remote-Site using predefined codecs and bit rate.

To assign a region to a device pool, so that it can be applied at a physical or logical site level, go to **System > Device Pool**, and under Roaming Sensitive Settings, select the region.

Before discussing the other bandwidth optimization mechanisms such as local conference bridge, transcoding, and multicast MOH, it is important to understand the CUCM media services stack construct (that is, the way these media resources can be assigned to endpoints [or device pools] in the CUCM cluster). The following section covers media resource group (MRG) and media resource group list (MRGL) basics.

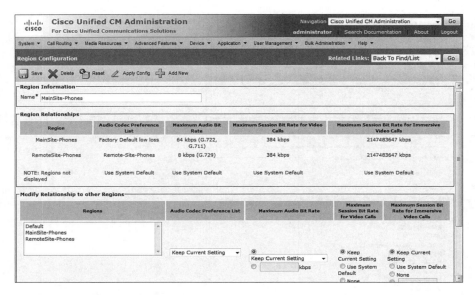

Figure 6-5   *Region Configuration*

Media Resource Group and Media Resource Group List

There are numerous types of media resources that CUCM has to offer for various functions, including the following:

- Annunciator

- Conference bridge

- Transcoder

- MOH server (MOH)

- Media termination point (MTP)

For details on these, refer to the *Implementing Cisco IP Telephony and Video, Part 1 (CIPTV1) Foundation Learning Guide*.

Once the function for these media resources and the various media resources are defined, they need to be managed and assigned to intended devices/endpoints/device pools. Media resource management is accomplished by two entities via the following:

- Media resource group (MRG)

- Media resource group lists (MRGLs)

Both are accessible from the Media Resources menu in CUCM Administration. Individual media resources can be assigned to MRGs, and in turn MRGs can be assigned to MRGLs.

Then, MRGLs can be assigned to endpoints/devices either directly or to device pools, which in turn get assigned to the endpoints or devices.

An MRG can contain one or more media resource and is used to aggregate all similar type of media resources into one MRG (for example, an MRG for annunciators, another MRG for MTPs), as shown in Figure 6-6.

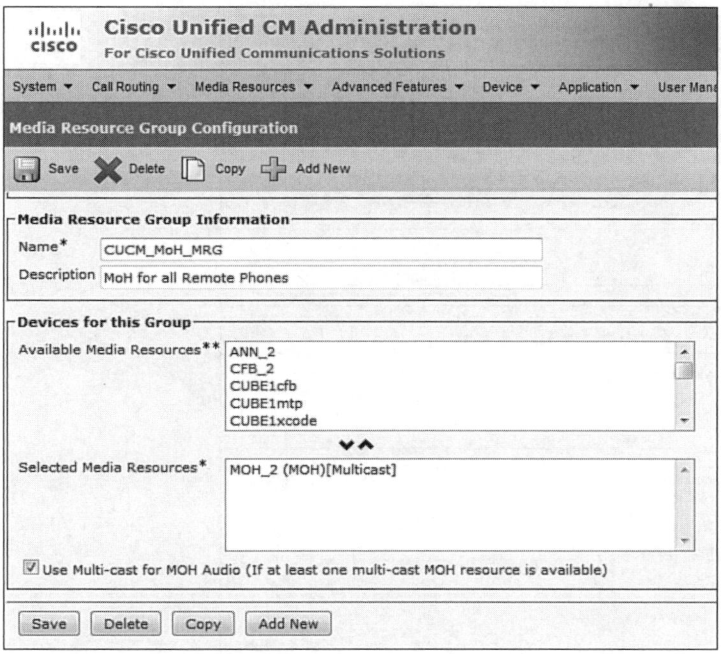

Figure 6-6 *Media Resource Group Definition*

An MRGL can aggregate multiple MRGs and forms a single point of convergence for media resources for endpoints/devices. In an MRGL, the order in which the MRGs are added determines which MRG is queried first for a requested resource, as shown in Figure 6-7.

To configure MRG and MRGL, follow these steps:

Step 1.   Go to **CUCM Administration GUI > Media Resources > Media Resource Group**. Click **Add New**.

Step 2.   Provide a name for the MRG and select a media resource from the available resources. Click **Save**.

Step 3.   To add an MRGL, go to **Media Resources > Media Resource Group List**. Click **Add New**.

Step 4. Enter the MRGL name and select one or more MRGs from available media resource groups.

Step 5. You can change the order in which MRGs are queried by highlighting the name and clicking the up and down arrows located to the right of Selected Media Resource Groups box. Click **Save**.

Figure 6-7 *Media Resource Group List Definition*

At this time, you can assign the MRGLs created to device pools or to endpoints. The next sections describe the various bandwidth conservation mechanisms.

Multicast Music on Hold

Similar to the traditional private automatic branch exchange (PABX), when a caller on hold gets music, MOH is streamed to the calling party's phone when the called party puts the call on hold. MOH requires an MOH server to be configured as well as an audio file that will play during the hold event.

Note   MOH is part of IP Voice Media Streaming Application Service (IPVMS), and an MOH server is automatically created when the IPVMS is activated on a CUCM server.

To configure or fine tune an MOH server, go to **Media Resources > Music On Hold Server** and select the server you want to configure. MOH audio can be played as unicast (default) or multicast streams. At the campus site, CUCM-based MOH, which is by default unicast, can be played out to local phones. This can be extended to remote phones as well, but not without bandwidth implications. To preserve bandwidth, the remote site survivable remote site telephony (SRST) gateway/router can play an audio file from its flash as a multicast stream. The following paragraphs explain the behind-the-scenes process of using multicast MOH.

Multicast MOH is based on MOH capabilities of the Cisco IOS router's feature (discussed in Chapter 5, "Remote Site Telephony and Branch Redundancy Options") (that is, Cisco Unified SRST [working in standby or fallback mode]). To leverage this feature, the Cisco IOS SRST gateway/router is configured for multicast MOH and continuously sends an MOH stream that to the endpoints appears as though a multicast MOH stream has been generated by the CUCM MOH server. In other words, the router sits as a proxy for MOH between CUCM and endpoints, as the logical next hop for endpoints to receive MOH from. This is done by configuring the CUCM MOH server for multicast MOH and setting the max-hops value in the MOH server configuration to 1.

Note   The max-hops parameter specifies the Time To Live (TTL) value used in the IP header of the RTP packets.

In the background, the CUCM MOH server at the main (campus) site and SRST gateway located at the remote site end up using the same multicast address and port number for their streams, as shown in Figure 6-8. Hence, the MOH packets generated by the CUCM MOH server at the campus site are dropped by the campus router as the TTL exceeds the hops. However, the SRST gateway continuously generates a multicast MOH stream so that the IP phones get this stream (which appears to be coming from the CUCM MOH server).

Enabling multicast MOH requires configuring CUCM server/cluster at the main site, campus router, and voice gateway at the remote site. The CUCM and remote site router multicast MOH configuration is covered in Chapter 5. In addition to the configuration covered in Chapter 5, the sections that follow describe the configuration required on both CUCM, campus router, and remote site router to support multicast MOH.

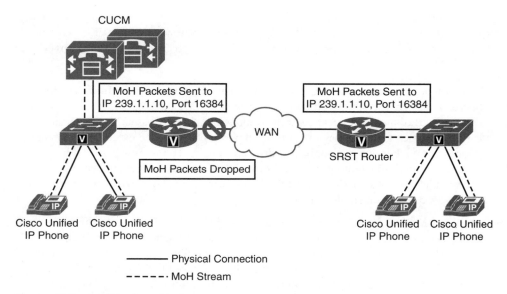

Figure 6-8   *Multicast MOH basics*

Figure 6-9 depicts the MOH server configuration for CUCM. As observed, the TTL is set to 1. The multicast-based IP and port have been defined. These must match the remote router's multicast configuration under call-manager-fallback.

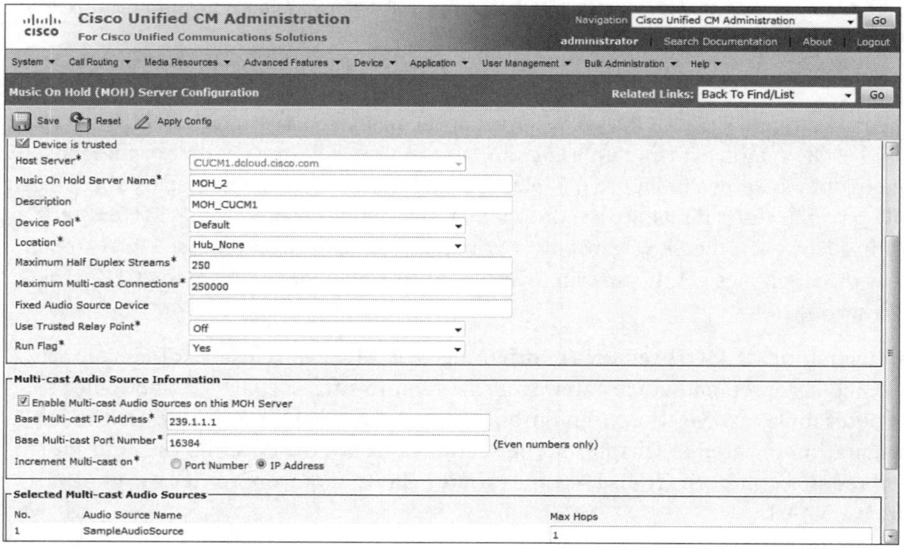

Figure 6-9   *CUCM MOH Server Multicast Configuration*

Note   A Cisco Unified SRST license is required irrespective of whether SRST functionality is used.

Because the MOH server (CUCM IPVMS-enabled server) is located at the campus or the main site is configured for multicast, the remote site router must be configured for multicast MOH. This implies that multicast routing has to be enabled to allow the multicast stream to be routed from the CUCM MOH server network to the phone networks. Example 6-1 covers the configuration to support multicast routing at the remote site router.

Example 6-1   *Multicast Configuration for IOS Router Supporting Multicast MOH*

```
Remote-Site(config)# ip multicast-routing
!
Remote-Site(config)# interface GigabitEthernet 0/0
Remote-Site(config-if)# description WAN
Remote-Site(config-if)# ip address 10.1.1.1 255.255.255.0
Remote-Site(config-if)# ip pim sparse-dense-mode
!
Remote-Site(config)# interface GigabitEthernet 0/1
Remote-Site(config-if)# description Local Site Phones
Remote-Site(config-if)# ip address 10.1.2.1 255.255.255.0
Remote-Site(config-if)# ip pim sparse-dense-mode
```

Two commands used in Example 6-1 are required to enable multicast routing:

- **ip multicast-routing** command in global configuration mode enables multicast routing on the Cisco IOS router.

- **ip pim sparse-dense-mode** command needs to be configured on each interface where multicast routing should be enabled (ingress/egress interfaces; from/to CUCM and IP phones).

Consequently, the main site router must be configured to drop any MOH multicast packets that are destined to a remote site, traversing WAN (considering that TTL was set to 1 in CUCM). Example 6-2 illustrates the configuration of the campus router.

Example 6-2   *Multicast Configuration for IOS Router Supporting Multicast MOH*

```
Campus-Site(config)# ip access-list extended MOH
Campus-Site(config-ext-nacl)# deny udp any host 239.1.1.10 eq 16384
Campus-Site(config-ext-nacl)# permit ip any any
!
Campus-Site(config)# interface GigabitEthernet 0/0
Campus-Site(config-if)# ip access-group MOH out
Campus-Site(config)# no ip pim sparse-mode
```

Multicast MOH IP Address and Port Considerations

CUCM MOH server can stream multiple multicast MOH files. This implies that a base multicast IP must be specified followed by a port. Moreover, the decision of choosing IP versus port increment must be made. To better understand the implication of IP or port increment, consider the following points:

- For each audio source enabled for multicast, four streams are enabled for the increment (one per codec): G.711 ulaw, G.711 alaw, G.729, and wideband.

- When incrementing on IP address, each stream consumes one IP address, which implies that each audio source requires four IP addresses.

- When incrementing on ports, two ports are reserved for each audio stream: one for RTP and the other for the Real-time Transport Control Protocol (RTCP). Hence, each stream consumes two ports, which further implies that it is two ports per codec and a total of eight ports per audio stream.

For example, an audio source enabled for multicast with base multicast IP as 249.1.1.10 will increment, taking 239.1.1.11 as the next IP (when using IP increment). However, if using port increment with initial port as 16384, the next port will be 16386 because ports get incremented in even numbers.

Local Conference Bridge

As discussed earlier in this chapter, having a local conference bridge has its merits. The phones local to the site leverage a local conference bridge instead of traversing a WAN link all the way to campus and converging in a conference on campus gateway resulting in savings of WAN bandwidth. This is particularly useful when the conferences are expected to occur more often and if mixed codecs (for example, G.711ulaw and G.729 conferences) have to take place. Figure 6-10 shows the campus-based CUCM software and IOS gateway-based hardware conference bridges and remote site conference bridge.

CUCM supports two types of conferences (Ad-Hoc and Meet-Me), and a conference bridge resource is required. This resource can be a hardware or software conference depending on the type of conference to be supported (for example, audio or video). A software conference bridge runs on the CUCM server on which IPVMS is activated, which also supports other media functions such as annunciator, MOH, media termination point (MTP), and so on. A software conference bridge is automatically created on that specific server, when IPVMS service is activated on the server. To change the default conference bridge settings, go to **Media Resources > Conference Bridge** and select the default (CFB_X, where X is any number assigned by CUCM). Figure 6-11 shows the default software conference bridge created on a CUCM server.

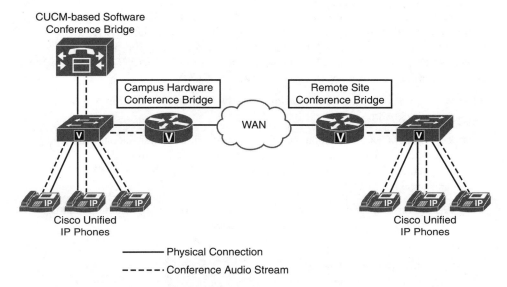

Figure 6-10 *Campus and Remote Site Conferencing*

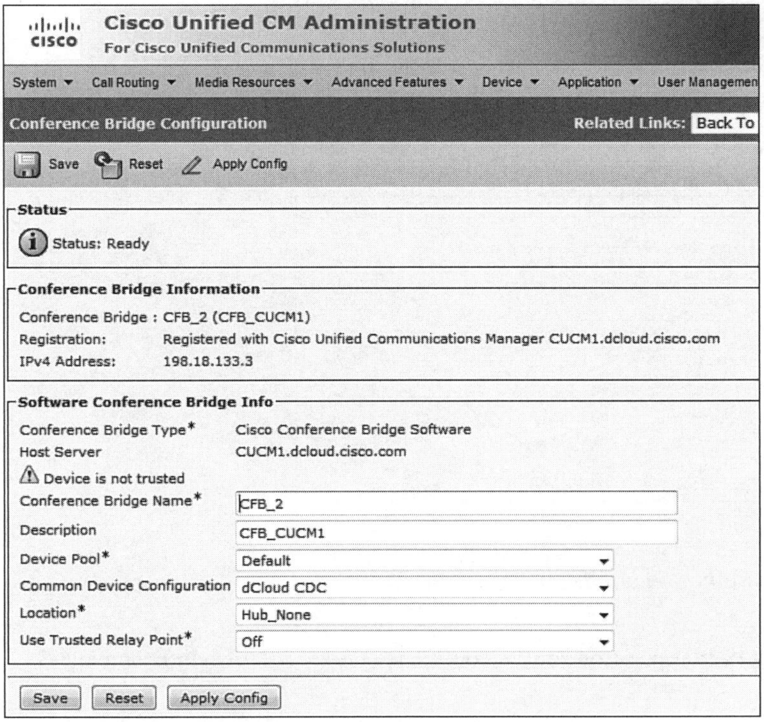

Figure 6-11 *CUCM Software Conference Bridge*

Cisco recommends using hardware conference bridges over software conference bridges. Software conference bridges have an impact on CUCM CPU and memory, and only support G.711 calls. Hardware conference bridges can be configured on IOS gateways leveraging DSPs on packet voice digital signal processor modules (PVDMs). Follow these steps to configure a hardware conference bridge on CUCM:

Step 1.   Go to **CUCM Administration > Media Resources > Conference Bridge.** Click **Add New.**

Step 2.   Select **Cisco IOS Enhanced Conference Bridge.** Enter the name matching the associated DSP profile name in the Cisco IOS router. In this case, the name of the conference bridge (as stated in Example 6-3) is HWCFB, as shown in Figure 6-12.

Step 3.   Enter other required details. Ensure that security mode matches what is configured on the IOS router. Click **Save.**

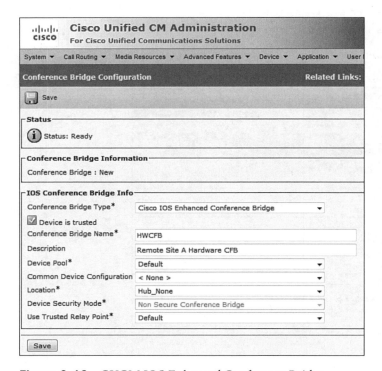

Figure 6-12   *CUCM IOS Enhanced Conference Bridge*

The subsequent IOS configuration for the conference bridge is shown in Example 6-3.

Example 6-3 *Hardware Conference Bridge Configuration*

```
CFBRouter(config)# voice-card 0
CFBRouter(config-voicecard)# dsp services dspfarm
!
CFBRouter(config)# sccp local GigabitEthernet 0/1
CFBRouter(config)# sccp ccm 10.76.108.146 identifier 1 priority 1 version 7.0+
CFBRouter(config)# sccp ccm 10.76.108.148 identifier 2 priority 2 version 7.0+
CFBRouter(config)# sccp
!
CFBRouter(config)# sccp ccm group 1
CFBRouter(config-sccp-ccm)# bind interface GigabitEthernet 0/1
CFBRouter(config-sccp-ccm)# associate ccm 1 priority 1
CFBRouter(config-sccp-ccm)# associate ccm 2 priority 2
CFBRouter(config-sccp-ccm)# associate profile 1 register HWCFB
CFBRouter(config-sccp-ccm)# switchback method graceful
!
CFBRouter(config)# dspfarm profile 1 conference
CFBRouter(config-dspfarm-profile)# codec g711ulaw
CFBRouter(config-dspfarm-profile)# codec g711alaw
CFBRouter(config-dspfarm-profile)# codec g729ar8
CFBRouter(config-dspfarm-profile)# codec g729abr8
CFBRouter(config-dspfarm-profile)# codec g729r8
CFBRouter(config-dspfarm-profile)# codec g729br8
CFBRouter(config-dspfarm-profile)# maximum conference-participants 30
CFBRouter(config-dspfarm-profile)# maximum sessions 50
CFBRouter(config-dspfarm-profile)# associate application SCCP
CFBRouter(config-dspfarm-profile)# no shut
```

Example 6-3 shows the following:

- **dsp services dsp-farm** under voice-card (slot number) enables the DSPs (DSP farm) on PVDM for supporting media functions.

- **sccp local** defines the local interface (physical, logical) that SCCP applications can use to register with CUCM servers.

- **sccp ccm** command adds CUCM servers to the list of available servers to which the voice gateway can register.

- **sccp** enables (administratively) SCCP function.

- **sccp ccm group (number)** command creates a CUCM group and enters SCCP configuration.

- **associate ccm** associates the previously defined CUCM servers to the **sccp ccm group.**

- **associate profile (number)** associates the dspfarm profile to **sccp ccm group.** (This profile can have multiple functions including conferencing, MTP, and transcoding.) This also defines the name of the conference bridge as it should be defined in CUCM (in this case, HWCFB; note that the name is case sensitive).

- **switchback method** defines the method to roll over to the next CUCM (per the priority defined in **sccm ccm group**) when the communication link to the active CUCM fails. It can be immediate or graceful.

- **dspfarm profile (number)** defines the type of profile (conferencing, transcoding) and defines the codecs to be supported within the profile.

- **maximum conference participants** and **maximum sessions** define the total concurrent participants and the total concurrent sessions, respectively.

- **associate application Skinny Client Control Protocol (SCCP)** associates the SCCP application so the conference bridge can register with CUCM servers and provide the required media services.

At this time, the remote site hardware conference resources register with CUCM using the SCCP application. The conference bridge can be added to an MRG, which in turn can be added to an MRGL, and finally the MRGL can be assigned to either a device pool or directly on a per-endpoint basis. This allows the remote site phones to leverage the local conference bridge instead of going over WAN to campus conferencing resources.

Transcoder

At times, it is required to have one codec converted to another (for example G.729 calls traversing over WAN from a remote site converted to G.711 at the data center or vice versa). This is done to save bandwidth over WAN and help ensure that "lossy" codecs are used. This does not imply that the call quality will be worse, as with favorable WAN parameters such as delay, jitter, packet loss, and so on. Under control, the mean opinion score (MOS) can be pretty decent. Transcoders are the hardware resources that enable a Cisco Collaboration network to convert calls from one codec to another. Transcoders can be enabled on Cisco IOS gateways. Like hardware conference bridges, they also leverage DSPs off PVDMs. Figure 6-13 shows a campus site transcoder in action. Here, the G.729 stream from a remote IP phone is being converted into G.711 for a local IP phone. Another use case is an application such as legacy voicemail or an IVR solution that only supports G.711.

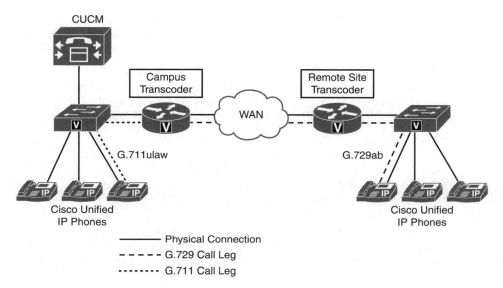

Figure 6-13 *IOS Transcoder for Campus Site*

The placement or design of a transcoder can be a daunting task. Follow these guidelines for the optimal usage of IOS transcoding resources:

- Implement an IOS-based transcoding resource. Because CUCM does not support software transcoding resources, the only option is to use a hardware-based transcoding resource by first configuring the transcoder at the Cisco IOS router and then adding the transcoder to CUCM. Remember that all IOS-based resources register with CUCM using SCCP.

- Implement CUCM-based regions so that only G.729 is permitted over the IP WAN, and an MRGL-based transcoder can be used that the IP phone has access to via a device pool or hard coded as a setting to the phone itself. To do so, all IP phones and G.711-only devices, such as third-party voicemail systems or software conference bridges that are located in the headquarters site, are placed in a region (such as campus region). Remote IP phones are placed in another region (such as remote branch region). The transcoding resource is put into a third (dedicated) region (such as XCODER region).

- Use the following relationships between regions:

  - **Within remote branch, set codec to G.711:** This allows local calls between remote IP phones to use G.711.

  - **Within campus region, set codec to G.711:** This allows local calls within headquarters to use G.711. These calls are not limited to calls between IP phones. This also includes calls to the G.711-only third-party voicemail system or calls that use the G.711-only software conference bridge. In addition, Unified Contact Center Express or Enterprise may also be configured to support G.711 only.

- **Within XCODER region, set codec to G.711:** Because this region includes only the transcoder media resource, this setting is irrelevant, because there are no calls within this region.

- **Between remote branch and campus region, set codec to G.729:** This ensures that calls between remote IP phones and headquarters devices, such as IP phones, software conference bridges, and voicemail systems, do not use the G.711 codec for calls that traverse the IP WAN.

> **Note** Calls between Unified IP phones at headquarters and remote Unified IP phones *do not require a transcoder*. They simply use the best allowed codec that is supported on both ends from the CUCM region settings (ideally, G.729). One important note: Cisco Unified IP phones and many video endpoints support *both G.711 and G.729 codecs.* They are essentially built in to the phone as part of its firmware and capabilities. The only time a transcoder needs to be invoked is for devices that *only support G.711*, such as Unified Contact Center Express (UCCX), Unified Contact Center Enterprise (UCCE), Unity Connection, and so on, depending on how these devices were configured/installed.

To configure a Transcoder on CUCM, follow these steps:

Step 1.   Go to **CUCM Administration > Media Resources > Transcoder**. Click **Add New**.

Step 2.   For transcoder type, select **Cisco Conference Enhanced IOS Media Termination Point** and enter a name matching the DSP profile configuration in the IOS router. Enter other required details. Click **Save**. Figure 6-14 shows the CUCM transcoder configuration.

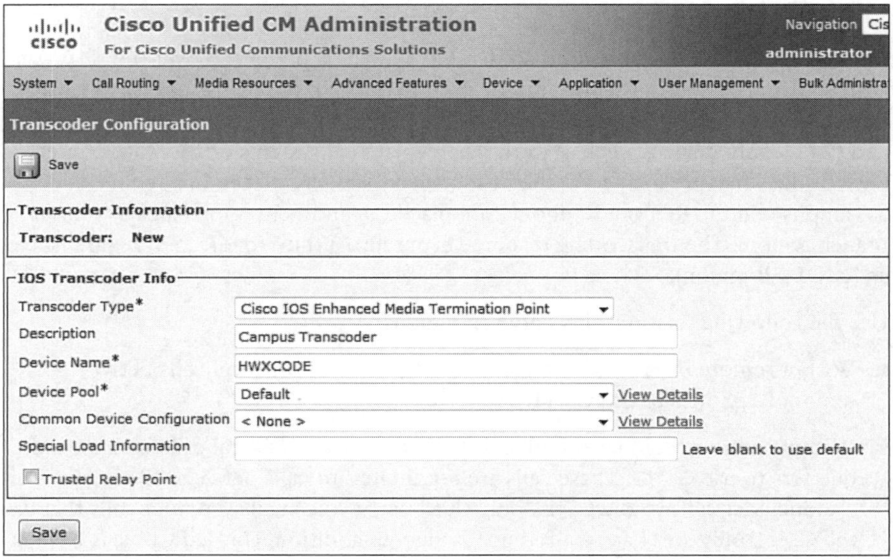

Figure 6-14   *IOS Transcoder for Campus Site*

Example 6-4 explains the configuration of SCCP group and DSP profile for enabling transcoding resources on a campus IOS gateway (building on the previous configuration Example 6-3 of the CFBRouter and conference bridge).

Example 6-4 *IOS Transcoder Configuration*

```
Campus-Router(config)# sccp ccm group 1
Campus-Router(config-sccp-ccm)# bind interface GigabitEthernet 0/1
Campus-Router(config-sccp-ccm)# associate ccm 1 priority 1
Campus-Router(config-sccp-ccm)# associate ccm 2 priority 2
Campus-Router(config-sccp-ccm)# associate profile 10 register HWXCODE
Campus-Router(config-sccp-ccm)# switchback method graceful
!
Campus-Router(config)# dspfarm profile 10 transcode
Campus-Router(config-dspfarm-profile)# codec g711ulaw
Campus-Router(config-dspfarm-profile)# codec g711alaw
Campus-Router(config-dspfarm-profile)# codec g729ar8
Campus-Router(config-dspfarm-profile)# codec g729abr8
Campus-Router(config-dspfarm-profile)# maximum sessions 20
Campus-Router(config-dspfarm-profile)# associate application SCCP
Campus-Router(config-dspfarm-profile)# no shut
```

Example 6-4 shows the following:

- **associate profile (number)** associates the dspfarm profile to **sccp ccm group** and also defines the name of the transcoder as it is defined in CUCM (in this case, HWXCODE).

- **dspfarm profile (number)** defines the type of profile (in this case, transcoding for profile 10) and defines the codecs to be supported within the profile.

- **maximum sessions** defines the total concurrent sessions the transcoder will support.

The rest of the commands are similar to that of conference bridge configuration. Also, similar to a conference bridge resource, a transcoder resource should be assigned to an MRG, which in turn gets assigned to an MRGL, which finally gets assigned at a device pool level or per-device level.

At this time, the campus router is ready to transcode the incoming or outgoing calls as per the defined region of originating and destination device and its relationship with the transcoder.

Summary

The following key points were discussed in this chapter:

- In an enterprise telephony solution, having calls over a WAN is implicit. However, saving bandwidth is a task that is explicit.

- To save WAN bandwidth, Cisco has empowered the UC administrators with various mechanisms and tools to optimize the WAN bandwidth consumption.

- Using MOH from within a router's flash instead of the WAN saves a considerable amount of bandwidth otherwise wasted in multiple unicast streams from campus site to remote site.

- Using a local conference bridge helps converge streams within the site instead of traveling all the way over to the campus site and back.

- The right choice of codecs and implementing transcoders helps save WAN bandwidth and improve the viability of the solution from the point of view of recurring costs and economies of scale.

References

For additional information, refer to these resources:

Cisco Systems, Inc. Cisco Collaboration Systems 10.x Solution Reference Network Designs (SRND), May 2014. http://www.cisco.com/c/en/us/td/docs/voice_ip_comm/cucm/srnd/collab10/collab10.html

Cisco Systems, Inc. Cisco Collaboration Systems 11.x Solution Reference Network Designs (SRND), July 2015. http://www.cisco.com/c/en/us/td/docs/voice_ip_comm/cucm/srnd/collab11/collab11.html

Review Questions

Use these questions to review what you have learned in this chapter. The answers appear in Appendix A, "Answers Appendix."

1. Which of the following are examples of bandwidth conservation mechanisms? (Choose three?)

   a. CAC

   b. LBM

   c. Campus site conference bridge

   d. Transcoder

   e. Remote site conference bridge

2. Which resources can be grouped in a media resource group? (Choose four.)

   a. MOH

   b. Annunciator

   c. RSVP

   d. Conference bridge (software and hardware)

   e. Transcoder

3. **Which statement about multicast MOH is true?**

   a. Multicast MOH can be implemented without SRST configuration.

   b. Multicast MOH from remote site router flash can also be used for unicast MOH when required.

   c. The remote site router with SRST 8.x can stream more than a single MOH file.

   d. CUCM streams Multicast MOH to remote site routers.

4. **How can a Cisco Collaboration solution administrator ensure that remote sites use local conference resources before failing over to a campus site?**

   a. Assign the remote site's phones in an MRG that links to a remote site conference bridge.

   b. Assign the main site's phones in an MRG that links to a remote site conference bridge.

   c. Assign the remote site's phones in an MRGL that links to a remote site conference bridge and has a secondary conference bridge available at the main site.

   d. Assign the remote site's phones in an MRGL that links to a remote site conference bridge.

5. **True or false: A transcoder can be deployed using CUCM's built-in IPVMS service.**

   a. True

   b. False

6. **True or false: The same DSP farm can be used for media functions such as conferencing and transcoding.**

   a. True

   b. False

7. **Where is the name of the IOS resource defined that should match the name configured in CUCM?**

   a. sccp ccm profile

   b. dspfarm profile

   c. voice-card

   d. voice services voip

8. **True or false: An IOS router can support both transcoding and conferencing at the same time.**

   a. True

   b. False

Chapter 7

Call Admission Control (CAC) Implementation

Cisco Unified Communications Manager (CUCM) and many other Cisco Collaboration applications use Skinny Client Control Protocol (SCCP), Session Initiation Protocol (SIP), H.323, Q-Signaling Protocol (QSIG), and many other forms of signaling protocols with endpoints and gateways. Cisco Unified IP phones leverage Real-time Transport Protocol (RTP) protocol (media) to communicate among themselves and with Cisco Collaboration applications such as Cisco Unity or Cisco Unity Express (CUE). All signaling and media traffic requires a certain degree of bandwidth to be transported within local-area network (LAN) or wide-area network (WAN).

Bandwidth provisioning within a LAN is much less challenging than over WAN. Although an organization can provision bandwidth, beyond a certain number of projected calls the bandwidth can run out pretty quickly and can cause poor quality or dropped calls. In such case, call admission control (CAC) can be used to regulate traffic volume to ensure, or maintain, a certain level of audio quality in voice communications networks. This chapter discusses the various CAC mechanisms Cisco offers to counter voice degradation due to bandwidth depletion.

Upon completing this chapter, you will be able to meet these objectives:

- Define CAC characteristics
- Define CUCM-based locations-based CAC (LCAC)
- Define CUCM-based enhanced CAC (E-LCAC)
- Define CUCM-based Resource Reservation Protocol (RSVP)
- Define automated alternate routing
- Define IOS-based CAC mechanisms

Call Admission Control Characteristics

As briefly discussed, the purpose of CAC is to avoid oversubscription of WAN link when voice/video calls are placed from one physical/logical site to another. In other words, there is a threshold as to how many calls are allowed across the network on a WAN link. Voice/video calls are admitted to the network only as long as the network can ensure sufficient quality of service (QoS). There are three types of CAC mechanisms:

- **Local CAC:** Based on a device's local determination (for example, state of an outgoing interface or link). Examples: Location-based CAC (LCAC) on CUCM and **max-connections** in IOS gateways.

- **Reservation-based CAC:** Based on the reservation or calculation of required resources before a call can be admitted on the network. Example: Resource Reservation Protocol (RSVP) in CUCM and IOS gateway.

- **Measurement-based CAC:** Measurement-based CAC techniques look ahead into the network to measure the state of the network in order to determine whether to allow a new call (for example, Advanced Voice BusyOut [AVBO] on IOS gateways).

The most common types of CAC mechanisms seen in a collaboration network are local CAC and reservation-based CAC.

CAC implementation can also be distinguished in terms of the following:

- **Topology-aware CAC:** RSVP-based CAC that does not require static values and can dynamically allow or limit the calls based on the resource reservation in the path of the call

- **Topology unaware CAC:** As any mechanism that is based on a static configuration (example, location-based CAC)

The following section describes CUCM implementation of CAC using various CAC mechanisms.

CUCM Call Admission Control

CUCM-based CAC mechanisms preserve call quality and prevent WAN bandwidth oversubscription by limiting the (concurrent) number of calls over WAN links. CUCM-based CAC mechanisms can be classified into three categories:

- Location-based CAC (LCAC)

- Enhanced-LCAC (E-LCAC)

- Resource Reservation Protocol (RSVP)

The following sections explain each of these CUCM-based CAC mechanisms.

Location-Based CAC

CUCM location defines the maximum audio/video bandwidth allowed into and out of a location (a virtual physical location/site). This helps protect a WAN link to a remote site from being oversubscribed (by only allowing a predefined [threshold] bandwidth for the calls traversing through the site associated with the location). CUCM LCAC is topology unaware.

Locations work in conjunction with regions to define the characteristics of a network link. Locations can be associated with device pools and devices themselves, such as phones, trunks, gateways, conference bridge, and music on hold (MOH) server (that is, basically any device that can be a source for media). Locations can be dynamically associated with endpoints via device mobility (IP subnets). Figure 7-1 represents location-based CAC configuration between a hub site and remote site.

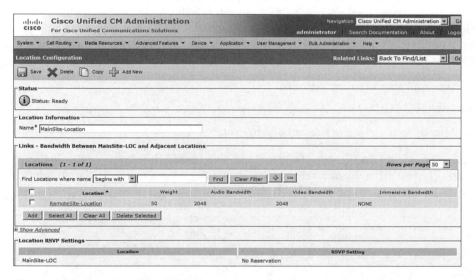

Figure 7-1 *Location-based CAC*

Note It is important to understand that a location provides bandwidth accounting "within" the location as well as "in" or "out" of the location. This can be referred to as intralocation and interlocation calls.

For example, when an endpoint (IP phone, analog phone, or a soft client) in location MainSite-Location places a call to an endpoint in location RemoteSite-Location; CUCM is aware that it is an intracluster call between two different locations and deducts bandwidth depending on the type of codec used for the call and whether the call was an audio call only, a video call, or an immersive TelePresence call. Calls are allowed until the total bandwidth between the locations (for a certain type of call) is consumed, then no more calls are allowed. It is at this time that automated alternate routing (AAR, explained later in this chapter) will kick in to allow further calls to be completed over public switched telephone network (PSTN) links.

As shown in Figure 7-1, LCAC has options for providing three types of bandwidth control:

- Audio
- Video
- Immersive (TelePresence)

Audio applies to Real-time Transport Protocol (RTP) audio-only calls between two locations. It is important however to understand the difference between video and immersive bandwidth. Location-based call admission control (LCAC) allows the administrator to set separate immersive bandwidth setting on locations and links. (Links are discussed in the next section.) Desktop video and TelePresence can reside in the same location, and the bandwidth can be deducted separately for either. SIP trunks are used to classify a device or system as follows:

- **Desktop:** Standard desktop video
- **Immersive:** High-definition immersive video
- **Mixed:** A mix of immersive and desktop video

This is achieved by configuring a SIP profile to set a specific video class (and assigning that profile to a SIP trunk or a TelePresence multipoint control unit [MCU], configured as a media resource), as shown in Figure 7-2.

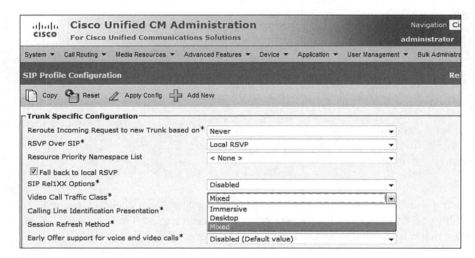

Figure 7-2 *SIP Profile for Video Traffic Class*

LCAC has some restrictions, such as the following:

■ It is not designed to work where customer networks are multitiered and multihops instead of simple hub-and-spoke topology.

■ It does not offer intercluster bandwidth management support. It is limited to intracluster calls.

These limitations are addressed by enhanced location-based CAC (E-LCAC) that is described in the next section. However, before E-LCAC is discussed, it is imperative to understand the CUCM service, which enables the UC administrators to configure and deploy E-LCAC.

Location Bandwidth Manager

E-LCAC leverages the Location Bandwidth Manager (LBM) service. LBM computes the effective path from source location to destination location. LBM is a CUCM service that is activated by going to **Cisco Unified Serviceability > Tools > Service Activation**. LBM can be enabled on multiple servers in a cluster so that LBM groups can be configured to provide LBM redundancy (like Cisco CallManager groups for call processing). Cisco CallManager service interacts with LBM service within a server, and the LBM service is full mesh replicated within a cluster, as shown in Figure 7-3.

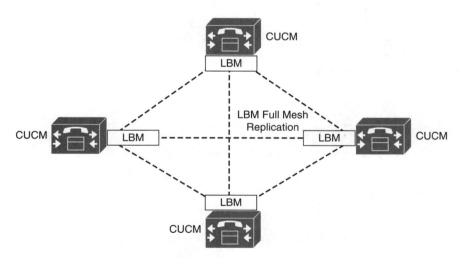

Figure 7-3 *LBM Replication (Intracluster)*

LBMs within the cluster create a fully meshed communications network via Extensible Markup Language (XML) over TCP for the replication of bandwidth change notifications between LBMs. A minimum of one instance of LBM must run in each cluster to enable E-LCAC (for intercluster CAC support). LBM groups can be configured within a single site as well as a cluster over a WAN. More than one subscriber may run the LBM service, providing for server redundancy, and up to two LBM servers may reside in an LBM group (one active and one standby). In large clusters, multiple LBM groups may be defined to distribute the load of CAC queries within the cluster. This arrangement is configured to control which CUCM uses which LBM groups in order to share the load.

As shown in Figure 7-4, to configure an LBM group, go to **System > Location Info > Location Bandwidth Manager Group**.

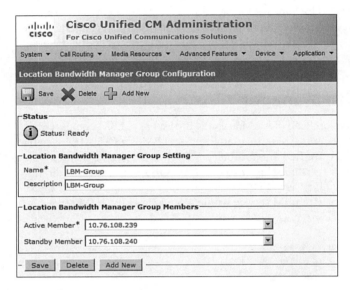

Figure 7-4   *LBM Group Configuration*

The main functions of LBM are as follows:

- Path assembly and calculation

- Servicing bandwidth requests from CUCM control (XML/TCP)

- Replication of bandwidth information to other LBMs within the cluster and between clusters

- Intercluster locations CAC

- Provides configured and dynamic information to serviceability

- Updates location Real-Time Monitoring Tool (RTMT) counters

It is important to understand that LBM should never be enabled on the Publisher. LBM should be enabled on a Subscriber that is also running the CallManager service. LBM may be enabled on a dedicated standalone subscriber.

The following are LBM group configuration recommendations:

- Use local LBM when available with redundant LBMs at each call processing site.

- In case of clustering over WAN or split data center (DC) designs, redundant LBMs should be configured for each DC.

- To enable load reduction on active (call processing) Subscribers, use dedicated LBMs (that is, enable LBM on standby Subscribers).

The following section describes the enhanced location-based CAC (E-LCAC).

Enhanced Location-Based CAC

E-LCAC enables network administrators to go beyond a traditional CAC modeling. It allows network modeling by applying locations and links to determine the best paths for a call to proceed between different sites. E-LCAC is a model-based static CAC mechanism and is topology unaware. The following are fundamentals of E-LCAC network modeling:

- **Location:** A location represents a *physical or logical site*, typically a LAN. It can contain endpoints or simply serve as a transit location between links for WAN network modeling.

- **Links:** Links represent the connection between two locations and is typically a WAN link. Links interconnect locations (to build the topology) and are used to define bandwidth available between locations.

- **Weights:** Weights are used on links to provide a *cost* to the *effective path*. Weights are applicable only when there is more than one path between any two locations.

- **Path:** A path is a sequence of links connecting locations (intermediary) that eventually leads from an originating location to a destination location.

- **Effective Path:** An effective path is the path with the *least cumulative weight*.

CUCM calculates shortest paths/least cost from all locations to all locations and automatically builds the effective paths. CUCM tracks bandwidth across any link that the network model represents, from the originating location to the terminating location. Figure 7-5 illustrates the network modeling with locations, links, and effective path.

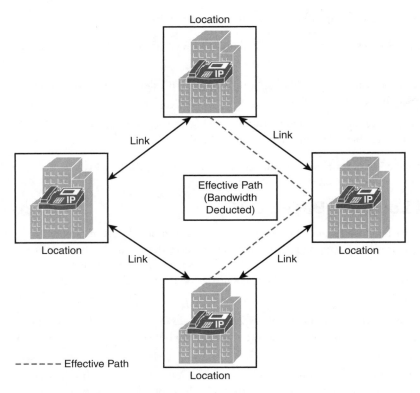

Figure 7-5   *E-LCAC-Based Network Modeling*

E-LCAC works in two modes:

■ Intracluster E-LCAC

■ Intercluster E-LCAC

Intracluster bandwidth limits are assigned to a location in order to CAC all calls made To/From/Within the location. Moreover, intralocation bandwidth values are unlimited by default (similar to Hub_None Location). The Locations Bandwidth Manager (LBM) service computes the effective path from source location to destination location (within the same cluster [intracluster]), as illustrated in Figure 7-6.

To calculate the Effective Path, E-LCAC does the sum of weight of links across each possible path from source to destination. The least cost value of the path's weight determines the effective path. For example, as shown in Figure 7-6, from Location A to Location D there are two paths via Location B (A>B>D) and via Location C (A>C>D). The sum of weights across A>B>D is 50 + 20, whereas the sum of weights across A>C>D is 50 + 10. Hence, A>C>D will be the effective path.

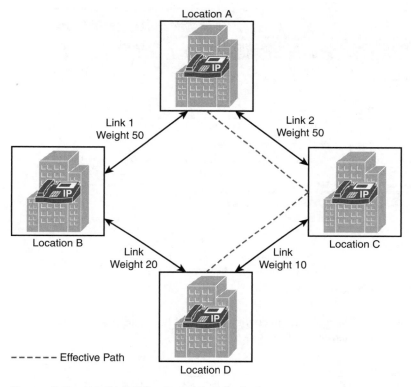

Figure 7-6 *E-LCAC Effective Path Calculation*

A tie-break of equally weighted paths is determined by LBM based on location name. (For example, if the weights from B>D and C>D are the same, the location name [alphabetic] will be used as a tie-breaker.)

Once the effective path is determined, all subsequent calls that have the same source and destination locations use the same effective path.

For intercluster CAC, each cluster manages its own topology. Consequently, each cluster propagates its topology to other clusters configured in the LBM intercluster replication network and creates a global topology (also known as assembled topology) piecing together each cluster's replicated topology.

In an enterprise environment, a Session Management Edition (SME) cluster, for example, can manage all the locations and links for the entire location replication network (all leaf clusters). All leaf clusters are required to only configure the locations that they require to associate with their local endpoints and devices. Figure 7-7 depicts the configuration of LBM hub group.

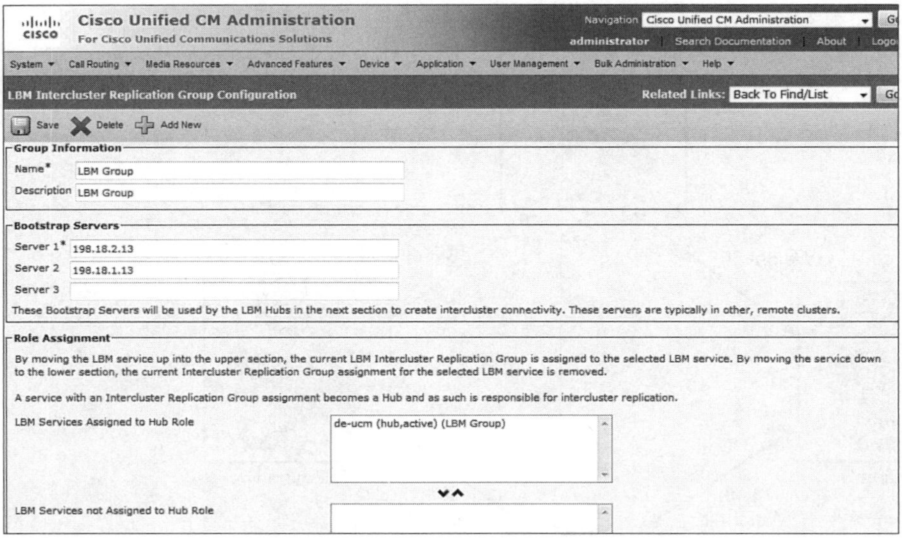

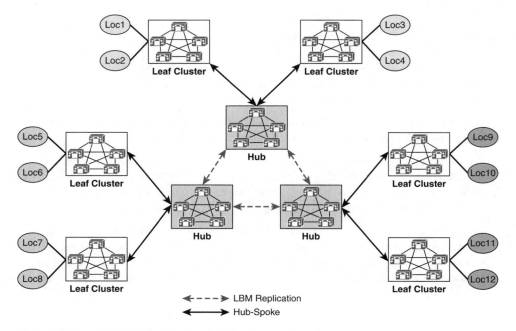

Figure 7-7   *LBM HUB Group Configuration*

Figure 7-8 depicts an LBM replication network with LBM hubs and spokes. Note that hubs replicate the information using the LBM service.

Figure 7-8   *LBM Replication in Hub-Spoke Topology*

An LBM is a hub when it is assigned to an LBM hub group. An LBM is a spoke when it is not assigned to an LBM hub group. Also, as shown in Figure 7-7, servers under bootstrap servers are responsible for informing the hub server network of the LBM hub servers.

Similar to LBM replication within a cluster, intercluster replication (SME <-> leaf) is set up by the hub-spoke relationship. Figure 7-9 shows the SME cluster acting as a hub and other clusters as spokes.

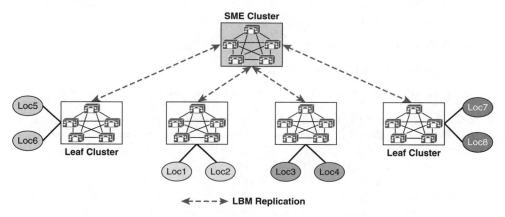

Figure 7-9 *LBM Replication in SME Topology*

The rules for establishing an LBM hub replication network are as follows:

■ If a cluster has multiple LBM hubs, the lowest IPv4 (full) address LBM hub functions as the sender of messages to other remote clusters.

■ The LBM hub that functions as the sender for messages in the cluster selects the first LBM hub of each cluster to send messages to.

■ The LBM hubs that receive messages from remote clusters forward the received messages to the LBM spokes in their local cluster.

■ Forwarded messages have a unique random string so that messages received twice will be dropped to avoid any replication storm/loop.

■ Other LBM hubs in the cluster that receive the forwarded message will not forward on to LBM spokes because the message is not from a "remote" cluster.

E-LCAC introduces two new location types:

■ **Shadow location:** A shadow location is used to enable a SIP trunk to pass E-LCAC information, such as location name and video traffic class, among other entities for E-LCAC for function across clusters. To pass this location information across clusters, the SIP intercluster trunk (ICT) needs to be assigned to the shadow location. Similar to the "phantom" location, a shadow location cannot have a link to other user/device locations, and therefore no bandwidth can be reserved between the shadow location and other user/device locations. Any device other than a SIP ICT that is assigned to the shadow location is treated as if it were associated to Hub_None.

■ **Shared location:** A shared/common location is a location that is configured with the "same name" on clusters participating in an LBM replication network. A shared location serves two purposes:

  ■ It enables clusters to collate their respective configured topologies to one another.

  ■ It provides the ability for multiple clusters to perform CAC for the same locations.

The following are some examples of E-LCAC-based audio only, video, and TelePresence (immersive) calls. The rule of thumb is as follows:

■ **Audio (audio-only calls):** RTP bit rate + IP and UDP header overhead

■ **Video (video calls):** RTP bit rate only

■ **Immersive (video calls by Cisco TelePresence endpoints):** RTP bit rate only

As shown in Figure 7-10, the audio-only call with Real-time Transport Control Protocol (RTCP) has an 80-kbps bandwidth (duplex). For this call, the Location Bandwidth Manager (LBM) deducts 80 kbps (bit rate + IP/UDP overhead) between location L1 and location L2 for a call established between desktop phones. Note that the actual bandwidth consumed at Layer 3 in the network with RTCP enabled is between 80 kbps and 84 kbps. However, the 4-byte RTCP overhead is not accounted for in E-LCAC and must be accounted for in network bandwidth provisioning.

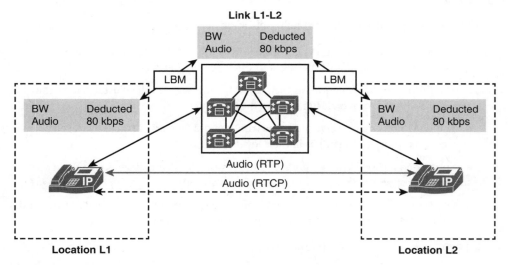

Figure 7-10 *LBM-Based E-LCAC in Audio Only Calls*

Next, Figure 7-11 illustrates that a video call is established between endpoints in location L1 and location L2. In this case, two audio streams, two audio-associated RTCP streams, two video streams, and two video-associated RTCP streams are established. This particular call is 2048 kbps, with 64 kbps of G.711 audio and 1984 kbps of video bandwidth. Note that LBM is centrally managing the bandwidth deduction for L1 and L2.

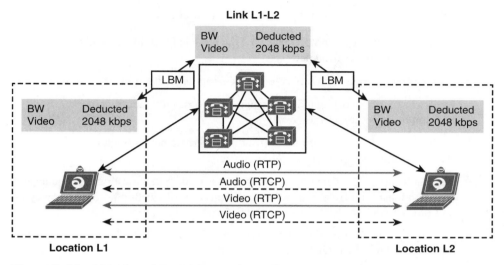

Figure 7-11 *LBM-based E-LCAC in Video Calls*

Finally, as illustrated in Figure 7-12, an immersive TelePresence call with Binary Floor Control Protocol (BFCP)-based screen sharing is established with audio stream and associated RTCP, video stream and its associated RTCP, and finally a presentation stream and BFCP signaling. In this scenario LBM deducts only the audio and video bandwidth, and the bandwidth for presentation and BFCP is to be provisioned outside of BFCP scope (in network bandwidth modeling).

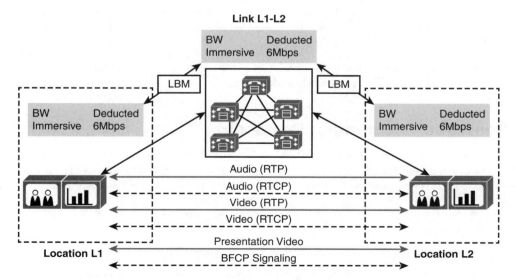

Figure 7-12 *LBM-Based E-LCAC in Immersive Calls*

The following are some important points relevant to E-LCAC:

■ Each cluster requires a complete view of the modeled topology, locally to calculate the end-to-end reservation path.

■ Each cluster is required to replicate the local topology that it manages using LBM.

■ Every cluster participating in LBM requires the location be locally configured for location to device association.

■ Each cluster should be configured with the immediate neighboring locations such that each cluster's topology can interconnect.

■ Naming locations consistently across clusters is vital. "Same location same name and different location different name" practice must be followed to have consistency in an LBM network.

■ The Hub_None location should be renamed to be unique in each cluster (or it will be shared by other clusters).

■ The cluster ID must be changed on each cluster participating in an LBM network so that it is unique on each cluster.

■ Links need to be configured to establish the interconnect points in the topology.

■ Discrepancies of bandwidth limits and weights on shared links and locations are resolved by using the minimum of the assigned values (names, actual values).

■ An LBM hub is responsible for communicating to hubs in remote clusters.

■ An LBM spoke does not directly communicate with other remote clusters. LBM spokes receive and send messages to remote clusters through the local LBM hub. If the LBM hub is not available, the associated spokes are dissociated from the LBM network.

■ E-LCAC cannot adjust its knowledge of effective paths due to link failures and reconfigurations (such as newly added links).

Resource Reservation Protocol

Resource Reservation Protocol (RSVP) is a topology-aware CAC signaling protocol that has been designed to work with any WAN topology. RSVP runs as a software feature on Cisco IOS routers and as RSVP agent on CUCM. It has the following characteristics:

■ Uses existing routing protocols and dynamically adjusts to link failures/topology changes.

■ Reservations are receiver initiated and are per-stream basis.

■ Operates transparently across non-RSVP routers, allowing for partial or gradual deployment. (Non-RSVP routers cannot guarantee bandwidth reservations, so end-to-end bandwidth is not guaranteed.)

- RSVP agents are dynamically inserted (in pairs) by CUCM based on location policy.

- Calls are accepted, re-marked, or rejected based on the outcome of RSVP reservation and configured policy (optional, mandatory).

Figure 7-13 depicts RSVP call flow including endpoints, CUCM RSVP agent, and RSVP-enabled IOS routers.

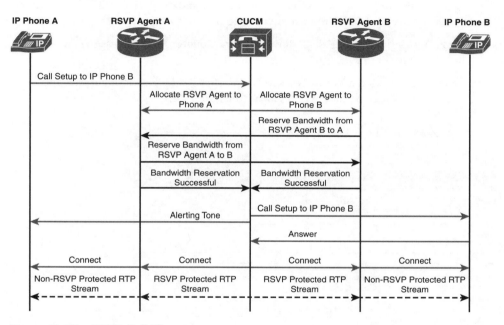

Figure 7-13   *RSVP Call Flow*

As shown in Figure 7-13, when IP phone A requests for call setup to IP phone B, all RSVP agents in the call confirm the resources and reserve bandwidth before the call setup can be sent to the destination phone.

CUCM uses two RSVP application IDs:

- **AudioStream:** Used for audio streams of voice calls

- **VideoStream:** Used for audio and video streams of video calls

These allow limiting the maximum number of audio or video calls across a link. RSVP accepts reservations or rejects them based on bandwidth pool limitations for each application ID.

Note   As per the leading practice, voice calls should be marked as EF (default) and video calls (both audio and video streams) are marked AF41 (default). Signaling traffic is marked as CS3 or AF31.

RSVP Configuration

RSVP configuration is done at CUCM and the IOS router. The following steps explain RSVP configuration on CUCM:

Step 1.   Configure a cluster-wide RSVP policy by browsing to **CUCM Administration GUI > System > Service Parameters > CUCM.** Policy for new call setup should be configured as **Mandatory.** (The call fails or reverts to AAR if RSVP reservation fails [equivalent to static location].)

Step 2.   Configure **Call Fails Following Retry Counter Exceeded** as the mid-call retry option.

Step 3.   Go to **Media Resources > MTP** and click **Add New.** Configure a media termination point (MTP) similar to Figure 7-14.

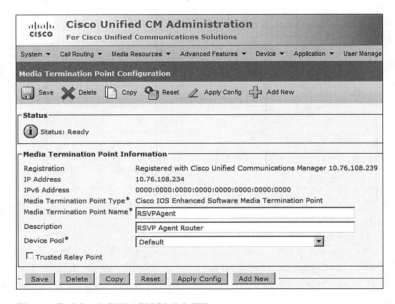

Figure 7-14   *RSVP (CUCM) MTP*

Step 4.   As a final step, configure the IOS gateway as an RSVP agent so that it can be inserted in the call between CUCM and a remote IP phone (provided the MTP is assigned to the IP phone's MRGL or the device pool MRGL). Example 7-1 outlines the configuration of a Cisco IOS router MTP as an RSVP agent. This RSVP agent will be inserted in a call setup when a new call request is initiated from an endpoint that has access to the MTP (via MRGL or device pool).

Example 7-1 *RSVP IOS Router Configuration*

```
RSVP-Router(config)# sccp local Loopback0
RSVP-Router(config)# sccp ccm 10.76.108.239 identifier 1 priority 1 version 8.0+
RSVP-Router(config)# sccp ccm 10.76.108.240 identifier 2 priority 2 version 8.0+
RSVP-Router(config)# sccp
!
RSVP-Router(config)# sccp ccm group 10
RSVP-Router(config-sccp-ccm)# associate ccm 1 priority 1
RSVP-Router(config-sccp-ccm)# associate ccm 2 priority 2
RSVP-Router(config-sccp-ccm)# associate profile 10 register RSVPAgent
!
RSVP-Router(config)# dspfarm profile 10 mtp
RSVP-Router(config-dspfarm-profile)# codec pass-through
RSVP-Router(config-dspfarm-profile)# rsvp
RSVP-Router(config-dspfarm-profile)# maximum sessions software 50
RSVP-Router(config-dspfarm-profile)# associate application SCCP
RSVP-Router(config-dspfarm-profile)# no shut
```

RSVP SIP Preconditions

SIP preconditions (as defined by RFC 3312 and RFC 4032) allow Cisco call control agents to synchronize RSVP requirements with one another. The SIP required header includes a precondition option tag and QoS precondition as part of the media attributes in the Session Description Protocol (SDP).

> **Note** SDP is a format for describing streaming media initialization parameters. SDP does not deliver media itself, but is used between endpoints for negotiation of media type, format, and all associated properties. SDP is defined in IETF RFC 4566.

In an SDP call flow, there can be multiple attributes. All of these attributes are beyond the scope of this text. The most commonly used call flow attributes are as follows:

- **Sendrecv:** Used to establish a two-way media stream.

- **Recvonly:** The SIP endpoint only receives (listen mode) and does not send media.

- **Sendonly:** The SIP endpoint only sends and does not receive media.

- **Inactive:** The SIP endpoint neither sends nor receives media.

Before RSVP initiation, an SDP can look like this:

m=audio 20000 RTP/AVP 0
c=IN IP4 10.10.100.201

a=curr:qos e2e **none**

a=des:qos mandatory e2e sendrecv

After exchange and handshake between RSVP agents, the SDP (result of UPDATE from initiator) can look like this:

m=audio 20000 RTP/AVP 0

c=IN IP4 10.10.100.201

a=curr:qos e2e **send**

a=des:qos mandatory e2e sendrecv

In reply, the other RSVP agent sends the following SDP (result of 200 OK UPDATE):

m=audio 30000 RTP/AVP 0

c=IN IP4 10.20.10.200

a=curr:qos e2e **sendrecv**

a=des:qos mandatory e2e sendrecv

With SIP preconditions, only one RSVP agent per cluster is required, and topology design with multiple clusters in a single data center or multisite distributed model is supported.

The following steps outline an RSVP SIP preconditions-based call processing, as illustrated in Figure 7-15.

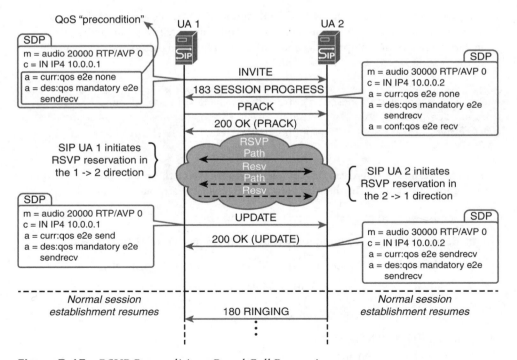

Figure 7-15   *RSVP Preconditions-Based Call Processing*

Step 1. The originating IP phone places a call to the destination IP phone, reachable through a SIP trunk. This invokes the RSVP functionality, as per the configuration of the calling IP phone's location.

Step 2. The originating CUCM server sends a SIP INVITE message with SDP. The IP address for the media stream in the SDP is set to the IP address of the originating RSVP agent.

Step 3. The destination CUCM server responds with a SIP SESSION PROGRESS message with SDP. It provides the IP address of the terminating RSVP agent and confirms the RSVP request for the forward direction, and at the same time sends an RSVP request for the reverse direction.

Step 4. The negotiation of SIP preconditions for RSVP CAC is completed by SIP Provisional Response Acknowledgement (PRACK), UPDATE, and OK messages. At this time, each of the two RSVP agents attempt an RSVP reservation for their respective forward directions (that is, toward the other RSVP agent) of the preconditioned bandwidth.

Step 5. If the RSVP reservation is successful, a standard call setup is performed by SIP RINGING, OK, and ACK messages, as described earlier in RSVP call setup process.

Step 6. When the call is answered, the destination CUCM requests a renegotiation of media capabilities by sending a SIP INVITE message without SDP, and the originating CUCM responds with a SIP OK message with SDP. The complete set of supported media capabilities is included in the SDP.

Step 7. The receiving CUCM sends a SIP OK with SDP message, including the selected codec. This codec is now used for the end-to-end call. If the selected codec has bandwidth requirements that differ from the requirements used during the SIP preconditions phase, the RSVP reservation is updated accordingly.

Step 8. The call is now established with three call legs (between the originating IP phone and its RSVP agent, the two RSVP agents, and the destination IP phone and its RSVP agent) (as with RSVP-enabled locations for calls within a cluster).

To configure RSVP SIP preconditions, follow these steps:

Step 1. Go to **CUCM Administration GUI > Device > Device Settings > SIP Profile**. Copy the default standard SIP profile and add trunk-specific configuration, as shown in Figure 7-16. Set **RSVP over SIP** to E2E. Ensure that the check box **Fall Back to Local RSVP** is checked. Set **SIP Rel1XX** options to **Send PRACK If 1XX Contains SDP**.

Step 2. Apply the new SIP profile with SIP preconditions to a SIP trunk. The next section details automated alternate routing (AAR) and its configuration.

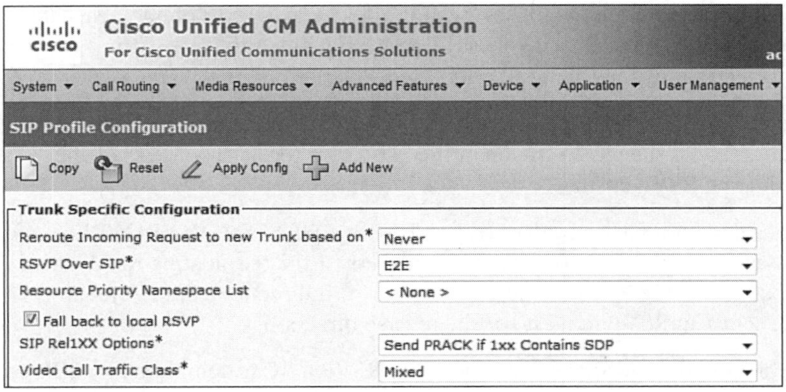

Cisco Unified CM Administration
For Cisco Unified Communications Solutions
System
Call Routing
Media Resources
Advanced Features
Device
Application
User Management
SIP Profile Configuration
Copy
Reset
Apply Config
Add New
Trunk Specific Configuration
Reroute Incoming Request to new Trunk based on*
Never
RSVP Over SIP*
E2E
Resource Priority Namespace List
< None >
Fall back to local RSVP
SIP Rel1XX Options*
Send PRACK if 1xx Contains SDP
Video Call Traffic Class*
Mixed

Figure 7-16 *SIP Profile Configuration for RSVP Preconditions*

Automated Alternate Routing

As a consequence of implementing CAC, calls beyond a specific threshold of bandwidth or link characteristic will be dropped to retain the quality on the existing calls. That directly translates to end users getting fast buy during peak hours. This outcome can be undesirable, especially when there are critical calls to be made. CUCM can automatically reroute calls through the PSTN or other networks when calls are blocked due to insufficient location bandwidth. This mechanism is known as *automated alternate routing* (AAR). Using AAR, the caller does not need to hang up and redial the called party. The AAR feature enables CUCM to establish an alternate path for the voice media when the preferred path (IP) between two intracluster endpoints runs out of available bandwidth, as determined by the locations mechanism for CAC. Figure 7-17 gives an overview of AAR-based call rerouting when the WAN link reaches a threshold bandwidth defined in location.

Main Site
CUCM Cluster
Remote Site (Branch Location)
WAN Bandwidth Full (CAC Threshold)
IP WAN
PSTN
Call Rerouted via PSTN
Voice Call Path (Media)

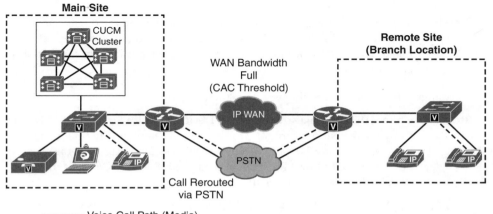

Figure 7-17 *AAR Call Flow*

To configure AAR, follow these steps:

Step 1.   Browse to **System > Service Parameters > CallManager** to set the cluster-wide service parameter **Automated Alternate Routing Enable** to **True** (False by default).

Step 2.   Browse to **Call Routing > AAR Group** and create AAR groups for local and remote sites. Define an AAR group dial prefix between a particular AAR group and other AAR groups (dial prefix is prepended to the external phone number mask of a line), as shown in Figure 7-18.

> **Note**   An AAR group represents the dialing area where the entities such as DN, voicemail port, trunks, and voice gateway are located.

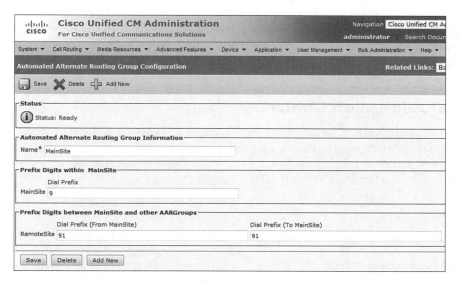

Figure 7-18   *AAR Group Definition*

Step 3.   Reuse existing or create new AAR partitions by browsing to **Call Routing > Class of Control > Partition**. Similarly, reuse existing or create a new AAR CSS by browsing to **Call Routing > Class of Control > Calling Search Space**. Assign the partitions created earlier to the respective CSS, and assign the CSS as AAR CSS at device pool and device level. AAR CSS allows the phone to access the PSTN partitions so that the calls can be completed over PSTN.

Step 4.   Go to **Call Routing > Route/Hunt > Route Patterns** and create a new route pattern. Select the external phone number mask as On, and discard digits as appropriate for the called number. Ensure that the route pattern is assigned to the appropriate AAR partition.

Step 5.   Assign the AAR setting for each phone in the line settings as shown in Figure 7-19.

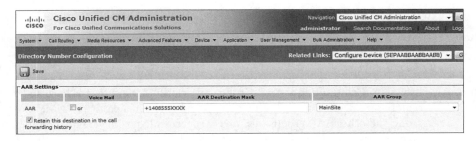

Figure 7-19   *AAR Phone Line Settings*

> **Note**   The AAR Destination Mask setting at phone line level supersedes the external phone number mask of the called phone.

When AAR is invoked, the message "Network congestion rerouting" is shown on the IP phone's display.

The next section covers the CAC mechanisms pertinent to Cisco IOS routers.

IOS Call Admission Control

Cisco IOS gateways also offer a host of CAC mechanisms to allow or disallow calls over WAN links. Although some of these are limited to just IOS, others work in conjunction with CUCM and other call controls within a Cisco Collaboration solution. The following sections describe a few commonly used IOS CAC mechanisms.

Local CAC

The examples of local CAC implementation on IOS gateway are as follows:

- **max-connections:** Allows a dial peer to enforce a maximum connection limit on a VoIP or POTS dial peer. The command **max-conn** puts limits on active connections through a dial peer. Example 7-2 gives an insight to the **max-conn** command, whereby a VoIP dial peer is limited to ten calls and has higher preference than a POTS dial peer.

Example 7-2   *Cisco IOS Voice Gateway max-connections*

```
CACRouter(config)# dial-peer voice 10 voip
CACRouter(config-dial-peer)# description dial-peer with CAC
CACRouter(config-dial-peer)# max-conn 10
CACRouter(config-dial-peer)# preference 1
CACRouter(config-dial-peer)# destination-pattern 4445551...
CACRouter(config-dial-peer)# session target ipv4:198.133.13.3
!
```

```
CACRouter(config)# dial-peer voice 110 pots
CACRouter(config-dial-peer)# description PSTN dial-peer
CACRouter(config-dial-peer)# preference 2
CACRouter(config-dial-peer)# destination-pattern 4445551...
CACRouter(config-dial-peer)# port 1/0:23
CACRouter(config-dial-peer)# prefix 9444555
```

- **Local Voice BusyOut (LVBO):** Allows private branch exchange (PBX) trunks connected to a voice gateway to be taken out of service when WAN conditions are not suitable for voice transport. LVBO allows a gateway to monitor up to 32 interfaces and provide the ability to monitor the state of network interfaces, including both LAN and WAN, and thereby busy out trunks to the PBX if any of the monitored links fail. LVBO can be implemented under voice ports (T1/E1) by using the command **busyout monitor** *<interface type>*.

- **Voice bandwidth:** It is a Voice over Frame Relay (VoFR)-only mechanism because it allows the **map class** to account for the bandwidth and deduct bandwidth on each active call. Voice bandwidth can be configured with the command **frame-relay voice-bandwidth** *<bandwidth in bps>*.

Reservation-Based CAC

Reservation-based CAC is based on the reservation or calculation of required resources before a call can be admitted. Example 7-3 shows configuration of a gatekeeper zone-based CAC.

Note   Gatekeeper-based CAC is used in legacy networks based on H.323. For newer Cisco Collaboration solutions based on SIP, previously described CAC mechanisms are used progressively.

Example 7-3   *Gatekeeper-Based RSVP*

```
GK-Router(config)# gatekeeper
GK-Router(config-gk)# zone local CUCMGK corp.local 10.10.10.180
GK-Router(config-gk)# zone remote CMEGK corp.local 10.10.100.180
GK-Router(config-gk)# zone prefix CUCMGK 1*
GK-Router(config-gk)# gw-type-prefix 1#* default-technology
GK-Router(config-gk)# bandwidth session zone CUCMGK 256
GK-Router(config-gk)# bandwidth total zone CUCMGK 2048
GK-Router(config-gk)# bandwidth interzone default 512
GK-Router(config-gk)# bandwidth remote 512
GK-Router(config-gk)# no shutdown
```

In Example 7-3, **bandwidth** commands are used to configure gatekeeper-based RSVP. The following points explain each of these commands:

- **bandwidth session:** Defines per-call bandwidth for calls placed from CUCMGK zone

- **bandwidth total:** Defines total bandwidth available for CUCMGK zone

- **bandwidth interzone:** Defines default bandwidth between CUCMGK and other zones

- **bandwidth remote:** defines the available bandwidth between local (CUCMGK) and remote (CMEGK) zones.

Measurement-Based CAC

Measurement-based CAC techniques involve the use of probes such as Cisco Service Level Agreement (SLA) or Service Assurance Agent (SAA) probes. Measurement-based CAC mechanisms include Advanced Voice BusyOut (AVBO) and PSTN fallback. AVBO is an advancement of the LVBO feature that adds the capability to probe destinations using Cisco SLA/SAA and gives the ability to busy out a trunk or voice ports based on network conditions. AVBO uses Impairment Calculated Planning Impairment Factor (ICPIF) or specific network delay, or loss values as configured. PSTN fallback is based on SAA and acts on ICPIF or delay, or loss values as configured. PSTN fallback does not busy out a trunk or voice ports, but works on a per-call basis.

Summary

CAC is an important component of every communications network and allows oversubscription of WAN links, and at the same time maintains call quality of ongoing calls. The following key points were discussed in this chapter:

- Various CAC mechanisms are available on CUCM and Cisco IOS routers.

- Some formats of CAC are topology unaware, and others are topology aware, thereby enabling proactive lookup for resources before committing the call setup/proceed.

- Numerous mechanisms can be deployed as per the technical/business requirements of an enterprise.

References

For additional information, refer to these resources:

Cisco Systems, Inc. Cisco Collaboration Systems 10.x Solution Reference Network Designs (SRND), May 2014. http://www.cisco.com/c/en/us/td/docs/voice_ip_comm/cucm/srnd/collab10/collab10.html

Cisco Systems, Inc. Cisco Collaboration Systems 11.x Solution Reference Network Designs (SRND), July 2015. http://www.cisco.com/c/en/us/td/docs/voice_ip_comm/cucm/srnd/collab11/collab11.html

Review Questions

Use these questions to review what you have learned in this chapter. The answers appear in Appendix A, "Answers Appendix."

1. Which of the following is an example of topology-aware CAC mechanism?

   a. Location-based CAC (LCAC)

   b. Enhanced location-based CAC (E-LCAC)

   c. max-connections

   d. RSVP

2. Which form of CAC leverages the Location Bandwidth Manager (LBM) service of CUCM?

   a. E-LCAC

   b. LCAC

   c. RSVP

   d. AAR

3. Will an AAR solution help if a WAN link is down? (Choose two.)

   a. Yes, calls will be rerouted over PSTN using AAR.

   b. No, AAR will not be invoked because it is a function available in case of WAN congestion, not WAN unavailability.

   c. Yes, AAR may be invoked if there is a route group with a PSTN gateway in the route list.

   d. No, AAR will not be invoked, and CUCM will reroute the calls using an alternate route.

4. Which bandwidth command defines the cumulative bandwidth for a zone on IOS gatekeeper?

   a. bandwidth session

   b. bandwidth total

   c. bandwidth remote

   d. bandwidth interzone

5. What is true about RSVP SIP preconditions? (Choose two.)

   a. Each local and remote endpoint has a direct call leg to respective local RSVP agents.

   b. All entities participating in the call maintain a call leg to all RSVP agents.

c. The CUCM cluster where the call originated maintains a call leg with local and remote endpoints.

d. Both local and remote endpoints have call legs to both local and remote RSVP agents.

6. **Which of the following statements is true about SIP preconditions?**

   a. SIP preconditions can be used to implement CAC.

   b. SIP preconditions cannot be used to implement CAC.

   c. SIP preconditions cannot be used in RSVP.

   d. SIP preconditions are not supported by CUCM.

7. **Which protocol(s) does the RSVP IOS RSVP agent uses to communicate with CUCM?**

   a. SCCP

   b. H.323

   c. SIP

   d. MGCP

   e. Both SIP and SCCP

8. **Which of the following is not part of E-LCAC network topology?**

   a. Region

   b. Location

   c. Link

   d. Effective path

   e. Weight

9. **Which of the following is not an option for LCAC configuration in CUCM?**

   a. Audio

   b. Video

   c. Immersive

   d. VCS

10. **True or false: The LBM service can only be activated on the Publisher node in a CUCM cluster.**

    a. True

    b. False

Chapter 8

Implementing Cisco Device Mobility

Cisco Collaboration solutions with multisite deployments can be challenging when users roam between buildings, sites, and campus environments. This is common in large commercial enterprises, governments, and educational institutions.

In multisite environments, it is usual for some users to roam between sites. When such users take their Cisco Unified Communications (UC) endpoints with them, devices such as Cisco Unified wireless IP phones, Cisco IP Communicator (softphone) phones, Cisco Jabber Client Service Framework (CSF) devices, or actual endpoints like Cisco 9900 series IP phones, the standard configuration of their endpoints needs to be adapted to suit the needs of the current physical location. It is important for a modern UC solution to recognize when users are roaming and adapt and change the settings on their endpoint for the current site.

This chapter describes device mobility. This feature of Cisco Unified Communications Manager (CUCM) allows CUCM endpoints to be dynamically reconfigured based on their actual location as determined by the IP address that is used by the device.

Upon completing this chapter, you will be able to meet these objectives:

- Identify the issues with devices roaming between sites
- Describe the device mobility feature
- Describe the device mobility configuration elements and their interaction
- Describe device mobility operation
- Describe the key facts that must be considered when implementing device mobility
- Describe device mobility interaction with globalized call routing
- Describe the steps required to implement and configure device mobility

Device Roaming Overview

In CUCM, much of the configuration data associated with an IP phone or video endpoint is assumed to be static data; the configuration values do not change. If a device remains stationary, the configuration values are accurate. In the event of a device moving sites or physical locations, the configuration values may be inaccurate. The configuration should change and be able to dynamically detect when a device moves or roams between sites, thus updating the underlying configuration values.

Figure 8-1 describes an IP phone or perhaps a small video endpoint roaming between two internal sites as a user brings along their device for business travel.

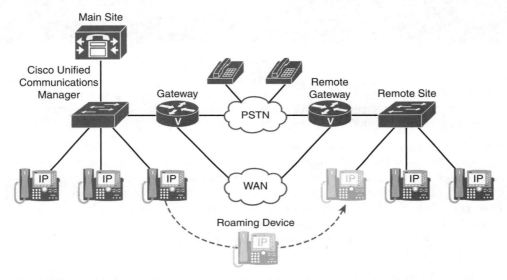

Figure 8-1   *Users and Devices Roaming Between Sites and Locations in a Unified Communications Environment*

When users roam between sites, they might take their devices with them. Although this typically does not apply to physical Cisco IP phones or video endpoints, it is still possible. In situations where employees are promoted or demoted, they may move physical office locations. An end user in UC is more likely to roam with softphones such as Cisco IP Communicator or a Cisco Jabber CSF device running on a laptop. Another common device that roams between sites often is Cisco Unified Wireless IP phones such as a Cisco 792x series endpoint.

Issues with Roaming Devices

When an endpoint moves between different CUCM sites, inaccurate device and line settings may occur. Settings such as regions, locations, calling search spaces (CSSs), and media resource groups (MRGs) all need to be updated. Imagine if a user roams between sites, if his device CSS is not updated, the user's call may use route patterns. The patterns will route the call out of the wrong site and affect the user's dialing habits. What a user may know to be a local call at one site may in fact require a long-distance call if he or she roams to another campus, building, or location.

The configuration of an IP phone includes personal settings and location-dependent settings that are all bound statically to the phone's MAC address or device name, hence to the device itself. Before the CUCM device mobility feature was introduced in CUCM Version 6.x, the physical device location was traditionally assumed to be constant.

If a CUCM registered device (most likely a softphone such as IP Communicator, Jabber CSF device, or wireless IP phone) is moved between sites, the location-dependent settings become inaccurate. Some of these settings and their errors are as follows:

- **Region:** Might cause wrong codec settings. Each device pool in CUCM can be associated with a region. The region defines the codec (or coder-decoder) to be used for audio. The "home" site of an IP phone may have a region relationship set up to use G.711 for calls internal to the home site. If a user roams across the WAN to another site, the IP phone is still using G.711 across the WAN link, as opposed to having its settings updated and reflected in the region to use G.729 or perhaps Internet Low Bitrate Codec (iLBC).

- **Location:** Might cause wrong call admission control (CAC) and bandwidth settings. Each device pool in CUCM can contain an associated Location or you can apply a Location directly to an endpoint. When devices roam it is important to have their CAC mechanisms updated to properly track bandwidth. This becomes important for newer CUCM concepts such as locations bandwidth manager and enhanced locations CAC.

- **Survivable remote site telephony (SRST) reference:** Might cause a malfunction of Cisco Unified survivable remote site telephony (SRST) by pointing to the wrong SRST Integrated Services Router (ISR).

- **Automated alternate routing (AAR) group:** Might cause a malfunction of the call redirection on no bandwidth.

- **Calling search space (CSS):** Might cause usage of remote gateways instead of local ones. This is perhaps the greatest issue with roaming devices in modern UC environments, because users who roam may be routed out the wrong gateway, trunk, or gatekeeper. This can potentially force the user to alter their dialing habits.

- **Media resource groups and media resource group lists:** Might cause allocation of wrong media resources, such as conference bridges or transcoders, media termination points, or video-conferencing resources.

To maintain the correct settings, CUCM needs to be aware of the physical location of all phones, including roaming devices within the CUCM cluster.

Using Device Mobility to Solve Roaming Device Issues

CUCM device mobility offers functionality that is designed to enhance the mobility of devices within an IP network. Table 8-1 summarizes the challenges and solutions of device mobility implemented in CUCM.

Table 8-1 *Device Mobility Solves Issues of Roaming Devices*

Issue Without Device Mobility	Device Mobility Feature to Solve the Issue
When the mobile user moves to a different location, CAC settings are not adjusted.	Location settings are dynamically assigned.
PSTN gateways to be used are fixed.	Dynamic phone CSS allows for site-independent local gateway access.
SRST reference is fixed.	SRST reference is dynamically assigned.
When the mobile user moves to a different region, codec settings are not adjusted.	Region settings are dynamically assigned.
AAR does not work for mobile users.	The AAR calling search space and the AAR group of the directory number (DN) are dynamically assigned.
Media resources are assigned location-independently.	The media resource list is dynamically assigned.
Extension mobility user device profiles are not dynamically updated based on location.	Extension mobility user device profile (UDP) settings are dynamically updated to reflect the roaming location.

Although devices such as IP phones, video endpoints, and IP Communicator/Jabber still register with the same CUCM cluster with Skinny Client Control Protocol (SCCP) or Session Initiation Protocol (SIP), they now dynamically adapt some of their behavior based on the actual site where they are located. Those changes are triggered by the IP subnet in which the phone is located. Table 8-1 shows the issues that are solved by configuring device mobility in the CUCM cluster.

Basically, all location-dependent parameters can be dynamically reconfigured by device mobility. So, the phone keeps its user-specific configuration, such as DN, speed dials, and call-forwarding settings. However, it adapts location-specific settings such as region, location, and SRST reference to the actual physical location.

Device mobility can also be configured so that dial plan-related settings, such as the device calling search space (CSS), automatic alternate routing (AAR) group, and AAR CSS, are modified. These dial plan-related settings only need to be modified if different sites require different CSSs. When you use device mobility in conjunction with local route groups and globalized call routing, the device-level CSS design can be greatly simplified. In certain situations, a single CSS may be used across all sites. Although this may not fit into every environment, it can dramatically simplify multisite deployments with dozens or hundreds of sites.

Device Mobility Overview

The following are key characteristics and features of CUCM device mobility:

■ Device mobility can be used in multisite environments with centralized call processing within a single CUCM cluster. It is important to note that there is no concept of multicluster device mobility. If a device needs to move between clusters, it must be added to each cluster, and the settings will remain cluster specific! Also, device mobility support both SCCP and SIP. It can track devices and dynamically adjust settings regardless of the registration protocol used.

■ Device mobility enables users to roam between sites with their Cisco IP phones Typically, these are Cisco Unified wireless IP phones, Cisco IP Communicator, or Jabber CSF devices. As mentioned previously, in certain situations, you may see users moving physical endpoints such as 8900 or 9900 series endpoints or perhaps SX or DX series video endpoints. IP phones are assigned with a location-specific IP address by Dynamic Host Configuration Protocol (DHCP) scopes specific to each location.

■ CUCM determines the physical location of the IP phone based on the IP address used by the IP phone.

■ Based on the physical location of the IP phone (tracked by subnet), the appropriate device configurations are applied.

Device Mobility: Dynamic Phone Configuration Parameters

Two types of IP phone or video endpoint configuration parameters can be dynamically assigned by device mobility:

■ Roaming-sensitive settings

■ Device mobility-related settings

Device mobility can reconfigure site-specific device configuration parameters based on the devices' physical location. It is important to remember that device mobility does not modify any user-specific phone parameters such as DNs or any IP phone button settings or phone services.

Figure 8-2 depicts both roaming sensitive and device mobility information settings that are available for an IP phone in CUCM.

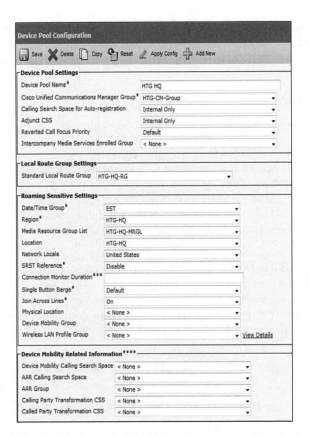

Figure 8-2   *Device Mobility Settings on a Device Pool in the CUCM Administration GUI*

The phone configuration parameters that can be dynamically applied to the device configuration are grouped in two categories:

- Roaming-sensitive settings:

  - Date/time group

  - Region

  - Location

  - Connection monitor duration

  - Network locale

  - SRST reference

  - Media resource group list (MRGL)

  - Physical location

- Device mobility group

- Local route group

Note The Date/time group, region, location, and connection monitor duration are configured at the device pool level only.

Note The Network locale, SRST reference, and media resource group list are overlapping parameters; that is, they can be configured at the phones or device pool level.

Note The physical location and device mobility group parameters determine which settings should be applied to a roaming phone. The options are none, the roaming-sensitive settings only, or the roaming-sensitive settings and the settings that are related to device mobility. They are not phone configuration parameters themselves, so they are not applied to the phone configuration like the other listed roaming-sensitive settings. Instead, they are used in the decision to change the phone configuration and determine which phone configuration values are changed. Consequently, they cannot be overlapping and can be configured only at device pools.

- Device mobility-related settings:

  - Device mobility CSS

  - AAR CSS

  - AAR group

  - Calling party transformation CSS

  - Called party transformation CSS

Note The device mobility CSS, AAR CSS, and AAR group are overlapping parameters. Therefore, they can be configured at phones and device pools. However, the device mobility CSS is called the CSS only in the Phone Configuration window. It does not overlap with the CSS configured at lines. It relates specifically to a phone's device CSS.

Roaming-sensitive settings are settings that do not have an impact on call routing. Device mobility-related settings, however, have a direct impact on call routing, because they modify the device CSS, AAR group, and AAR CSS. Depending on the implementation of device mobility, roaming-sensitive settings only, or both roaming-sensitive settings and device mobility-related settings, can be applied to a roaming phone.

The CUCM Administration GUI does not show the local route group in the roaming-sensitive settings pane. However, the local route group is a roaming-sensitive setting and is updated when the physical locations of the home device pool and the roaming device pool are different. The called-party transformation CSS is shown in the device mobility-related settings pane of the GUI, but this setting does not apply to IP phones and so it is not a device mobility-related setting, despite being listed.

Device Mobility Dynamic Configuration by Location-Dependent Device Pools

Location-dependent configuration settings are stored in the CUCM IBM Informix database once applied and saved as part of the device pool for configuration on the endpoint.

Figure 8-3 depicts an IP phone roaming between a main site and a remote site. Note the phone will move physical sites as well as be in different IP subnet schemes.

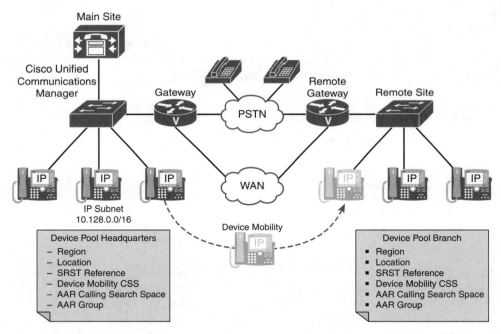

Figure 8-3   *Dynamic Configuration Settings Applied by Location-Dependent Device Pools*

As shown in Figure 8-3, the location-dependent parameters such as roaming-sensitive settings and device mobility-related settings are configured at the device pool level. Based on the IP subnet that is used by the phone and associated with a device pool, CUCM can choose the appropriate device pool and dynamically apply the location-dependent parameters.

With the introduction of device mobility, CUCM is aware of the physical location of a device based on its IP address within its IP subnet and applies the appropriate location-specific configuration by selecting the corresponding device pool in the branch site.

An IP phone can only be manually configured to be in one device pool at any one time. In Figure 8-3, the manually configured device pool is referred to as *Device Pool Main*, and the dynamically assigned device pool from device mobility is referred to as *Device Pool Remote*. Because these sites are independent, each site can have associated settings, such as the following:

- Regions

- Locations

- SRST references

- Device mobility CSS

- AAR CSS

- AAR group

These independent settings play key roles in how calls route depending on where an actual device is located. This forms the basis of device mobility.

Device Mobility Configuration Elements

Inside CUCM, there are various configuration elements used by device mobility. Table 8-2 lists the device mobility-related configuration elements and describes their functions. The newly introduced elements are device mobility info (DMI), physical location (PL), and device mobility group (DMG).

Table 8-2 *Device Mobility Configuration Element Functions*

Configuration Element Name	Configuration Element Function
Device pool (DP)	Defines a set of common characteristics for devices. It contains only device- and location-related information. One device pool has to be assigned to each device.
Device mobility info (DMI)	Specifies an IP subnet and associates it with one or more device pools. DMIs can be associated with one device pool.
Physical location (PL)	A tag assigned to one or more device pools. It is used to identify whether a device is roaming within a physical location or between physical locations. A great example is a building with multiple floors where a user can take a wireless phone and roam to a different floor in the building, but still be in the same physical site.

Table 8-2   *continued*

Configuration Element Name	Configuration Element Function
Device mobility group (DMG)	A tag assigned to one or more device pools. It is used to identify whether a device is roaming within a device mobility group or between device mobility groups. An easy way to think of a device mobility group is having a DMG per building, site, or perhaps geographic region such as United States versus Great Brittan. All devices located in this DMG are then inside a PL, which has device pools that track phones based on DMI.

The DMI is configured with a name and an IP subnet and is associated with one or more device pools. Multiple DMIs can be associated with the same device pool.

The PL and the device mobility group are just tags. They are configured with a name only and do not include any other configuration settings. Both are nonmandatory device pool configuration parameters. Therefore, at the device pool, you can choose no physical location or one physical location and one or no device mobility group. They are used to determine whether two device pools are at the same physical location or in the same device mobility group.

Relationship Between Device Mobility Configuration Elements

Figure 8-4 shows an example of the different device mobility configuration elements and how they relate to each other.

As Figure 8-4 shows, there are five DMIs for three physical locations: San Jose, New York, and London. They are configured as follows:

- **SJ1_dmi:** The IP subnet of this device mobility info is 10.1.1.0/24. This DMI is used at Building A of the San Jose campus and is associated with DP SJ_A_dp.

- **SJ2_dmi:** The IP subnet of this DMI is 10.1.2.0/24. This DMI is used at Building B1 of the San Jose campus and is associated with device pool SJ_B1_dp.

- **SJ3_dmi:** The IP subnet of this DMI is 10.1.3.0/24. Like SJ2_dmi, this DMI is used at Building B1, which is associated with device pool SJ_B1_dp, but it is also used at Building B2 and is associated with device pool SJ_B2_dp.

- **NY_dmi:** The IP subnet of this DMI is 10.3.1.0/24. This DMI is used at the New York campus and is associated with device pool NY_dp.

- **LON_dmi:** The IP subnet of this DMI is 10.10.1.0/24. This DMI is used at the London campus and is associated with device pool LON_dp.

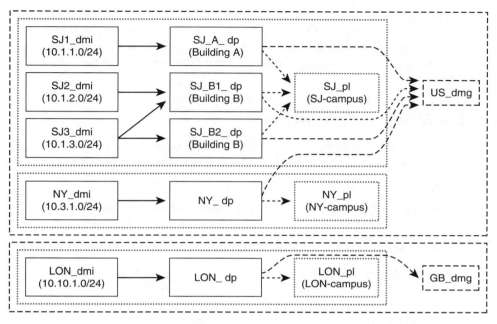

Figure 8-4 *Relationship Between Device Mobility Configuration Elements*

Device pools SJ_A_dp, SJ_B1_dp, and SJ_B2_dp are all configured with the same physical location (SJ_pl) because they are all used for devices located at the San Jose campus. Keep in mind that a PL could represent a campus with multiple buildings or a building with multiple floors.

Device pool NY_dp, serving the New York campus, is configured with physical location NY_pl. Device pool LON_dp, serving the London campus, is configured with physical location LON_pl.

All device pools that are assigned with a U.S. physical location (that is, SJ_A_dp, SJ_B1_dp, SJ_B2_dp, and NY_dp) are configured with device mobility group US_dmg. This setting means that all U.S. device pools are in the same DMG. The London campus is in a different DMG: GB_dmg. The DMG configured represents a container of sites or locations. If a phone is in the same DMG but a different physical location, settings will be updated. An example is a phone traveling from San Jose to New York; it is in the same DMG, but different PLs.

In summary, the U.S. DMG consists of two physical locations: San Jose and New York. At San Jose, IP subnets 10.1.1.0/24, 10.1.2.0/24, and 10.1.3.0/24 are used; New York uses IP subnet 10.3.1.0/24, and London is configured with IP subnet 10.10.1.0/24. DMGs are used to recognize differences in dial patterns in different geographic locations.

Based on the IP address of an IP phone, CUCM can determine one or more associated device pools and the physical location and DMG of the device pool or pools. If an IP phone uses an IP address of IP subnet 10.1.3.0/24, the device pool has two candidates. However, in this example, the physical location and DMG are the same for these two device pools.

Device Mobility Operation

As discussed earlier, each endpoint is configured with a device pool, similar to previous versions of CUCM. This device pool is considered the endpoints' home device pool.

IP subnets are associated with device pools by configuring DMIs.

The following occurs when a device mobility-enabled endpoint registers with CUCM with an IP address that matches an IP subnet configured in a DMI:

- **The current device pool is chosen as follows:**

  - If the DMI is associated with the endpoint's main or home device pool, the endpoint is considered to be in its home location. Therefore, device mobility does not reconfigure the endpoint.

  - If the DMI is associated with one or more device pools other than the phone's main or home device pool, one of the associated device pools is chosen based on a round-robin load-sharing algorithm.

- **If the current device pool is different from the home device pool, the following checks are performed:**

  - If the physical locations are not different, the phone's configuration is not modified.

  - If the physical locations are different, the roaming-sensitive parameters of the current roaming device pool are applied.

  - If the device mobility groups are the same, in addition to different physical locations, the device mobility-related settings are also applied, along with the roaming-sensitive parameters.

In summary, the roaming-sensitive parameters are applied when the physical location of the current device pool is different from the physical location of the main or home device pool. The device mobility-related settings are also applied in addition to the roaming-sensitive parameters when the physical locations are different and the device mobility groups are the same. This occurs when roaming between physical locations within the same device mobility group.

As a consequence, physical locations and device mobility groups should be used as follows:

- **Physical locations:** Configure physical locations in such a way that codec choice and CAC truly reflect the device's current location. Also, local SRST references and local media resources at the roaming site should be used instead of those located at the currently remote home network. Depending on the network structure, IP subnets, and allocation of services, you can define physical locations based on a city, enterprise campus, building, or floor/department.

■ **Device mobility groups:** A DMG should define a group of sites with similar dialing patterns or dialing behavior. DMGs represent the highest-level geographic entities in your network. Depending on the network size and scope, your DMGs could represent countries, regions, states or provinces, cities, or other geographic entities. Device mobility-related settings that are applied only when roaming within the same DMG impact call routing. Therefore, different DMGs should be set up whenever a roaming user should not be forced to adapt his dialing behavior. In this case, when roaming between different DMGs, the phone device mobility-related settings that impact call routing are not modified.

Note   When using globalized call routing and local route groups, DMGs are irrelevant. There is no need to change the device-level CSS, the AAR CSS, and the device-level AAR group. The section, "Device Mobility Interaction with Globalized Call Routing" provides more information about the interaction of globalized call routing and device mobility.

Device Mobility Operation Flowchart

Figure 8-5 illustrates the flow of device mobility operation. Again, DP is the device pool, and DMI is the device mobility info (IP subnet). DMG is device mobility group, and PL is the physical location.

The process of a phone registration with device mobility enabled is shown in the following steps:

Step 1.   A phone device attempts to register with CUCM. Phones that do not register with CUCM cannot be part of the CUCM cluster and therefore do not have any device mobility configuration. Endpoints can register with CUCM using either SCCP or SIP protocols; device mobility has no impact on the protocol used. If the phone successfully registers with a CUCM server, the process continues.

Step 2.   CUCM checks whether device mobility is enabled for the device. If device mobility is not enabled, the default registration process behavior applies; proceed to Step 10. Otherwise, if device mobility is enabled, the registration and device mobility selection process continues.

Step 3.   Assuming device mobility is enabled, CUCM checks whether the IP address of the endpoint is found in one of the DMGs. If the IP address is not found in any DMG, the default behavior applies; proceed to Step 10. Otherwise, CUCM will attempt to pick a device pool based on the associated DMI and continue the process.

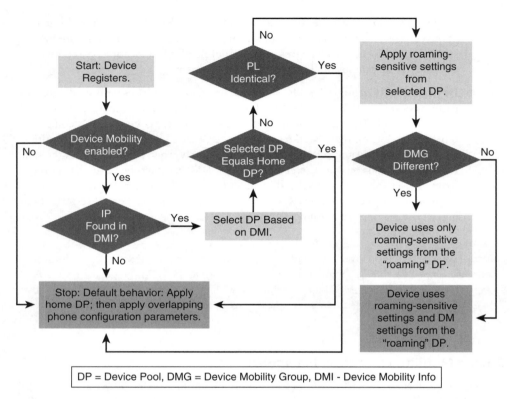

DP = Device Pool, DMG = Device Mobility Group, DMI - Device Mobility Info

Figure 8-5 *Device Mobility Operational Flowchart*

Step 4.   If the endpoint's home device pool is associated with the DMI in which the phone's IP address was found, the home or main device pool is chosen. If the home or main device pool is not associated with the DMI in which the phone's IP address was found, the device pool is chosen based on a load-sharing algorithm. The load-sharing algorithm applies if more than one device pool is associated with the DMI.

Step 5.   If the chosen device pool is the home or main device pool, the default behavior applies; proceed to Step 10. Otherwise, CUCM will evaluate the PLs, thus continuing the process.

Step 6.   If the PLs of the chosen device pool and the home or main device pool are the same, the default behavior applies; proceed to Step 10. Otherwise, CUCM will evaluate roaming-sensitive settings and continue.

Step 7.   The roaming-sensitive settings of the chosen device pool of the roaming or remote device pool are used to update the phone's configuration.

Note   In this case, there is a potential to have overlapping settings at the device pool and device levels in CUCM. Should this occur, the device pool (namely, media resource group list, location, and network locale of the roaming or remote device pool) has priority over the corresponding settings at the device level on the phone. This behavior is different from the default behavior in Step 10.

Step 8.   If the DMG of the chosen device pool and the home or main device pool are different, the device uses only the roaming-sensitive settings from the "roaming" or "remote" device pool. If they are not different, the device uses the roaming or remote settings and the device mobility (DM) settings from the "roaming" or "remote" device pool.

Note   In this case, all settings are overlapping settings that are device mobility-related settings that exist at the phone and device pool. Therefore, the parameters of the roaming or remote device pool have priority over the corresponding settings at the phone.

Step 9.   Next, where the phone configuration has been updated with either the roaming-sensitive settings only, or with the roaming-sensitive settings and the device mobility-related settings, the phone is reset for the updated configuration to be applied to the phone.

Note   This is the end of the process. Step 10 applies only in the conditions outlined in the previous steps.

Step 10.   The default behavior is the settings of the home device pool, which is the device pool configured on the phone. Some configuration parameters of the device pool can also be set individually at the phone. These overlapping phone configuration parameters are as follows:

- Media resource group list
- Location
- Network locale
- Device mobility CSS (which is just called CSS at the phone)
- AAR CSS
- AAR group

If these are configured at the phone, implying they are not set to [None], the phone configuration settings have priority over the corresponding setting at the device pool.

Device Mobility Considerations

In multisite deployments with device mobility enabled, there are special considerations.

Roaming-sensitive settings ensure that the roaming device uses local media resources and SRST references. In addition, they confirm the correct use of codecs and CAC between sites. Typically, this is always desired when a device roams between different sites. It is not required when the device moves only between IP subnets within the same site. Therefore, the recommendation is to assign all device pools that are associated with IP subnets (DMI) that are used at the same site to the same physical location. This results in phone configuration changes only when the phone roams between sites (physical locations) and not in a situation where a phone is only moved between different networks of the same site.

Device mobility-related settings impact call routing. By applying the device CSS, AAR group, and AAR, CSS calls are routed differently. The settings at the roaming device pool determine which gateway will be used for public switch telephone network (PSTN) access and AAR PSTN calls based on the device CSS and AAR CSS. They also determine how the number to be used for AAR calls is composed based on the AAR group.

Such changes can result in different dialing behavior. For instance, when you roam between different countries, the PSTN access code and PSTN numbering plans might be different. For example, to dial the Austrian destination +43 699 18900009, users in Germany dial 0.0043 699 18900009, whereas users in the United States have to dial 9.01143 699 18900009.

For example, European users who roam with their softphones to the United States might be confused when they have to use U.S. dialing rules access code 9 instead of 0 and 011 instead of 00 for international numbers. If you want to avoid this confusion, suppress the application of device mobility-related settings. You do this by assigning device pools that are to be used at sites with different dialing rules to different device mobility groups and different physical locations. Now, when a user roams with a device from, for example Germany, to the United States, all the roaming-sensitive settings are applied such as local media resources, SRST gateways, codecs, and CAC settings. The device mobility-related settings are not applied. The phone now uses the PSTN gateway and dial rules of its home location even though the user moved to another site. The user does not have to adapt to the dial rules of the local site to which the phone was moved.

Note The preceding statements regarding call routing and dial behavior that are based on device mobility-related settings do not apply when globalized call routing is used. The section "Device Mobility Interaction with Globalized Call Routing" presents more information about the interaction of globalized call routing and device mobility.

Review of Line and Device CSSs

An IP phone can be configured with a line CSS and a device CSS. If both exist, the partitions configured to the line CSS are considered before the partitions of the device CSS when routing a call. (This is another example of "more specific overrides more general.") Remember line trumps device, and typically a line CSS contains restrictions or partitions with block patterns, whereas a device CSS contains route patterns for PSTN access.

These two CSSs allow the use of the line/device approach for implementing calling privileges and the choice of a local gateway for PSTN calls. With the line/device approach, all possible PSTN route patterns exist once per location, which is configured with a site-specific partition. This partition is included in the device CSS of the phones and therefore enables the use of a local gateway for PSTN calls. To implement class of service (CoS), PSTN route patterns that should not be available to all users (for example, international calls, long-distance calls, or all toll calls) are configured as blocked route patterns and are assigned to separate partitions. The line CSS of a phone now includes the partitions of the route patterns that should be blocked for this phone. Because the line CSS has priority over the device CSS, the blocked pattern takes precedence over the routed pattern that is found in a partition listed at the device CSS.

Device Mobility and CSSs

The CSSs assigned at both the line and device level on an endpoint play a crucial role in device mobility operations. This section examines the impact of CSS selection in the device mobility process. There are a few rules and design considerations for CSS and device mobility.

Device mobility never modifies the line CSS of a phone. It does, however, change the device CSS and AAR CSS of a phone when the phone is roaming between different physical locations within the same device mobility group.

The line CSS implements CoS configuration by permitting internal destinations, such as phone DNs, call park, and Meet-Me conferences, but blocks PSTN destinations. Because the line CSS is not changed by device mobility, CoS settings of the device are kept when the device is roaming.

The device CSS is modified when roaming within the same device mobility group. In this case, the device CSS that is used at the home location is replaced by a device CSS that is applicable to the roaming location. This device CSS typically contains translation patterns, route patterns, or perhaps called-party transformation patterns that reference a site-specific gateway. When device mobility is invoked, the device CSS now refers to the local gateway of the roaming site instead of the gateway that is used at the home location, thus routing the call out of the proper site specific gateway.

If the traditional approach of using only one CSS combining CoS and gateway choice is used, the device CSS must be used. This is because device mobility cannot modify the line CSS, and the line CSS has priority over the device CSS. These settings can be modified by device mobility.

The AAR CSS can be configured only at the device level. Therefore, it is always correctly replaced when roaming between physical locations within the same device mobility group.

> **Note**   When using globalized call routing and local route groups, there is no need for site-specific device-level CSS. More information about the interaction of globalized call routing and device mobility is provided in the section "Device Mobility Interaction with Globalized Call Routing."

Examples of Different Call-Routing Paths Based on Device Mobility Groups and Tail-End Hop-Off

Table 8-3 shows how calls are routed in different device mobility scenarios.

Table 8-3   *Examples of Different Call Routing Paths Based on Device Mobility Groups and TEHO*

Scenario	Result
Same DMG, call to PSTN destination close to home location, no TEHO	The call uses the local PSTN gateway at the roaming location to place a long-distance PSTN call.
Same DMG, call to PSTN destination close to home location, TEHO	The call uses the IP WAN to the gateway at the home location to place a local PSTN call.
Same DMG, call to PSTN destination close to roaming location	The call uses the local PSTN gateway at the roaming location to place a local PSTN call.
Different DMG, call to PSTN destination close to home location	The call uses the IP WAN to the gateway at the home location to place a local PSTN call.
Different DMG, call to PSTN destination close to roaming location, no TEHO	The call uses the IP WAN to the gateway at the home location to place a long-distance PSTN call.
Different DMG, call to PSTN destination close to roaming location, TEHO	The call uses the local PSTN gateway at the roaming location to place a local PSTN call.

Calls are routed differently depending on the configuration of device mobility groups. Call routing factors depend on whether device mobility-related settings are applied, the dialed destination, and the use of tail-end hop-off (TEHO). In some scenarios, calls might take suboptimal paths.

For example, assume that a user from London roams to the U.S. office with Cisco IP Communicator. For simplicity, assume that there is only one U.S. office.

For the following three scenarios, the home device pool and the roaming device pool are assigned to the same DMG, which means that device mobility applies device mobility-related settings. As a result, PSTN calls placed from the roaming device are treated like PSTN calls of standard U.S. phones:

1. If a call is placed to a PSTN number that is close to the home location of the phone, such as a U.K. number, assuming TEHO is not configured, the call uses the local U.S. PSTN gateway to place an international PSTN call. From a toll perspective, this is a suboptimal solution, because the IP WAN is not used as much as it could be when implementing TEHO. This factor applies not only to the roaming user, but also to U.S. users who place calls to PSTN destinations in Great Britain.

2. If the same call is placed to a U.K. PSTN number and TEHO is configured, the call uses the IP WAN to the London site and breaks out to the PSTN at the London gateway with a local call. This solution is the optimal one from a toll perspective.

3. If a call to a U.S. destination number is placed, the U.S. gateway is used for a local or national call. This event is optimal from a toll perspective.

Note In all the examples shown that are based on Table 8-3, the user has to dial PSTN destinations by following the North American Numbering Plan (NANP) dial rules.

For the next three scenarios, the home or main device pool and the roaming or remote device pool are assigned to different DMGs. This means that the device mobility-related settings are not applied. Therefore, calls placed from the roaming device are routed the same way as they are when the device is in its home location:

1. If a call from the U.S. to a U.K. PSTN destination is placed, the call uses the IP WAN to the London site and breaks out to the PSTN at the London gateway with a local or national call. This solution is the optimal one from a toll perspective.

2. If a call is placed from the U.S. to a PSTN destination close to the roaming or remote location and TEHO is not configured, the call uses the IP WAN from the U.S. office to the London site and breaks out to the PSTN at the London gateway to place an international call back to the United States. From a toll perspective, this is the worst possible solution, because the call first goes from the United States to London over the IP WAN, wasting bandwidth, and then goes back from London to the United States via a costly international call.

3. If a call is placed from the U.S. to a PSTN destination close to the roaming or remote location and TEHO is configured, the U.S. gateway is used for a local or national call. This event is optimal from a toll perspective.

Note In these three examples, the user has to dial PSTN destinations by following the dial rules of the home or main location (the United Kingdom).

In summary, when allowing the device mobility-related settings to be applied by using the same DMG, calls to the home location use a local PSTN gateway to place a long-distance or international call when not implementing TEHO. All other calls are optimal.

When the device mobility-related settings are not applied by using different device mobility groups and by not using TEHO, calls to the roaming location first use the IP WAN to go from the roaming location to the home location and then use the home gateway to place a long-distance or international call back to the roaming location. All other calls are optimal.

The discussed scenarios assume that globalized call routing and local route groups are not used. The impact of globalized call routing and local route groups is discussed in the next section.

Device Mobility Interaction with Globalized Call Routing

This section examines the interactions of globalized call routing and local route groups with CUCM device mobility enabled for IP phones and video endpoints.

Local route groups were introduced in CUCM Version 7. When local route groups and globalized call routing (which use local route groups) are not used or supported, device mobility is typically implemented as follows:

- **Roaming-sensitive settings:** These settings are updated when the device roams between different physical locations. These settings are location, region, SRST reference, MRGL, and other parameters that do not affect the selection of the PSTN gateway or the local rules.

- **Device Mobility-related settings:** These settings can be applied in addition to the roaming-sensitive settings (meaning that a phone has to roam between different physical locations). The device mobility-related settings are device CSS, AAR CSS, and AAR group. The configuration of the device mobility group should determine your decision about whether to apply the device mobility-related settings. Here are two scenarios:

  - If the device roams between different device mobility groups, the device mobility-related settings are not updated with the values that were configured at the roaming or remote device pool. This configuration has the advantage that users do not have to adapt to different dial rules between home and roaming location (if they exist). The disadvantage is that all PSTN calls use the home gateway, which can lead to suboptimal routing.

  - If the device roams within the same device mobility group, the device mobility-related settings are updated with the values of the roaming device pool. This configuration has the advantage that all PSTN calls will use the local (roaming or remote) gateway, which is typically desired for roaming or remote users. However, the users have to adhere to the local dial rules.

Note If TEHO is used, there are no suboptimal paths when using device mobility with different device mobility groups. When local route groups and globalized call routing are not being used, TEHO implementation can be complex, especially when local PSTN backup is desired and when TEHO is implemented in international deployments.

In summary, unless TEHO is used, the implementation of device mobility without globalized call routing leads to this situation: Either the home gateway has to be employed (when allowing the home dial rules), or the user is forced to adhere to the dial rules of the roaming site (to use the local gateway of the roaming site).

Advantages of Using Local Route Groups and Globalized Call Routing

Device mobility has extensive benefits when paired or designed with globalized call routing and local route groups, especially when implemented in international environments.

When device mobility with globalized call routing is used, there are no changes when implementing the roaming-sensitive settings. Their application remains the same when roaming between sites. One setting that is important is the application of the local route group, which is a roaming-sensitive setting.

The dial plan-related part of device mobility, however, changes substantially with globalized call routing. It allows a roaming user to follow the home dial rules for external calls and use the local gateway of the roaming site.

This situation is possible because globalization of localized call ingress occurs at the phone. This function is provided by the line CSS of the phone. It offers access to phone-specific translation patterns that normalize the localized input of the user to global format. The device CSS that was used for gateway selection is obsolete because gateway selection is now performed by the local route group feature.

The AAR CSS and AAR group that are configured at the device level can be the same for all phones as long as the AAR number is always in global format. (Ensure that it is always in global format by configuring either the external phone number mask or the AAR transformation mask to E.164 format.) In this case, no different AAR groups are required because there is no need for different prefixes that are based on the location of the two phones.

Furthermore, there is no need for different AAR CSS, because the gateway selection is not based on different route lists (referenced from different route patterns in different partitions). Instead, it is based on the local route group that was configured at the device pool of the calling phone.

In summary, when globalized call routing is used, device mobility allows users to utilize local gateways at roaming sites for PSTN access (or for backup when TEHO is configured) while adhering to their home dial rules. There is no need to apply different device CSS, AAR CSS, and AAR groups, so device mobility groups are no longer required.

An Example of Globalized Call Routing That Is *Not* Configured with a Different Device Mobility Group

Figure 8-6 shows an example of device mobility with different device mobility groups in an environment where globalized call routing is not implemented. Also, gateway selection is performed by the device CSS of the IP phone.

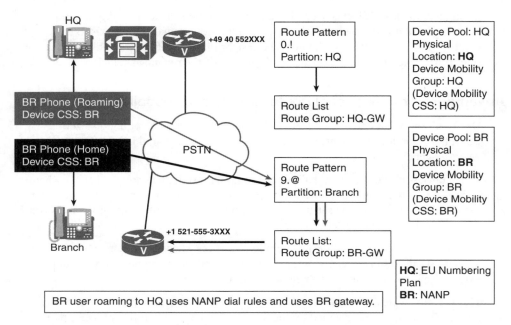

Figure 8-6   *Call Routing Without Globalized Call Routing Using Different Device Mobility Groups*

In Figure 8-6, there are two sites: the main site (HQ) is in Europe, and the branch site (BR) is in the United States. Separate route patterns (representing the different dial rules) are configured in different partitions. The CSS of HQ phones provides access to the HQ gateway; the CSS of BR phones provides access to the BR gateway.

Device mobility is configured with different device mobility groups. This configuration allows BR users who are roaming with their phones to the HQ to use the home dial rules. The device CSS is not updated by device mobility, and therefore the CSS still provides access to the BR route pattern (9.@). However, as a consequence, the BR gateway is used for all PSTN calls. In the configuration shown in Figure 8-6, a nonoptimal route is chosen for path selection. Ideally, the call should route out the headquarters gateway, but because the device CSS is not updated by device mobility, the wrong gateway is chosen for path selection.

An Example of Globalized Call Routing That Is *Not* Configured with the Same Device Mobility Group

Figure 8-7 shows an example of device mobility with identical device mobility groups in an environment where globalized call routing is not implemented. Also, gateway selection is performed by the device CSS of the IP phone.

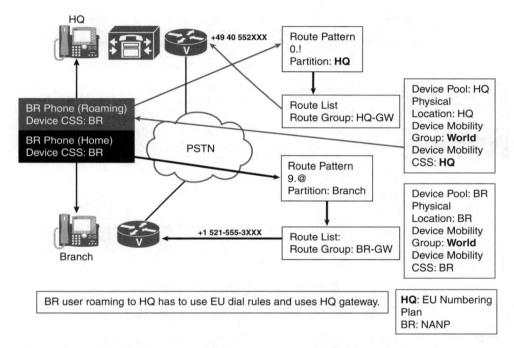

Figure 8-7 *Call Routing Without Globalized Call Routing Using the Same Device Mobility Group*

This example is identical to the previous example with one exception: This time, the device mobility group of the home and the roaming device pool are the same.

When a BR user roams to the HQ, the device CSS of the phone is updated with the device CSS of the roaming device pool. In the example, CSS BR is changed to HQ. As a consequence, the phone has access to the HQ partition that includes PSTN route patterns in EU dialing format (0.!).

Therefore, the roaming user has to follow EU dial rules. Calls to 9.@ are not possible anymore. However, this configuration allows the BR user to use the HQ gateway when roaming to the HQ. Figure 8-7 shows optimal call routing configuration and path logic, but it uses route patterns that may be unfamiliar to some readers.

An Example of Globalized Call Routing

Figure 8-8 shows an example of device mobility in an environment where globalized call routing is implemented. Also, gateway selection is performed by the local route group feature.

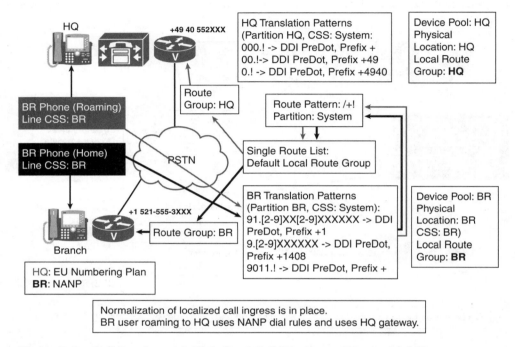

Figure 8-8   *Call Routing with Globalized Call Routing and Device Mobility*

This example is based on the previous scenario: HQ is in Europe, and BR is in the United States. A BR user roams to Europe.

However, in this example, globalized call routing has been implemented. Therefore, the (line) CSS of BR phones provides access to translation patterns that convert localized call ingress at the phone (NANP format) to global E.164 format. EU phones have access to translation patterns that convert EU input to global E.164 format.

A single PSTN route pattern (\+!) is configured; it is in a partition that is accessible by all translation patterns.

When a BR user roams to the HQ, the line CSS is not modified; no device CSS is configured at the phone or at the device pool. The device mobility groups are also not set (or are set differently).

As a result, there is effectively no change in matching the translation patterns: The BR user still uses NANP dial rules (like at home). The number is converted to international format by translation patterns and matches the (only) PSTN route pattern. The route pattern refers to a route list that is configured to use the default local route group. The default local route group is taken from the roaming device pool. Therefore, if the phone is physically located in the BR office, the local route group is BR. If the phone is roaming to the HQ site, the local route group is HQ. As a result, the local gateway is always used for a PSTN call.

If TEHO was configured, there would be a TEHO route pattern in E.164 format with a leading + sign. The TEHO pattern would refer to a site-specific route list in order to select the correct gateway for PSTN egress. The backup gateway would then again be selected by the local route group feature. In the configuration shown in Figure 8-8, local dialing habits and rules optimally route the call by choosing the local route group for path selection.

Device Mobility Configuration

The following steps describe how to configure device mobility on CUCM:

Step 1.   To configure physical locations and device mobility groups, as shown in Figure 8-9, navigate to CUCM Administration and choose **System > Physical Location.** For each physical location, a name and a description are configured.

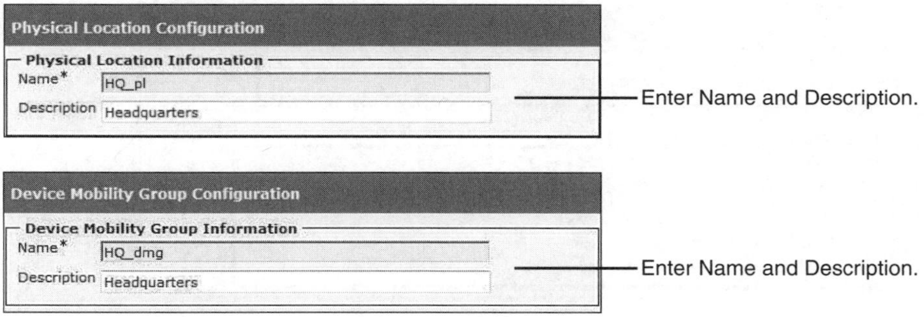

Figure 8-9   *Steps 1 and 2: Configure Physical Locations and Device Mobility Groups*

Note   Device mobility groups are not necessary when there is no need to change the device level CSS, AAR CSS, and AAR group. This principle applies also when local route groups are used in an environment where all sites share the same dial rules or in an environment where globalized call routing is implemented.

Step 2.   To configure a device pool when device mobility is used, as shown in Figure 8-10, in CUCM Administration, go to **System > Device Pool.** A device pool is configured with a name and a CUCM group. It includes roaming-sensitive settings and device mobility-related settings configured under Device Mobility Related Information. The physical location and the device mobility group, both configured under Roaming Sensitive Settings, are used to decide what settings should be applied to a phone.

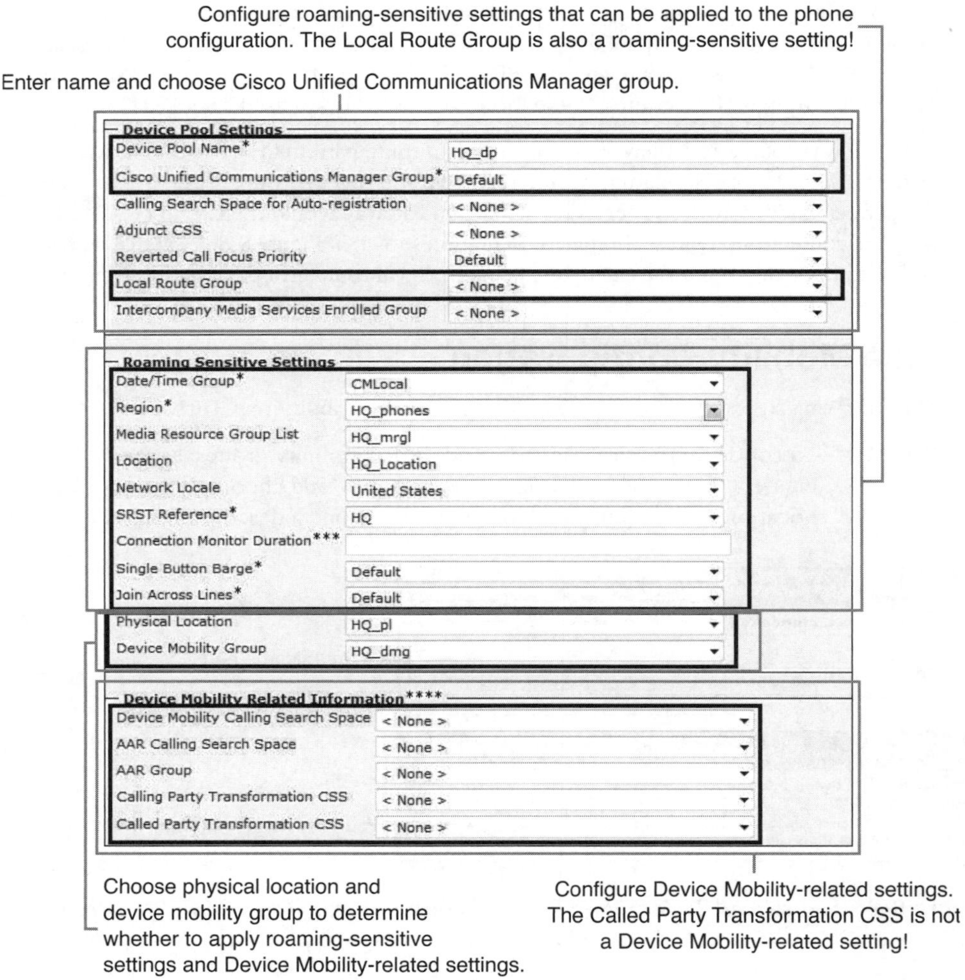

Configure roaming-sensitive settings that can be applied to the phone
configuration. The Local Route Group is also a roaming-sensitive setting!

Enter name and choose Cisco Unified Communications Manager group.

Device Pool Settings

Device Pool Name*	HQ_dp
Cisco Unified Communications Manager Group*	Default
Calling Search Space for Auto-registration	< None >
Adjunct CSS	< None >
Reverted Call Focus Priority	Default
Local Route Group	< None >
Intercompany Media Services Enrolled Group	< None >

Roaming Sensitive Settings

Date/Time Group*	CMLocal
Region*	HQ_phones
Media Resource Group List	HQ_mrgl
Location	HQ_Location
Network Locale	United States
SRST Reference*	HQ
Connection Monitor Duration***	
Single Button Barge*	Default
Join Across Lines*	Default
Physical Location	HQ_pl
Device Mobility Group	HQ_dmg

Device Mobility Related Information**

Device Mobility Calling Search Space	< None >
AAR Calling Search Space	< None >
AAR Group	< None >
Calling Party Transformation CSS	< None >
Called Party Transformation CSS	< None >

Choose physical location and
device mobility group to determine
whether to apply roaming-sensitive
settings and Device Mobility-related settings.

Configure Device Mobility-related settings.
The Called Party Transformation CSS is not
a Device Mobility-related setting!

Figure 8-10 *Configuring Device Pool Device Mobility Settings*

Note   The physical location and the device mobility group themselves are not applied to
the configuration of a phone but are used only to control which settings to apply.

Figure 8-10 does not show the local route group in the roaming-sensitive settings pane.
Nevertheless, the local route group is a roaming-sensitive setting and is updated when the
physical locations of the home device pool and the roaming device pool are different.
The called-party transformation CSS is shown in the device mobility-related settings pane,
but this setting does not apply to IP phones so there is no device mobility-related setting.

Step 3. To configure a device pool when device mobility is used, as shown in Figure 8-11, in CUCM Administration, choose **System > Device Mobility > Device Mobility Info**. The DMIs are configured with a name, a subnet, and a subnet mask. Then they are associated with one or more device pools.

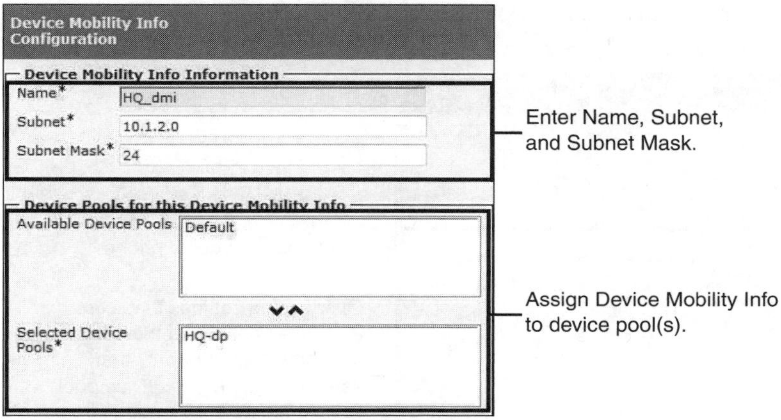

Figure 8-11 *Configuring Device Mobility Information for Device Pools*

Step 4. Device mobility is off by default and can be configured per phone. The default for the device mobility mode, if it is not set differently at the phone, is set under **System > Service Parameter**, as shown in Figure 8-12. Choose the **Cisco CallManager** service, and set the **Device Mobility Mode** to **On** or **Off**. Note that Off is the default. The parameter is found in the Clusterwide Parameters (Device - Phone) section.

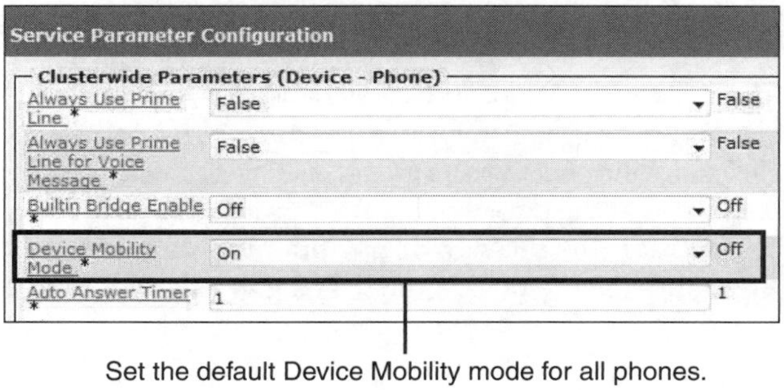

Set the default Device Mobility mode for all phones.

Figure 8-12 *Device Mobility Mode CCM Service Parameter Settings*

Photo credits: charcomphoto, Vladimir Agapov

Note There are many entries in CUCM under **System > Service Parameter** after you choose Cisco CallManager. Press **Ctrl-F** in Internet Explorer and search for the specific title or entry.

Step 5.   As shown in Figure 8-13, set the device mobility mode per phone in CUCM Administration by choosing **Device > Phone**. In the Phone Configuration window, you enable or disable device mobility for the phone by either setting the **Device Mobility Mode** to **On** or **Off** or leaving the default value **Default**. If the device mobility mode is set to Default, the device mobility mode is set at the Cisco CallManager service parameter.

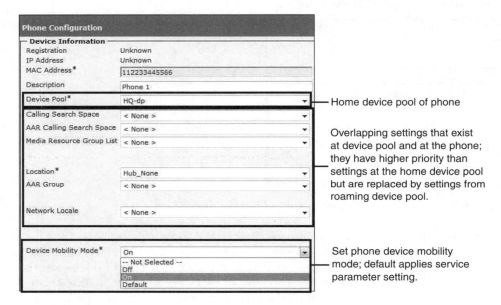

Figure 8-13   *Configuring Device Mobility Settings on IP phones*

Note   Figure 8-13 also shows the configuration of the overlapping parameters, which can be configured at both the phone and the device pool. The overlapping parameters for roaming-sensitive settings are media resource group list, location, and network locale. The overlapping parameters for the device mobility-related settings are calling search space (called device mobility calling search space at the device pool), AAR group, and AAR CSS. Overlapping parameters configured at the phone have higher priority than settings at the home device pool and lower priority than settings at the roaming device pool.

Summary

The following key points were discussed in this chapter:

- Issues with roaming devices include inappropriate region, location, time zone, and SRST reference configuration. PSTN calls use the home gateway instead of the local gateway at the roaming site.

- Device mobility allows roaming devices to be identified by their IP addresses. It also allows configuration settings to be applied that are suitable for the device's current physical location.

- Device mobility configuration elements are device mobility groups, physical locations, device pools, and DMIs.

- Roaming-sensitive settings are applied to devices that roam between physical locations. In addition, device mobility-related settings are applied to devices that roam within the same device mobility group.

- After you configure device mobility groups, physical locations, device pools, and DMIs, you need to enable device mobility either cluster-wide as a service parameter or individually per phone.

References

For additional information, refer to the following:

Cisco Systems, Inc. Cisco Collaboration Systems 10.x Solution Reference Network Designs (SRND), May 2014. http://www.cisco.com/c/en/us/td/docs/voice_ip_comm/ cucm/srnd/collab10/collab10.html

Cisco Systems, Inc. Cisco Collaboration Systems 11.x Solution Reference Network Designs (SRND), July 2015. http://www.cisco.com/c/en/us/td/docs/voice_ip_comm/ cucm/srnd/collab11/collab11.html

Cisco Systems, Inc. Features and Services Guide for Cisco Unified Communications Manager, Release 10.0(1), May 2014. http://www.cisco.com/c/en/us/td/docs/voice_ ip_comm/cucm/admin/10_0_1/ccmfeat/CUCM_BK_F3AC1C0F_00_cucm-features-services-guide-100/CUCM_BK_F3AC1C0F_00_cucm-features-services-guide-100_ chapter_0110011.html

Review Questions

Use these questions to review what you have learned in this chapter. The answers appear in Appendix A, "Answers Appendix."

1. **Which setting is not modified for a laptop with IP Communicator when a user roams between sites when device mobility is enabled?**

    a. Region

    b. Directory number

    c. Location

    d. SRST reference

2. **Which two statements about the relationship between device mobility configuration elements are true? (Choose two.)**

   a. Device mobility infos refer to one or more device pools.

   b. Device pools refer to one or more physical locations.

   c. Device pools refer to one device mobility group.

   d. Device pools refer to one device mobility info.

   e. Physical locations refer to device mobility groups.

3. **Which statement about device mobility operation is false?**

   a. A device pool is selected based on the roaming device's IP address.

   b. If the selected device pool is the home device pool, no changes are made.

   c. If the selected device pool is in a different device mobility group than the home device pool, the device mobility-related settings of the roaming device pool are applied.

   d. If the selected device pool is in a different physical location than the home device pool, the roaming-sensitive settings of the roaming device pool are applied.

4. **True or false: If no device pool settings are configured on the phone, the settings configured on the phone are used.**

   a. True

   b. False

5. **Which two of the following are valid issues that device mobility fixes by changing the settings for roaming (mobile) phones in CUCM? (Choose two.)**

   a. SRST references for mobile phones are dynamically assigned.

   b. Phone speed dials for mobile phones are dynamically assigned.

   c. Phone services for mobile phones are dynamically assigned.

   d. Region settings for mobile phones are dynamically assigned.

   e. Call forward settings for mobile phones are dynamically assigned.

6. **Which of the following statements about device mobility is the most accurate?**

   a. Device mobility is essentially the same as extension mobility.

   b. Device mobility has been available in CUCM and CCM since Version 3.0.

   c. Device mobility is a feature added in CUCM v7.x and later.

   d. Device mobility greatly enhances Cisco IP phone services for mobile users.

7. **What is the correct configuration relationship between device pools and device mobility in CUCM?**

   a. IP phones may optionally be configured to use device pools when enabling device mobility.

   b. IP phones must be configured to use device pools when enabling device mobility.

   c. Device pools can optionally be configured to work with device mobility.

   d. Device pools are not used when enabling device mobility.

8. **Which statement about configuring device mobility in a CUCM cluster is true?**

   a. Device mobility is enabled by default in the cluster and must be enabled for devices in CUCM v6.

   b. Device mobility is disabled by default and must be enabled for the CUCM cluster and then configured for individual devices in CUCM.

   c. Device mobility is enabled by default in the cluster and must be enabled for devices in all versions of CCM and CUCM.

   d. Device mobility is disabled by default and must be enabled for the CUCM cluster and then configured for individual devices in all versions of CCM and CUCM.

9. **Which three of the following are valid device mobility configuration elements?**

   a. Device pool

   b. Region

   c. Physical location

   d. Device mobility group

   e. CUCM server

   f. Cisco IP phone

Chapter 9

Cisco Extension Mobility

In Cisco Unified Communications Manager (CUCM) multisite environments, it is common for some users to roam between desks or sites on a regular basis. This is sometimes referred to as *hoteling* or *hot desking*. This technique is prevalent in areas where traditional desk space is limited and in call centers. When people use phones that are provided at the desks or sites they visit, they want to, but traditionally cannot, use their personal phone settings. These settings include directory numbers (DNs), speed dials, calling privileges, and message waiting indicator (MWI). A Unified Communications (UC) solution solves this problem.

This chapter describes extension mobility, a feature of CUCM that allows CUCM users to log in to an IP phone and have their personal profile applied, regardless of the device and physical location that they are using.

Upon completing this chapter, you will be able to meet these objectives:

■ Identify issues with users roaming between sites

■ Describe the Cisco extension mobility feature

■ Describe the Cisco extension mobility configuration elements and their interaction

■ Describe Cisco extension mobility operation

■ Describe the issues that must be considered when implementing Cisco extension mobility

■ List the steps that are required to configure Cisco extension mobility

Overview of Roaming Between Sites

Chapter 8, "Implementing Cisco Device Mobility" addressed the situation when users travel to other sites within their organization and bring their phone with them, such as IP Communicator or Cisco Jabber on their laptop. This chapter focuses on issues that can occur if users temporarily change their workplace and roam between different sites and use an available phone at the current location Some organizations call this hoteling, and often provide offices or cubicles for traveling employees. Figure 9-1 illustrates the concept of a roaming user.

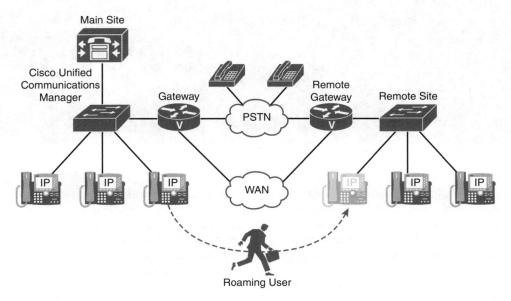

Figure 9-1   *Roaming Users*

Challenges with Roaming Users

Without extension mobility, roaming users may be subject to the following challenges:

- Extensions are traditionally bound to a user's permanent devices. When roaming, these extensions are not available.

- Users' wrong extension on the phone they are using while roaming.

- Users get the wrong calling privileges on the phone they are using while roaming.

- Users do not have personal speed dials available on the phone they are using while roaming.

- Users have the wrong IP phone services assigned to the phone they are using while roaming.

- MWI status and voicemail profiles do not work with a different extension.

To effectively address these issues, the user would require the CUCM administrator to reconfigure the roaming phone that is used with user-specific configuration, instead of having device-specific settings applied to the phone.

CUCM Extension Mobility Overview and Characteristics

All the stated issues are addressed by extension mobility. Extension mobility offers functionality that is designed to enhance the mobility of users within a CUCM cluster, as described in Table 9-1.

Table 9-1   *Extension Mobility Solves Issues of Roaming Users*

Issue Without Extension Mobility	Extension Mobility Feature to Solve the Issue
Extensions are bound to physical devices.	Extensions are bound to device profiles.
Speed dials are assigned to physical devices.	Speed dials are assigned to device profiles.
Services are assigned to physical devices.	Services are assigned to device profiles.
MWI status is defined for physical devices.	MWI status is updated during Extension Mobility login.
Calling privileges are defined for physical devices and locations.	Calling privileges result from merging line settings (device based) and physical device settings (location related).

Although the device is not the user's home device, it is reconfigured with user-specific settings that are stored in profiles. This action lets you separate user-specific parameters configured in user profiles from device-specific parameters that are still stored at the phone configuration along with default values for user-specific settings. The phone adapts some of its behavior based on the individual who is currently using the phone.

The configuration changes are triggered by a user login where the user is identified by a user ID and PIN. The phone configuration adapts to the individual user. When the user stops utilizing the phone and logs out (or is auto-logged out) the default configuration is reapplied.

The following are some key characteristics of extension mobility:

■ CUCM extension mobility allows users to log in to any phone and get their individual user-specific phone configuration applied to that phone. So, users can be reached at their personal directory number (DN), regardless of their location or the physical phone they are using.

■ Extension mobility is implemented as a phone service and works by default within a single CUCM cluster. With CUCM 8.x or later, extension mobility cross cluster (EMCC) can be enabled. EMCC is not covered in this book, but you can find additional information in the *CUCM Features and Services Guide* at http://www.cisco.com/c/en/us/td/docs/voice_ip_comm/cucm/admin/10_0_1/ ccmfeat/CUCM_BK_F3AC1C0F_00_cucm-features-services-guide-100/ CUCM_BK_F3AC1C0F_00_cucm-features-services-guide-100_chapter_011001. html#CUCM_TP_C040B6F7_00.

- The user-specific configuration is stored in device profiles. After successful login, the phone is reconfigured with user-specific parameters, and other device-specific parameters remain the same.

If a user logs in with a user ID that is still logged in at another device, one of the following options can be configured under CUCM service parameters:

- **Allow multiple logins:** When this method is configured, the user profile is applied to the phone where the user is logging in, and the same configuration remains active at the device where the user logged in before. The line number or numbers become shared lines because they are active on multiple devices.

- **Deny login:** In this case, the user gets an error message. Login is successful only after the user logs out of the previous device.

- **Auto-logout:** Like the preceding option, this ensures that a user can be logged in on only one device at a time. However, it allows the new login by automatically logging out the user at the other device.

On a phone configured for extension mobility, either another device profile that is a logout device profile can be applied or the parameters as configured at the phone are applied. The logout itself can be triggered by the user or enforced by the system after expiration of a maximum login time.

Extension Mobility: Dynamic Phone Configuration Parameters

Two types of configuration parameters are dynamically updated when extension mobility is used:

- **User-specific device-level parameters:** These are user-specific phone configuration parameters such as user music on hold (MOH) audio source, phone button templates, softkey templates, user locales, do not disturb (DND), privacy settings, and phone service subscriptions. All these parameters are configured at the device level of an IP phone.

- **Configuration of phone buttons, including lines:** All phone buttons, not only the button types as specified in the phone button template but also the complete configuration of the phone buttons, are updated by extension mobility. This update includes all configured lines, with all the line-configuration settings, speed dials, service URLs, call park buttons, and any other buttons that are configured in the device profile that is to be applied.

Extension Mobility with Dynamic Phone Configuration by Device Profiles

Figure 9-2 illustrates how user-specific settings roam with the user when the user logs out of one phone at the main site and then logs in to an extension mobility-enabled phone located anywhere on the CUCM cluster. Users can log in to any phone either at their home location or a remote location using extension mobility.

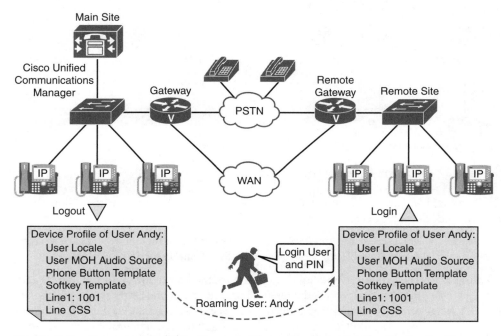

Figure 9-2 *Extension Mobility Dynamic Phone Configuration by Device Profiles*

As shown in Figure 9-2, the user-specific parameters, such as device-level parameters and all phone button settings (including line configurations), are formed in device profiles. Based on the user ID entered during login, CUCM can apply the user's personal device profile and reconfigure the phone with the configuration profile of the individual user who logs in.

With extension mobility, CUCM is aware of the end user sitting behind a device and applies the appropriate user-specific configuration based on a device profile associated with the logged-in user.

CUCM Extension Mobility Operation

The following steps, illustrated in Figure 9-3, describe how extension mobility works, how phone model mismatches are handled, and how calling search spaces (CSSs) and partitions are updated when CUCM extension mobility is used.

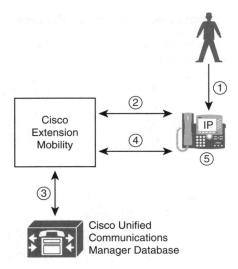

Figure 9-3 *Extension Mobility Login Process*

When a phone user wants to log in to a phone with extension mobility, the following
sequence of events occurs:

Step 1. The user presses the IP phone services button on the phone and chooses the
extension mobility service from the list of phone services available.

Step 2. The extension mobility service requires the user to log in using a user ID and
PIN. The user enters the required data on the phone by pressing each phone
button as many times as needed to select the alphanumeric characters for the
user ID and PIN. Therefore, it is suggested to use numerical usernames and
PINs to make login easier.

Step 3. If the entered user ID and PIN are correct, extension mobility chooses the
device profile that is associated with the user.

Note If a user is associated with more than one device profile, all associated profiles are
displayed, and the user has to choose the desired profile. Assigning multiple profiles to a
user means that the user is provided with a separate device profile for each site or phone
type available at a site.

Step 4. CUCM updates the phone configuration with the settings of the chosen
device profile. User-specific device-level parameters, lines, and other phone
buttons are updated with user-specific settings.

Step 5. The IP phone is reset and loads the updated configuration.

At this point, the phone can be used as if it was in the home location. From the user's phone experience, DNs, speed dials, and MWI are all correct regardless of the location and the IP phone that is used.

Users can log out of extension mobility by pressing the Services button again and choosing Logout in the extension mobility service. If users do not log out themselves, the system can be set to automatically log them out after the maximum login time expires. The administrator can configure the CUCM service parameter for the maximum login time.

Alternatively, UC administrators can set up CUCM such that users are also automatically logged out of a phone when they log in to another phone. Another option is that the next user of the phone logs out a previous user to be able to log in and have the phone updated with the settings of that new user. After logout, CUCM reconfigures the phone either with the standard configuration of the IP phone or by using another device profile, as specified in the Phone Configuration window.

Cisco Extension Mobility and CSSs

Cisco Extension Mobility does not modify the device calling search space (CSS) or the automated alternate routing (AAR) CSS (both of which are configured at the device level). Cisco extension mobility replaces the line CSS/CSSs that are configured at the phone with the line CSS/CSSs that are configured at the device profile of the logged-in user.

In an implementation that uses the line/device approach, the following applies:

■ The line CSS of the login device is updated with the line CSS of the user's chosen device profile.

■ The device CSS of the login device is not updated, and the same gateways (those that were initially configured at the phone before the user logged in) are used for external route patterns. Because the phone did not physically move, the same local gateways should be used for PSTN calls, even when a different user is currently logged in to the device.

> **Note**   When using globalized call routing (including local route groups), the implication of Cisco extension mobility CSSs is similar to the previously described scenario with the line-device approach. The only difference is that the gateway selection is done by the local route group and not by the site-specific device CSS.

If the traditional device-only CSS approach is used to implement partitions and CSS, the following applies:

■ If only device CSSs are used, the CSS is not updated, and no user-specific privileges can be applied. The user inherits the privileges that are configured at the device that is used for logging in.

- If only line CSSs are used, the line CSS that is configured at the device profile of the user replaces the line CSS of the login device. In a multisite environment, this configuration can cause problems in terms of gateway choice because the same gateway is always used for external calls. To avoid gateway selection problems in such an environment, use local route groups.

To avoid issues with mismatching IP phone models or with calling privileges when the traditional approach for implementing partitions and CSSs is used, multiple device profiles can be configured per user.

When different phone model series are used, issues can arise when the settings of the default device profile are applied. Different users might require various settings. This problem can be solved by creating multiple device profiles per user. When you configure and associate one device profile (per phone model) with a username, CUCM displays this list of profiles after successful login. The user can choose a device profile that matches the phone model of the login device. However, if many users need to employ Cisco extension mobility and many different phone models are used, this solution does not scale well.

The same concept can be used as an alternative to local route groups or the line/device approach for implementing CSSs. A separate device profile can be created per site and is configured with the appropriate CSS to allow local gateways to be used for external calls. Again, the user chooses the corresponding device profile after logging in, and the correct class of service (CoS) and gateway choice are applied without depending on a separate line and device CSS. The recommendation, however, is to use the line/device approach in a multisite environment, because that approach simplifies the dial plan and scales better.

> **Note**   When using the traditional CSS approach with only one CSS applied at the line, use local route groups to prevent gateway-selection problems.

CUCM Extension Mobility Device Profile Overview

The device profile is configured with all user-specific settings that are found at the device level of an IP phone, such as the following:

- User MOH audio source
- Phone button templates
- Softkey templates
- User locales
- DND and privacy settings

- Phone service subscriptions

- All phone buttons (including lines and speed dials)

One or more device profiles are applied to an end user in the End User Configuration window.

The default device profile stores default device configuration parameters that are applied by extension mobility when there is a mismatch between the actual phone model where the user logs in and the phone model configured in the user's device profile. The default device profile exists once per phone model type and per protocol for Session Initiation Protocol (SIP) and Skinny Client Control Protocol (SCCP). All the parameters that cannot be applied from the user's device profile are taken from the default device profile.

For example, a user is associated with a device profile for a Cisco Unified IP Phone 7945 running SCCP. If this user logs on to a Cisco Unified IP Phone 7965 running SIP, some features exist on the target phone that cannot be configured at the Cisco Unified IP Phone 7945 device profile. In this case, the configuration parameters that are unavailable on the user's device profile are taken from the default device profile of the Cisco Unified IP Phone 7945 SCCP.

If a device profile includes more parameters than are supported on the target phone, the additional settings are ignored when the target phone with the user-specific settings is reconfigured.

Note   The default device profile is not applied if a device profile of a user and the phone on which the user tries to log in are of the same phone model series (for example, Cisco Unified IP phone 7960, 7961, or 7965). CUCM automatically creates a default device profile for a specific phone model and protocol as soon as Cisco extension mobility is enabled on any phone configuration page for this phone model.

As shown in Figure 9-3, an end user is associated with one or more device profiles. For each possible IP phone model, whether SCCP or SIP, a default device profile is configured. Because extension mobility is implemented as an IP phone service, all phones that should support extension mobility *must* be subscribed to the extension mobility phone service to allow a user to log in to the phone. In addition, each device profile *must* be subscribed to the extension mobility phone service. This subscription is required to allow a user to log out of a phone.

Relationship Between Extension Mobility Configuration Elements

The Cisco IP phone models 7940 and 7965 are shown as an example in Figure 9-4, but they are not the only Cisco IP phones supported in CUCM for extension mobility. Almost all Cisco Unified IP phones support extension mobility.

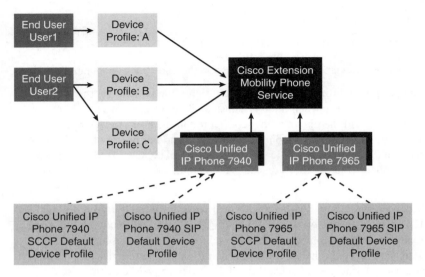

Figure 9-4   *Relationship of Cisco Extension Mobility Configuration Elements*

When different IP phone models are implemented in a CUCM cluster where extension mobility is enabled, an end user may log in to an IP phone that is a different model than the one configured in the user's device profile, as shown in Figure 9-5.

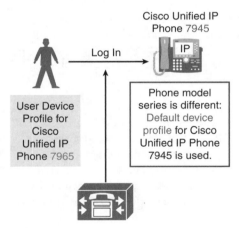

Figure 9-5   *Logging In to a Different Phone Model with Extension Mobility*

Because different phones support different features, when a user logs in to a phone that supports features different than the model associated with the user, the default device profile is used to apply parameters that are supported by the target phone but that are not included in the user's device profile. The default device profile includes phone configuration parameters such as phone button templates, softkey templates, phone services, and other phone configuration settings, but it does not include line or feature button configurations.

The result is that some phone features available on the user's home desk are unavailable on their remote phone when logging in with extension mobility. The next section describes the methodology to overcome this issue.

Default Device Profile and Feature Safe

As described in the previous section, a user with a different device profile can try to log in to a phone that is a different model than the user's default device profile. This can cause not all features to appear on the phone as would be desirable. Figure 9-6 depicts the user logging in to a phone with a matching device profile.

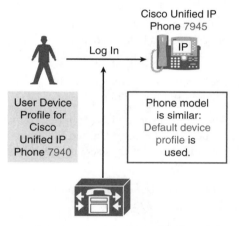

Cisco Unified IP
Phone 7945

Log In

User Device
Profile for
Cisco
Unified IP
Phone 7940

Phone model
is similar:
Default device
profile is
used.

Figure 9-6 *Feature Safe Functionality of the Default Device Profile*

When the phone model series of the physical phone and the user device profile are the same, the feature safe function allows different phone models to be used for user device profiles and physical phone models. For example, a user with an associated device profile for a Cisco Unified IP Phone 7940 phone can log into a Cisco Unified IP Phone 7945 phone without having the default device profile applied. No administrative tasks are required to enable feature safe. Feature safe is independent of the used signaling protocol (SIP or SCCP).

So, when logging in to a completely different model than configured for the user of the device profile, the following sequence of events happen:

Note   All of these events occur only after successful authentication of the user on the device and if the phone model series of the device profile does not match the phone model series.

Step 1.   Device-dependent parameters, such as phone button template and softkey template, from the default device profile are applied to the phone.

Step 2.   The system copies all device-independent configuration settings (user hold audio source, user locale, speed dials, and line configuration, except for the parameters that are specified under Line Setting for This Device) from the device profile to the login device.

Step 3.   The applicable device-dependent parameters of the device profile of the user are applied. These parameters include buttons (such as line and feature buttons) that are based on the phone button template applied from the default device profile.

Step 4.   If supported on the login device, phone service subscriptions from the device profile of the user are applied to the phone.

Step 5.   If the user's device profile does not have phone services configured, the system uses the phone services that are configured in the default device profile of the login device.

For example, the following events occur when a user who has a device profile for a Cisco Unified IP Phone 7960 logs in to a Cisco Unified IP Phone 7905:

Step 1.   The personal user hold audio source, user locale, speed dials (if supported by the phone button template that is configured in the Cisco Unified IP Phone 7905 default device profile), and DN configuration of the user are applied to the Cisco Unified IP Phone 7905.

Step 2.   The phone button template and the softkey template of the default device profile are applied to the Cisco Unified IP Phone 7905.

Step 3.   The user has access to the phone services that are configured in the Cisco Unified IP Phone 7905 default device profile.

CUCM Extension Mobility Configuration

The following steps describe the extension mobility login process:

Step 1.   Activate the Cisco extension mobility feature service by choosing **Tools > Service Activation** in CUCM Serviceability.

Note   Starting with CUCM Version 6.x, Cisco extension mobility is considered a user-facing feature and can be activated on any server in a CUCM cluster to provide a redundant Cisco extension mobility environment.

Step 2.   The Cisco extension mobility service has several configurable service parameters. Choose **System > Service Parameters** in CUCM, as shown in Figure 9-7. Select your CUCM server, and select **Cisco Extension Mobility (Active)**. Press the **Advance** button on the bottom to see all the options shown in Figure 9-7.

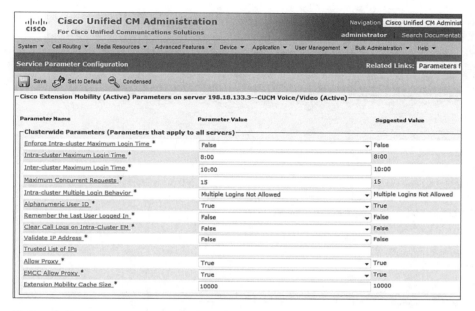

Figure 9-7 *Cisco Extension Mobility Service Parameters*

Note   All these parameters are cluster-wide service parameters of the Cisco extension mobility service. You can access them from CUCM Administration by choosing **System > Service Parameters**.

Step 3.   Add the Cisco extension mobility phone service, as shown in Figure 9-8. Browse to **Device > Device Settings > Phone Services** and select **Add New**.

Note   The service URL of http://*IP address of server*:8080/emapp /EMAppServlet?device=#DEVICENAME# is case sensitive and must be entered exactly as worded, except that you replace the *IP address of server* in the URL with the actual IP address of your CUCM server with extension mobility enabled. Be sure not to change anything else when entering this URL.

Step 4.   To configure a device profile, as shown in Figure 9-9, choose **Device > Device Settings > Device Profile**. First, you must choose the product type, which is the phone model. Then click **Next**, and then select the device protocol and click **Next** again. Then, you can configure the settings on the device profile.

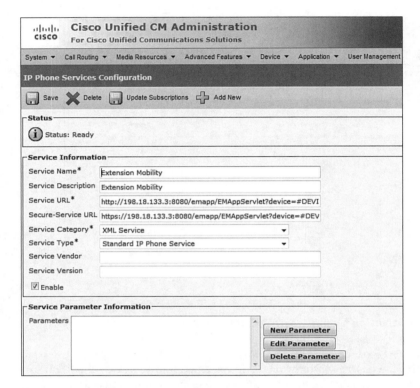

Figure 9-8   *Cisco Extension Mobility Phone Service*

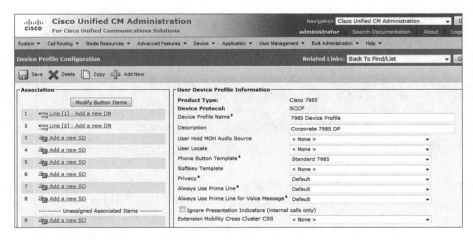

Figure 9-9   *Device Profile Configuration*

Note   The available configuration options depend on the phone model and protocol you choose. Each device profile name should be unique, and best practice is to reference the user in the device profile name.

Step 5.   Subscribe the configured device profile to the Cisco extension mobility phone service, as shown in Figure 9-10. Within the device profile you just created, in the upper right under Related Links, select **Subscribe/ Unsubscribe Services** and follow the wizard. Repeat this step for all the device profiles you have made.

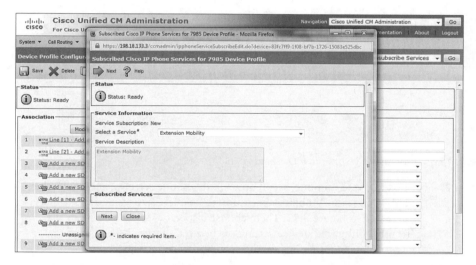

Figure 9-10   *Subscribing the Device Profile to the Cisco Extension Mobility Phone Service*

Note   If the device profile is not subscribed to the Cisco extension mobility service, users do not have access to Cisco extension mobility phone service after they log in and their device profile has been applied. As a result, users can no longer log out of Cisco extension mobility at the phone. Do not forget to subscribe to the Cisco extension mobility phone service. Since CUCM Version 7, an enterprise subscription can be enabled at each phone service. If an enterprise subscription is enabled, the corresponding phone service applies to all phones and device profiles, and per device configuration is not required.

Step 6.   Go to **User Management > End User** and choose the device profile or profiles that you want to associate with the user in the list of available profiles, as shown in Figure 9-11. You can also set the default profile.

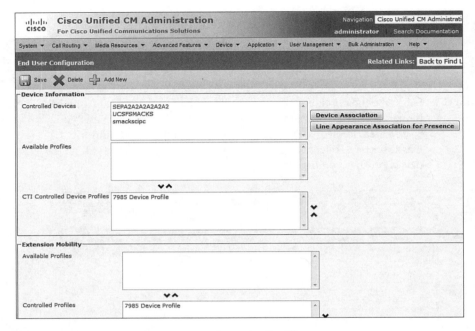

Figure 9-11   *Associating Users with Device Profiles*

Step 7.   Choose **Device > Phone**, scroll down in the phone settings, and check the **Enable Extension Mobility** check box to enable Cisco extension mobility, as depicted in Figure 9-12. Then, choose a specific device profile or the currently configured device settings to be used during the logout state. The recommendation is to use the current device settings.

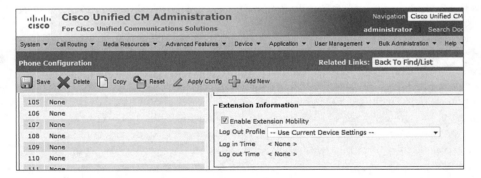

Figure 9-12   *Configure Phones for Cisco Extension Mobility*

Note   If you choose a device profile as a logout profile, you will not be able to delete it until it is unassigned from the phone.

At this time, the extension mobility configuration is complete.

Summary

The following key points were discussed in this chapter:

- The device mobility and extension mobility features of CUCM allow users to roam between sites.

- Extension mobility enables users to log in to IP phones and apply their profiles, including extension number, speed dials, services, MWI status, and calling privileges.

- The user's device profile is used to generate the phone configuration in the login state.

- Seven steps are needed to configure extension mobility.

References

For additional information, refer to the following:

Cisco Systems, Inc. Cisco Collaboration Systems 10.x Solution Reference Network Designs (SRND), May 2014. http://www.cisco.com/c/en/us/td/docs/voice_ip_comm/cucm/srnd/collab10/collab10.html

Cisco Systems, Inc. Cisco Collaboration Systems 11.x Solution Reference Network Designs (SRND), July 2015. http://www.cisco.com/c/en/us/td/docs/voice_ip_comm/cucm/srnd/collab11/collab11.html

Cisco Systems, Inc. Features and Services Guide for Cisco Unified Communications Manager, Release 10.0(1), May 2014. http://www.cisco.com/c/en/us/td/docs/voice_ip_comm/cucm/admin/10_0_1/ccmfeat/CUCM_BK_F3AC1C0F_00_cucm-features-services-guide-100/CUCM_BK_F3AC1C0F_00_cucm-features-services-guide-100_chapter_0110011.html

Review Questions

Use these questions to review what you have learned in this chapter. The answers appear in Appendix A, "Answers Appendix."

1. **Which of the following is not a problem when users roam between sites and use guest phones in a hoteling office where extension mobility is not enabled?**

   a. The phone they use uses the wrong location and region settings.

   b. The user gets the wrong extension on that phone.

   c. The user gets the wrong calling privileges from his or her home desk phone.

   d. The user does not have his or her speed dials available.

2. **Which two settings cannot be updated when you use extension mobility?**

   a. Phone button template

   b. Softkey template

   c. Device CSS

   d. Network locale

   e. Phone service subscriptions

   f. Phone lines and speed dials

3. **Which three of the following are not configuration elements relevant to extension mobility configuration? (Choose three.)**

   a. Location

   b. Phone

   c. End user

   d. Device security profile

   e. Device pool

   f. Device profile

   g. Phone service

4. **Which two of the following are Cisco-recommended approaches to implementing calling privileges when using extension mobility? (Choose two.)**

   a. Configure the lines of the user's device profile with a CSS that includes blocked route patterns for the destinations the user should not be allowed to dial.

   b. Do not configure a device CSS in this case.

   c. Do not configure a line CSS in this case.

   d. Configure the device with a CSS that includes all PSTN route patterns pointing to the local gateway.

   e. Configure the lines of the physical phone with the CSS that includes blocked route patterns for the destinations the user should not be allowed to dial.

5. **Which two of the following happen if the user logs into a device but is still logged into another device? (Choose two.)**

   a. If the multiple login behavior service parameter is set to Not Allowed, the second login fails.

   b. If the multiple login service parameter is set to Auto-Logout, the user is automatically logged out of the other device.

c. If the multiple login behavior enterprise parameter is set to allowed, the login succeeds, and the user is logged out at the other device.

d. If the multiple login behavior enterprise parameter is set to prompt, the user is asked whether he wants to be logged out at the other device first.

e. The login fails independent of any CUCM extension mobility configuration.

6. **Which of the following best describes a user using CUCM extension mobility prior to CUCM 8.x?**

a. An employee travels to a remote site in the same company CUCM cluster and logs in on a Cisco IP phone to make a PSTN call from the remote location.

b. An employee travels to a remote site in the same company CUCM cluster and uses IP Communicator on her laptop to make a PSTN call from the remote location.

c. An employee travels to a remote site in the same company on a different CUCM cluster and logs in on a Cisco IP phone to make a PSTN call from the remote location.

d. An employee travels to a remote site in the same company on a different CUCM cluster and uses IP communicator on his or her laptop to make a PSTN call from the remote location.

7. **What role do device profiles play in a CUCM cluster?**

a. It is optimal to configure for users who do not travel but who want to best use their IP phone services.

b. It is optimal to configure for users who travel to other sites but who want to best use their IP phone services.

c. It is optimal to configure for users who do not travel but who want to use the maximum number of their phone features.

d. It is optimal to configure for users who travel to other sites and who want their home site phone settings to be available remotely.

8. **Which statement about configuring extension mobility in a CUCM cluster is true?**

a. Extension mobility is enabled by default on the CUCM cluster but needs to be configured for each user.

b. Extension mobility is not enabled by default on the CUCM cluster and must be enabled on a CCM server and configured for each user.

c. Extension mobility is not enabled by default on the CUCM cluster and must be enabled on all CUCM servers and configured for each user.

d. Extension mobility is enabled by default on the CUCM cluster but needs to be enabled for all CUCM servers and configured for each user.

9. **Which implementation example of extension mobility would result in the simplest CUCM administrator configuration while maintaining maximum user Cisco IP phone features for roaming users?**

   a. Allow different users to own any Cisco IP phone model they choose for their homes site and for hoteling.

   b. Standardize on different Cisco IP phone models for each location to best suit the business models of different sites.

   c. Standardize on the 7965 Cisco IP phone model to be used for all sites.

   d. Standardize on the 7905 Cisco IP phone model to be used for different sites.

10. **True or false: In CUCM 8.x and later, extension mobility cross cluster (EMCC) is a supported feature.**

   a. True

   b. False

Chapter 10

Implementing Cisco Unified Mobility

The growing use of mobile devices, such as cell phones, tablets, and laptops, allows users to enjoy the efficiencies and speed of Cisco Collaboration. Many of these devices allow Cisco Unified Communications (UC) clients such as Cisco Jabber, Cisco Jabber Movi, XLite, and even IP Communicator to be installed on various devices from smartphones to tablets to laptops. With the list of applications and softphones or soft clients growing on a yearly basis, some users can experience "gadget overload." *Gadget overload* is a term used to define users in the organization who own multiple devices, ranging from office phones to home office phones, laptop computers, to mobile phones. These users often spend vast amounts of time managing their communications across different phone numbers and voice mailboxes, which can limit their ability to work efficiently. This is true whether users are on a retail floor, at an airport, or at a Wi-Fi hotspot in a local coffee shop.

Cisco Unified Mobility is a Collaboration application that comes as part of Cisco Collaboration suite and allows users to be reached at a single number regardless of the device they use. An example is a user who has a desk phone, a tablet with Cisco Jabber, and perhaps a work-issued cell phone. Unified Mobility allows them to receive calls simultaneously across all devices. This chapter describes the features of Cisco Unified Mobility and how they work and how to configure them.

This chapter describes the concepts, configuration, and implementation relevant to Cisco Unified Mobility for single number reach (SNR) and mobile voice access (MVA). Upon completing this chapter, you will be able to meet these objectives:

- Describe the purpose of Cisco Unified Mobility, how it works, and when to implement it

- Analyze call flows of Cisco Unified Mobility applications

- List the requirements for implementing and installing Cisco Unified Mobility

- Describe how to implement mobile voice access when using MGCP or SCCP PSTN gateways

■ Describe how calling search spaces are implemented in Cisco Unified Mobility

■ List the steps that are required to configure Cisco Unified Mobility

Cisco Unified Mobility Overview

Cisco Unified Mobility consists of two main components:

■ Mobile Connect, which is sometimes referred to as single number reach (SNR)

■ Mobile voice access (MVA)

Both of these components are illustrated in Figure 10-1.

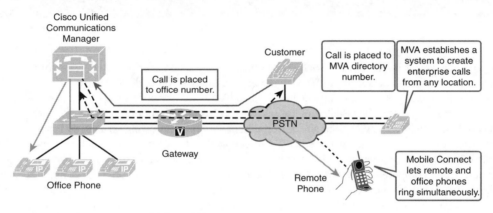

Figure 10-1 *Cisco Unified Mobility*

The Mobile Connect feature allows an incoming call to the enterprise phone number of a user to be offered to the user's office phone, and up to ten configurable remote destinations. Typically, such remote destinations are mobile phones and home office phones. This feature is referred to as SNR. Note that Mobile Connect and SNR refer to the same mechanisms, but their names can be used interchangeably.

Built on top of the Mobile Connect feature, mobile voice access (MVA) provides features similar to those of Mobile Connect for outgoing calls. With MVA enabled, users who are outside the enterprise can make calls as if they were directly connected to CUCM, in effect masking their caller ID appear as if it were originating from their IP phone or video endpoint. This functionality is commonly known as direct inward system access (DISA) in traditional telephony environments.

One common issue that arises with users in many organizations is the ability for users to mask their caller ID or automatic number identification (ANI). I am sure a few readers of this book can appreciate a scenario where a customer or client has obtained your cell phone number when you called them by accident. With Unified Mobility enabled, many users can receive and place calls from their mobile phones. With MVA enabled, the caller ID appears as if it was coming or originating from their primary device. MVA can also benefit the enterprise by limiting toll charges and consolidating phone billing directly to the enterprise rather than billing to each mobile user.

The next section explains the characteristics of MVA.

Mobile Connect and Mobile Voice Access Characteristics

A major benefit of Unified Mobility is allowing active calls to be switched between the IP phone and the remote phone. For example, if a user initiates a call from a mobile phone while traveling to the office, the call can be switched to the office phone after arriving at the office. Another example is a call initiated from the office phone that could be switched to a mobile phone if the user needs to leave the office. This means that users can receive business calls at a single phone number, regardless of the device that is used to receive a call. Mobile Connect allows users to answer incoming calls on the office phone or at a remote destination and pick up in-progress calls on the office phone or remote destination without losing the connection. When the call is offered to the desk phone and remote destination phone or phones, the user can answer at either of those phones and hand off the call to another phone.

For example, if a user receives a call placed to a business number or DID, the user's office phone and cell phone ring. If the user is traveling to the office, he can accept the call on his cell phone. After arriving at work, the in-progress call can be picked up at the office IP phone just by pressing a single key on the office IP phone. The call continues on the office IP phone without interruption. The person on the other end of the call does not notice the handover from the cell phone to the IP phone.

Note   When MVA is used, users can invoke mid-call features after the call is connected, or pick up the call on their desk phones just like they can with received Mobile Connect calls. This is possible because the call is anchored at the enterprise gateway.

SNR and MVA enable flexible management of enterprise and remote destinations. These two techniques provide several features and benefits:

- **Single enterprise number:** Regardless of the device that is used (whether an IP phone, cell phone, or home phone), calls can be received at a single enterprise phone number, meaning one number is dialed but multiple destinations ring simultaneously. The caller ID of the enterprise phone is also preserved on outgoing calls, regardless of the phone from which the call is initiated. This means that the enterprise voice mailbox can serve as a single, consolidated voice mailbox for all business calls.

- **Access lists:** Cisco Unified Mobility users can configure access lists to either permit or deny calling numbers to ring remote destinations. Users may not wish to have the mobility feature and SNR activated for every caller who dials their enterprise number, or in other words, users can set up their access lists: permit lists or deny lists (analogous to an IOS router's access control list [ACL]). If a permit access list is used, unlisted callers or callers who do not appear on the permit list are not allowed to ring remote destinations. Permit lists are created using the ANI of the outside caller. If a deny access list is used, only unlisted callers are allowed to ring remote destinations. In other words, anyone whose caller ID does not appear on the list is allowed to simultaneously ring the remote destination.

■ **User interfaces for enabling and disabling Cisco Unified Mobility:** Users can turn Cisco Unified Mobility on and off by using a telephone user interface (TUI) provided by MVA. A graphical user interface (GUI) for Cisco Unified Mobility user configuration is also available through the CUCM user web pages.

■ **Access to enterprise features:** CUCM features can be accessed by using dual-tone multifrequency (DTMF) feature access codes (much like traditional private automatic branch exchanges [PABXs]): The supported features include hold (the default is *81), exclusive hold (the default is *82), resume (the default is *83), transfer (the default is *84), and conference (the default is *85). A remote destination that initiates an outbound call can dial these numbers with an active call to initiate the features listed above, such as hold, resume, transfer, and conference.

> **Note**   The feature codes can be configured as CUCM service parameters.

■ **Smart client support:** On phones that have smart clients installed, softkeys can be used to access features such as hold, resume, transfer, and conference. Users can also enable or disable Cisco Unified Mobility from a smart client.

■ **Call logging:** Enterprise calls are logged regardless of the device that is used, whether it is an enterprise phone or a remote phone.

Cisco Unified Mobility Call Flow

This section discusses the call flows for Mobile Connect and MVA.

Mobile Connect Call Flow

Figure 10-2 illustrates the call flow when Mobile Connect is used, with an IP phone at extension 2001 and a mobile phone that belongs to the user of the IP phone.

A caller using the outside public switched telephone network (PSTN) calls the user's office number, 1-511-555-2001. Because Mobile Connect is enabled on CUCM, both the desktop phone 2001 and the configured remote destination mobile phone number 408-555-1001 ring simultaneously. The call is presented to the remote phone with the original caller ID of 479-555-1555. As soon as the call is accepted on one of the two phones, the other phone stops ringing. The user can switch the call between the office phone and the mobile phone and vice versa during the call without losing the connection.

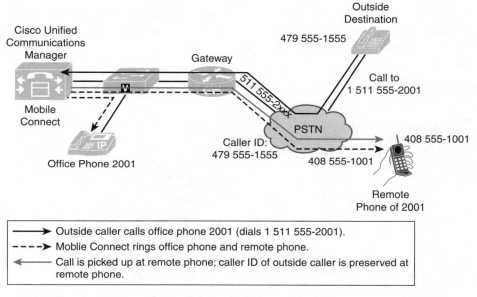

Figure 10-2 *Mobile Connect Call Flow of Incoming Calls to an Office Phone*

It is important to understand that Mobile Connect influences the calling number presentation. If a call is received from a recognized remote destination, the corresponding internal directory number, not the E.164 number of the remote device, is used for the calling number. As shown in Figure 10-3, internal extension 2001 has a Mobile Connect remote destination 408-555-1001, which is the cell phone of the user of 2001. If the user places a call from the mobile phone to an enterprise PSTN number of another internal colleague by dialing 1-511-555-2002, the called colleague sees the call coming from the internal directory number 2001 instead of the external mobile phone number.

The same process applies to calls placed to other internal destinations, such as voice mail. If the user of extension 2001 places a call to Cisco Unity Connection from the cell phone, Cisco Unity Connection sees directory number 2001 as the source of the call and not the PSTN number of the cell phone, which is 408-555-1001. Cisco Unity Connection can identify the user by the directory number and provide access to the appropriate mailbox instead of playing a generic welcome greeting.

To recognize Mobile Connect remote destinations, the Mobile Connect remote destination number has to match the ANI of the incoming call. Typically, Mobile Connect remote destinations include an access code (for example, 9-1 in the number

9-1-408-555-1001). Therefore, access code 9 and the long-distance code 1 have to be prefixed to the incoming ANI 408-555-1001 for the call source to be recognized as a Mobile Connect remote destination. Alternatively, the Cisco CallManager service parameter Matching Caller ID with Remote Destination can be set to Partial Match, and the number of digits for Caller ID Partial Match can be specified. This number specifies how many digits of the incoming ANI have to match a configured remote destination number (starting with the least-significant digit).

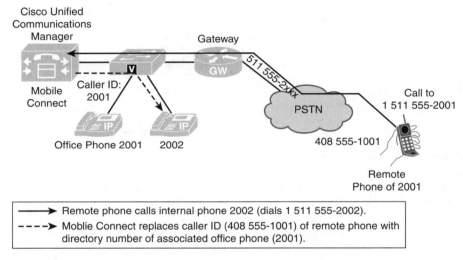

Figure 10-3 *Mobile Connect Call Flow of Internal Calls Placed from a Remote Phone*

If the source of the call is not recognized as a Mobile Connect remote destination, the PSTN number of the remote destination is used for the calling number, and it is not changed to the internal directory number.

Mobile Voice Access Call Flow

With MVA, users can place calls to the outside from a remote destination as if they were dialing from the desktop phone. In Figure 10-4, the user of the IP phone with directory number 2001 uses his cell phone 408-555-1001 to dial the MVA PSTN number of headquarters, extension 2999, by dialing 1-511-555-2999.

The gateway is configured to start an interactive voice response (IVR) call application for calls placed to that number. The Voice Extensible Markup Language (VXML)-based call application offers a prompt and asks for the remote destination number and the user's PIN.

Note If the source of an MVA call is a defined remote destination, the IVR only asks for a PIN, because it recognizes the remote destination.

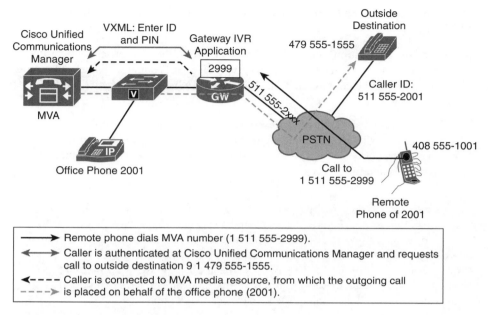

Figure 10-4 *Mobile Voice Access Call Flow*

After login, besides activating and deactivating MVA, the user can initiate a call from the enterprise network. The call is set up with the E.164 PSTN calling number of directory number 2001 instead of 408-555-1001. This action allows the called party to identify the caller by the user's enterprise number.

After the user has initiated a call from a remote destination by using MVA, the user can switch the call to the office phone during the call without losing the connection and can switch back again as needed.

> **Note** In Figure 10-4, the caller dials from the cell phone into the voice gateway, and backs out the same voice gateway to the outside destination phone number that answers, and thus the call is a trombone/hair-pinned call and uses two lines or channels on the gateway.

Cisco Unified Mobility Implementation Requirements

Cisco Unified Mobility requires CUCM Release 6.0 or later. At least one CUCM server in the cluster needs the MVA service to be started. As a best practice, it is recommended to start the MVA service on a CUCM Subscriber, not the CUCM Publisher. The MVA service also interacts with a Cisco Integrated Services Router (ISR) or Aggregation Services Router (ASR) and the default call application running on a Cisco IOS gateway.

The following are the key requirements for implementing Cisco Unified Mobility and MVA:

- The remote destination cannot be an IP phone within the enterprise. The remote destination has to be an external device, typically a PSTN number. Up to ten remote destinations can be configured. Class of service (CoS) can be configured to limit access to the PSTN.

- MVA requires an H.323 or SIP gateway to provide a VXML call application to remote callers who are dialing a certain number. VXML call application is special configuration added to the voice gateway to forward calls for a specific number onto CUCM and the MVA service.

Note   The call application actually runs in CUCM and is referenced in the gateway configuration and does not require files in a router flash like TCL scripts. These scripts can be loaded onto a gateway's flash to play an auto-attendant, or in the case of Unified Contact Center Enterprise (UCCE) it can be used to instruct calls to an IVR or SIP proxy server or Customer Voice Portal (CVP). Because the application resides on CUCM, you bypass the need for running these specialized scripts and simply have a dial peer that points the caller to CUCM and the MVA application dial-in number.

Note   Media Gateway Control Protocol (MGCP) is not supported because it does not maintain call applications and all the signaling is backhauled to the CUCM server.

- DTMF has to be sent out-of-band for MVA to work. When a user dials the MVA number, the user needs to enter a series of digits for the call to proceed. The user may need to authenticate to the service, and the digits entered for verification will arrive out-of-band of the media or Real-time Transfer Protocol (RTP) stream.

Cisco Unified Mobility Configuration Elements

The following are configuration elements of Cisco Unified Mobility and their functions.

- **End user:** Each end user must first be created and have a PIN assigned, which is used for authentication when MVA is used. Then, three important Cisco Unified Mobility-related settings can be configured for the end user:

- **Enable Mobility:** Check this option to allow the user to use the Mobile Connect feature.

- **Enable Mobile Voice Access:** Check this option to allow the user to place MVA-based outgoing calls using enterprise network.

■ **Remote Destination Limit:** This setting is used to limit the number of remote destinations that can be configured. The maximum is ten.

■ **IP phone:** The office phone of a Cisco Unified Mobility user must refer or establish a link to the end-user name. This linking is achieved by setting the Owner ID field on the Phone Configuration window to the user ID of the end user.

Figure 10-5 illustrates various configuration elements of Cisco Unified Mobility features and their respective interactions.

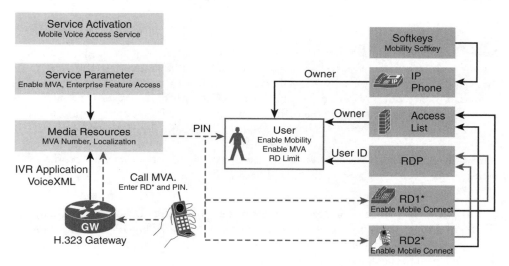

Figure 10-5 *Interaction of Cisco Unified Mobility Configuration Elements*

Note   In the End User Configuration window, the end user can be associated with one or more devices, such as IP phones. Such an association allows the end user to configure the device from the CUCM user web pages, but it is not relevant for Cisco Unified Mobility. The IP phone must be mapped to the end user by setting the owner in the Phone Configuration window.

■ **Remote destination profile (RDP):** This setting creates a virtual phone that is linked to the end user that represents all remote destinations associated with the user. It includes phone device-level configuration settings such as user and network music on hold (MOH) audio sources and calling search spaces (CSSs). For each office phone that an end user should be able to use for Cisco Unified Mobility, a shared line with the lines of the office phones has to be added to the RDP.

- **Remote destination:** A remote destination represents a user's cell phone and any number where the user may be reachable and is associated with one or more shared lines of a remote destination profile. For each remote destination, the remote destination number has to be specified as dialed from within the enterprise. The rerouting CSS of the specified remote destination profile is used to look up the configured remote destination number.

Note   The organization utilized outside access codes such as 9 need to be included in remote destination.

Note   The remote destination profile information has two CSSs that are used for call routing called the Calling Search Space (CSS) and the Rerouting Calling Search Space (RCSS). The CSS is used for outgoing calls that are initiated by using MVA. The RCSS is then used when a call is placed to the remote destination either when receiving a call to the number of the line shared by the office phone and the remote destination profile or when a call is handed over from the office phone to the remote destination. Therefore, the remote destination number has to be reachable by the RCSS. For MVA calls, the CSS is the CSS configured at the shared line and the CSS of the remote destination profile, with priority given to the CSS of the shared line.

- **Access list:** As described earlier, access lists can be configured to permit or deny calls to be placed to a remote destination when the shared line is called; the filter is based on the calling number. If no access list is applied, all calling numbers are allowed to ring the remote destination.

- **MVA media resource:** This media resource interacts with the Voice XML (VXML) call application running on the Cisco IOS gateway. It is required for MVA only. The number at which the Cisco IOS router can reach the media resource has to be specified, a partition can be applied, and one or more locales have to be chosen.

Note   The CSS of the gateway that runs the VXML call application has to include the partition that is applied to the number of the MVA media resource.

Figure 10-6 illustrates how a remote destination profile shares its line or lines with the associated Cisco IP phone or phones.

In Figure 10-6, a call is placed to directory number 2002. Line 1 at Office Phone 2 and all remote destinations associated with Line 2 of the remote destination ring, because the line is a shared line. For the call to the remote destination number, the rerouting CSS is used.

Suppose the remote phone with number 9-1-479-555-1555 (Remote Destination 2) calls in to the mobile voice application and requests an outgoing call to be placed. The CSS of Line 2 and the CSS of the RDP are used for the outgoing enterprise call initiated by Remote Destination 2.

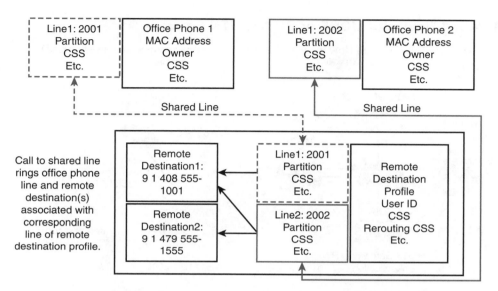

Figure 10-6 *Shared Lines Between Phone and Remote Destination Profiles*

Note To allow an active call to be handed over from an IP phone to a remote destination, the IP phone needs to have the Mobility softkey configured for the Connected call state. If the Mobility softkey is also added to the On Hook call state, it can be used to check the status of Cisco Unified Mobility (Mobile Connect on or off).

Cisco Unified Mobility MGCP or SCCP Gateway PSTN Access

On a Cisco ISR or ASR voice gateway, the MVA application is configured and triggered as part of a voice dial peer application. An incoming call enters one of the router's PSTN ports; this could be a primary rate interface (PRI) or plain old telephone system (POTS) port. Once the call signaling has been established, dial peer matching determines how to route the call. In terms of MVA, a dedicated dial peer will send the call to CUCM and the MVA application on the CUCM server. As described earlier, MGCP and SCCP gateways do not support MVA because the call control is dependent on CUCM.

In case an MGCP or SCCP gateway is used, CUCM must forward calls that are received from an MGCP- or SCCP-controlled interface to an H.323 or SIP gateway, to start the MVA application. From then on, the call treatment is like any H.323 or SIP environment, where dial peers are matched to kick off the MVA application and process. Essentially, CUCM "brokers" the call, acting as a middle man or agent, sending the incoming call from an MGCP or SCCP gateway to an H.323 or SIP gateway to activate the MVA process.

MVA Call Flow with MGCP or SCCP PSTN Gateway Access

Figure 10-7 shows a call flow in an MGCP or SCCP gateway environment where MVA is deployed.

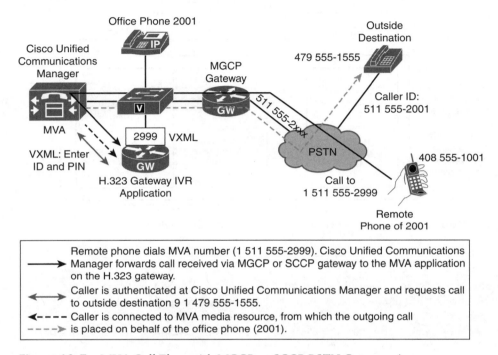

Remote phone dials MVA number (1 511 555-2999). Cisco Unified Communications Manager forwards call received via MGCP or SCCP gateway to the MVA application on the H.323 gateway.

Caller is authenticated at Cisco Unified Communications Manager and requests call to outside destination 9 1 479 555-1555.

Caller is connected to MVA media resource, from which the outgoing call is placed on behalf of the office phone (2001).

Figure 10-7 *MVA Call Flow with MGCP or SCCP PSTN Gateway Access*

In Figure 10-7, the incoming PSTN call is received on an MGCP-controlled interface. CUCM forwards the call to an H.323 gateway. On the H.323 gateway, the MVA application is started. The caller can be authenticated and can specify the final destination for the call. The caller is then connected to the MVA media resource on CUCM, from which the outgoing call is placed on behalf of the caller office phone (2001). CUCM establishes the outgoing call via the MGCP gateway.

Note   The H.323 gateway functionality can be combined on the gateway that receives the PSTN call on the MGCP-controlled interface. In this case, only one gateway that provides MGCP and H.323 signaling is required. In essence, you combined signaling protocols on the router, and it runs MGCP to terminate the PSTN circuit and H.323 to process MVA simultaneously.

Calling Search Space Handling in Cisco Unified Mobility

One important topic to consider when implementing Cisco Unified Mobility is calling search spaces (CSSs). Depending on the origin of a call that uses the Mobile Connect feature, different CSSs are used:

- For an incoming PSTN call to an office phone that is associated with a remote destination, the rerouting CSS at the RDP needs access to the mapped remote destination number. What this means is the RDP uses the rerouting CSS to access partitions and ultimately route patterns that could dial remote destinations. When an outside caller dials a desk phone with SNR activated, the rerouting CSS dictates which route patterns and remote destinations can be reached.

- For an incoming call from a remote phone (remote destination) to an internal destination, the CSS of the receiving device (trunk or gateway) needs access to the called internal number. If a user with a cellular phone as a remote destination calls back into the enterprise to reach another internal user, the CSS on the trunk or gateway terminating the call is used to determine if the caller can reach the internal destination or phone.

CSS Handling in Mobile Voice Access

With MVA, the incoming and outgoing call legs of the MVA call are treated independently. The incoming call leg is the call leg from the gateway where the MVA call application is running to the MVA media resource in CUCM. In effect, this call leg is from the router or gateway to CUCM. The CSS that is used for this call leg depends on a Cisco CallManager service parameter called Inbound CSS for Remote Destination. The parameter can be set to one of these values:

- **Trunk or Gateway Inbound CSS:** This value is the default value in CUCM. If this option is chosen, CUCM uses the CSS of the trunk or gateway from which the MVA call arrived. The CSS of the shared line and the CSS that is configured at the remote destination profile are not considered for the incoming call leg of an MVA call.

- **Remote Destination Profile + Line CSS:** If this option is selected, the CSS of the shared line and the CSS that is configured at the remote destination profile are combined (with priority given to the partitions of the shared-line CSS).

The outgoing call leg of an MVA call is the call leg from the MVA media resource to the PSTN destination that is called from the MVA call application. The CSS that is used for this call leg is always the combination of the CSS of the shared line and the CSS that is configured at the RDP (with priority given to the partitions of the shared-line CSS).

Cisco Unified Mobility Access List Functions

This section describes how to implement call filtering for Cisco Mobile Connect calls. In CUCM, the end user and the administrator can control access to remote destinations based on the time of day and the day of the week.

To support time-of-day access to remote destinations, the remote destination configuration page allows the configuration of a ring schedule. This schedule applies to the remote destination configuration page on both the administrator and user web pages. In essence, the ring schedule is both user configurable via the end user self-care portal and configurable via the call manager administration web page.

The remote destination can be generally enabled (enabled all the time) or explicit time ranges can be configured. The default is to enable the remote destination all the time. This often leads many users to have "SNR fatigue," meaning their remote destination rings all hours of the day. Many Cisco Collaboration engineers reading this chapter can possibly relate to SNR fatigue.

When an explicit time range is configured, each day of the week can be disabled, enabled for the entire day (24 hours), or configured with a From and To time range. This allows users to limit their SNR functionality to only business hours.

Access lists can limit caller IDs. These lists are applied at the remote destination configuration page as follows:

- The Allowed Access List default behavior or Access List setting is called Ring This Destination Only If Caller Is In <Access List>.

- The Blocked Access List default behavior or Access List setting is called Do Not Ring This Destination If Caller Is In <Access List>.

Operation of Time-of-Day Access Control

Figure 10-8 shows the operation of time-of-day access control on remote destinations.

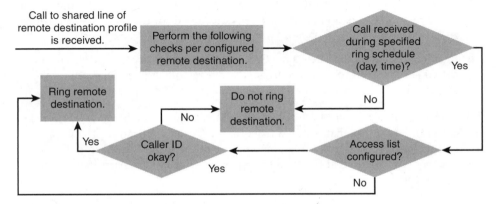

Figure 10-8   *Operation of Time-of-Day Access Control*

When considering the use of time-of-day access control for remote destinations, take two factors into consideration:

■ The remote destination rings only when the call is received during the specified ring schedule. This first decision is independent of the access list configuration. Basically, if the call falls outside of the ring schedule permitted hours, regardless of the access list the call is not allowed to the remote destination. This prevents after-hours callers from reaching end-user remote destinations.

■ If no access list is configured, all callers are permitted. However, this permission applies only after the first check (the call received during the specified ring schedule). If a caller is permitted according to an access list configuration but the call is received outside the configuration ring schedule, the call is not extended to the remote destination.

Figure 10-8 shows how a call is processed when it is received at a shared line that is configured at an RDP. For each remote destination that is associated with the called line, the ring schedule that is configured at the remote destination is checked in the following way:

■ If the call is received outside the configured ring schedule, the remote destination does not ring.

■ If the call is received within the configured ring schedule, the access list configuration of the remote destination is checked. If the caller ID is permitted, the remote destination rings. If the caller ID is not permitted, the remote destination does not ring.

The caller ID is permitted in the following scenarios:

■ The Always Ring the Destination parameter is selected.

■ An access list is applied by using the Ring This Destination Only If Caller Is In <Access List> parameter, and the caller ID is found in the specified access list.

■ An access list is applied by using the Do Not Ring This Destination Only If Caller Is In <Access List> parameter, and the caller ID is not found in the specified access list.

Cisco Unified Mobility Configuration

The following steps describe how to configure Cisco Unified Mobility by configuring the first piece of Mobile Connect.

Configuring Mobile Connect

Step 1.   Configure a softkey template that includes the Mobility softkey, as shown in Figure 10-9. This must be done for On Hook and Connected call states. Go to **Device > Device Settings > Softkey Template**.

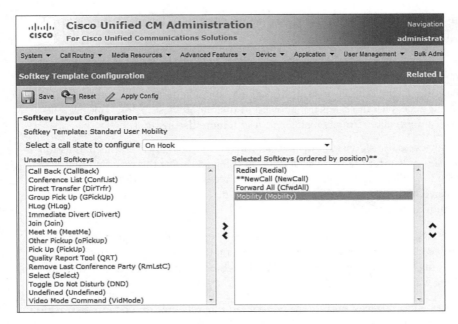

Figure 10-9 *Configuring a Softkey Template for Unified Mobility*

Step 2.   As shown in Figure 10-10, choose **User Management > End User** to
configure end users. Check the options **Enable Mobility, Enable Mobile
Voice Access,** and set the **Remote Destination Limit** as per the requirements.

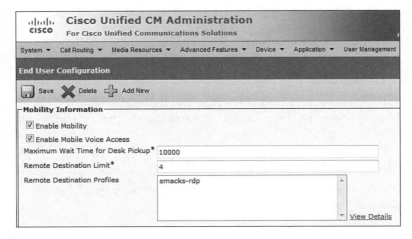

Figure 10-10 *Configuring End User*

Note Be sure to associate the end user with the IP phone in the Device Associations
pane lower in the End User Configuration page.

Step 3.   Configure the user's office IP phone for Cisco Unified Mobility, as shown in Figure 10-11, by choosing **Device > Phone**.

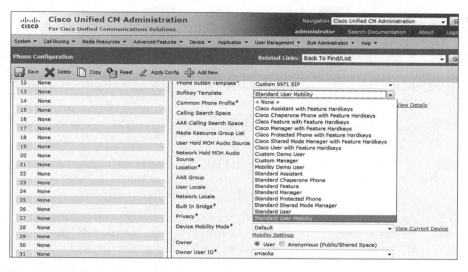

Figure 10-11   *IP Phone Configuration*

Step 4.   Configure the RDP, as shown in Figure 10-12, by choosing **Device > Device Settings > Remote Destination Profile**.

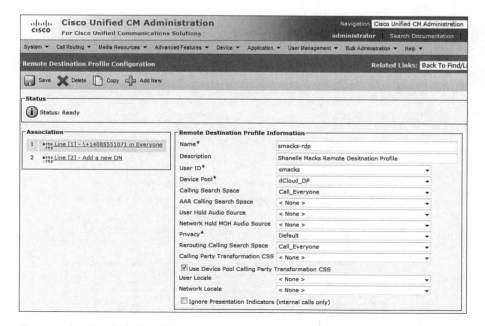

Figure 10-12   *RDP Configuration*

The RDP contains the parameters that apply to all the user's remote destinations. Enter a name, description, device pool, CSS, rerouting CSS, and network and user MOH audio sources for the RDP. The CSS (combined with the line CSS) is used for outgoing enterprise calls placed from a remote destination by using MVA. Partitions and CSSs are discussed in detail in *Implementing Cisco Unified Communications Manager, Part 1 (CIPTV1)*.

The rerouting CSS should allow sending calls placed to the user's enterprise phone number to the specified remote destinations. This CSS is also used when an active call is handed over from the office phone to a remote phone.

Check the Ignore Presentation Indicators option to ignore the connected line ID presentation. This is recommended for internal calls.

After an RDP is created, one shared line has to be configured for each DN used at the user's office phone or phones. To do so, navigate in CUCM to **Device > Phone** and choose the user's phones by clicking **Add a New DN** at the appropriate link.

Step 5. As shown in Figure 10-13, configure remote destinations by choosing **Device > Remote Destination**. Alternatively, you can click the **Add a New Remote Destination** link in an RDP. Define the name of RDP, destination number, and check **Enable Mobile Connect, Enable Single Number Reach and Enable Move to Mobile**. Optionally, you can also set the **Single Number Reach Voicemail Policy** and check **Enable Extend and Connect** (for Jabber clients). Lastly, the time-of-day policies can be defined within the remote destination using **Ring Schedule** combined with access lists.

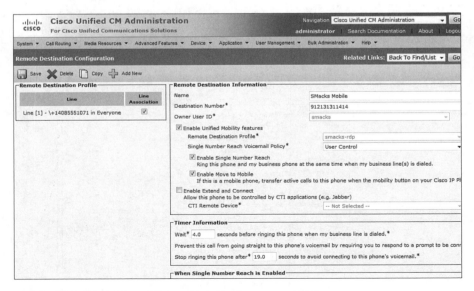

Figure 10-13 *Adding Remote Destinations to an RDP*

Note End users can create their own remote destinations on the CUCM user web pages, in addition to how the CUCM administrator does so in this step.

The remote destination has to be associated with one or more shared lines of the specified RDP.

Note After remote destination is associated, the remote destination rings if a call is placed to the appropriate shared line of an office phone. When a call is placed from a recognized remote destination to an internal destination, the calling number is modified from the remote phone number to the office phone DN. The remote DN usually has a PSTN access code (for example, 9 in NANP) and then an 11-digit number (trunk prefix 1 followed by the 10-digit number). If the incoming calling number is not prefixed with 91, internal phones see the call coming from the E.164 number of the remote phone instead of the associated internal DN. Step 6 shows how to resolve such issues.

Step 6. Set partial matches so that a calling number is recognized as a remote destination by configuring Cisco CallManager service parameters by browsing to **System > Service Parameters**, as shown in Figure 10-14.

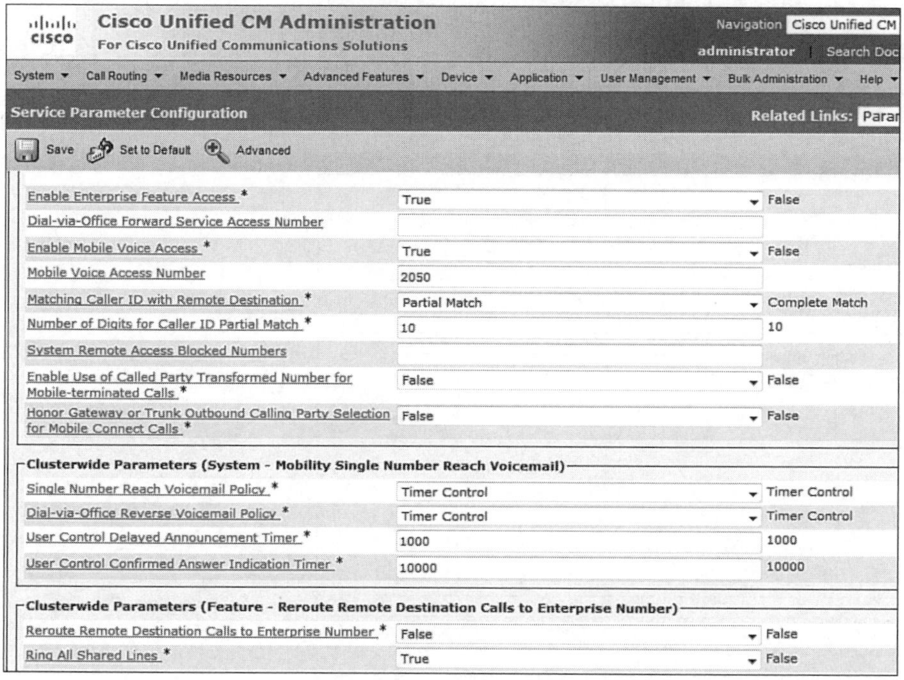

Figure 10-14 *Service Parameters Configuration*

Set **Matching Caller ID with Remote Destination** parameter to Partial Match. (The default is Complete Match.)

Set **Number of Digits for Caller ID Partial Match** parameter to the number of digits that have to match (beginning with the least-significant digit) when comparing the incoming calling number to the configured remote destination number.

> **Note**   Alternatively, you can choose **Call Routing > Transformation Pattern** to configure caller ID transformations. Each pattern can be assigned with a partition. The Calling Party Transformation CSS, which is configured in the RDP, is used to control access to the configured transformation patterns.

Step 7.   Access lists can be configured (optional) as shown in Figure 10-15, to control which callers are allowed to ring a remote destination. To configure access lists, choose **Call Routing > Class of Control > Access Lists**. Enter a name and description for the access list. Choose the user to whom the access list applies in the Owner drop-down list. Then check the **Allowed** option to create a list of phone numbers that should be allowed or check **Blocked** to create a list of phone numbers that should be blocked.

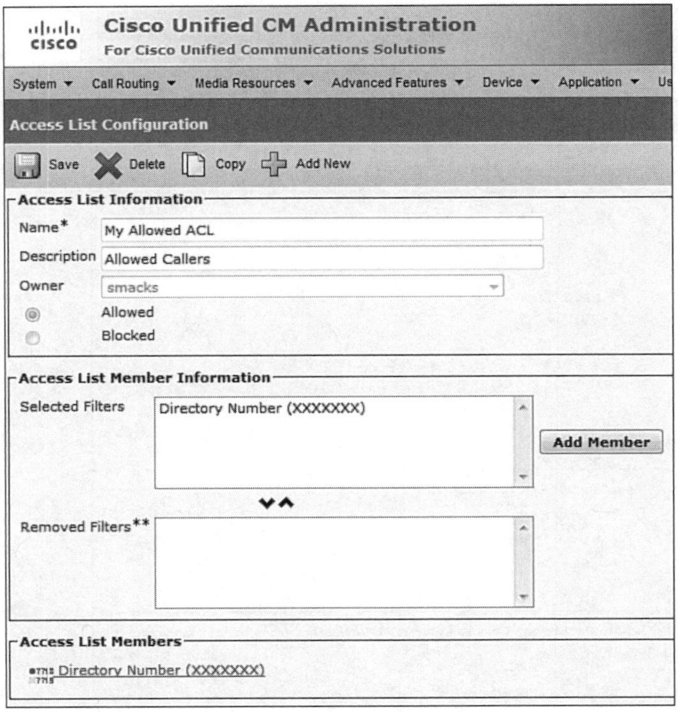

Figure 10-15   *Configuring the Access List*

Note   After you save the access list, the window reopens to display the Access List Member Information area. Click **Add Member** to add a member, and then choose one of the three options from the Filter Mask drop-down list. Choose to allow or disallow a specific DN (Directory Number), filter calls that do not have caller ID (Not Available), or filter calls that do not display their caller ID (Private).

Step 8.   Apply the access list on a remote destination by browsing to **Device > Remote Destination** and selecting the access list from the drop-down list under Allowed Access List or Blocked Access List.

Step 9.   Configure appropriate partitions and CSSs to align with the existing dial plan. It is important to note that the shared line directory number can be assigned with a partition. Also note that (standard) CSSs can be configured at the shared directory number (the line CSS) and at the device level (the IP phone and RDP). In addition, for Mobile Connect calls (that is, calls to a remote destination), a rerouting CSS can be configured in the RDP.

The next section describes MVA configuration.

Configuring Mobile Voice Access

Step 1.   Activate the **Cisco Unified Mobile Voice Access Service** by browsing to the Cisco Unified Serviceability. Choose **Tools > Service Activation** to activate the Cisco Unified MVA service.

Step 2.   Go to **System > Service Parameters**. Then choose the **Cisco CallManager** service and configure the service parameters applicable to MVA, as shown in Figure 10-14.

Note   Enable access to enterprise features by setting the **Enable Enterprise Feature Access** (described previously in this chapter) parameter to **True**.

Step 3.   Enable MVA for an end user, as shown in Figure 10-16, by browsing to **User Management > End User**. In the End User Configuration window, check the **Enable Mobile Voice Access** check box to allow the end user to use MVA.

Step 4.   The MVA media resource is automatically added when the Cisco Unified MVA service (see Figure 10-17) is activated. You can configure it by choosing **Media Resources > Mobile Voice Access**.

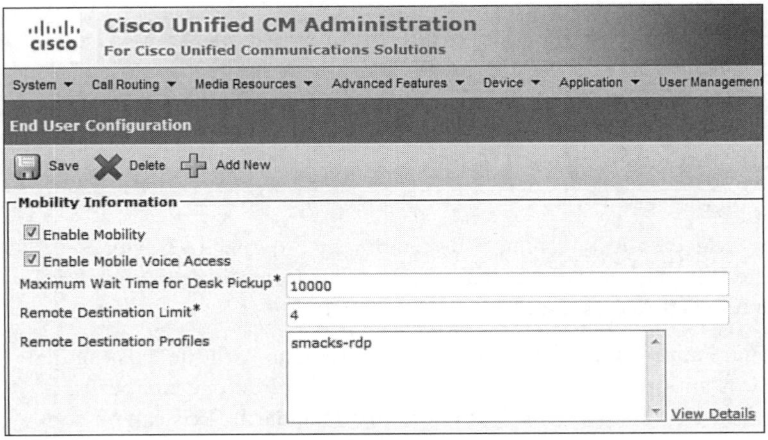

Figure 10-16   *Enabling MVA for End User*

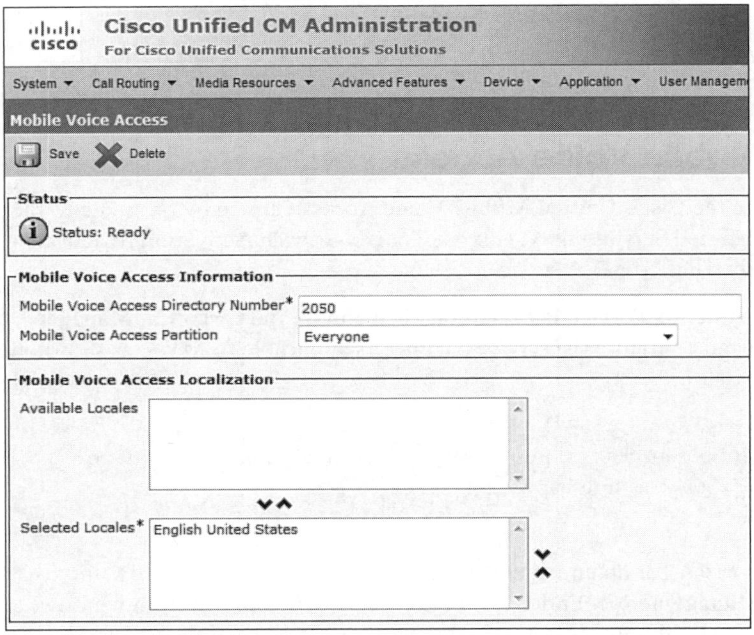

Figure 10-17   *MVA Media Resource*

Note   By default, only U.S. English is available. Both Mobile Connect and MVA support a maximum of nine locales, so Cisco Unified Communications Manager Administration blocks you from configuring ten or more locales for Cisco Unified Mobility. In Mobile Voice Access Localization, more than nine locales can appear in the Available Locales pane if they are installed for CUCM, but you can save only nine locales in the Selected Locales pane.

Example 10-1 shows the configuration of an H.323 gateway that provides access to the
Cisco Unified MVA feature.

Example 10-1   *Configure Mobile Access at the Cisco IOS Gateway*

```
MVARouter(config)# application
MVARouter(config-app)# service mva http://10.1.1.1:8080/ccmivr/pages/IVRMainpage.vxml
!
MVARouter(config)# voice-port 0/0/0:23
MVARouter(config-voiceport)# translation-profile incoming pstn-in
!
MVARouter(config)# voice translation-profile pstn-in
MVARouter(cfg-translation-profile)# translate called 1
!
MVARouter(config)# voice translation-rule 1
MVARouter(cfg-translation-rule)# rule 1 /.*5552\(...$\)/ /2\1/
!
MVARouter(config)# dial-peer voice 29991 pots
MVARouter(config-dial-peer)# service mva
MVARouter(config-dial-peer)# incoming called-number 2999
MVARouter(config-dial-peer)# direct-inward-dial
MVARouter(config-dial-peer)# port 0/0/0:23
!
MVARouter(config)# dial-peer voice 29992 voip
MVARouter(config-dial-peer)# destination-pattern 2999
MVARouter(config-dial-peer)# session target ipv4:10.1.1.1
MVARouter(config-dial-peer)# dtmf-relay h245-alphanumeric
MVARouter(config-dial-peer)# codec g711ulaw
MVARouter(config-dial-peer)# no vad
!
MVARouter(config)# dial-peer voice 1 pots
MVARouter(config-dial-peer)# destination-pattern 9T
MVARouter(config-dial-peer)# incoming called-number 2...
MVARouter(config-dial-peer)# direct-inward-dial
MVARouter(config-dial-peer)# port 0/0/0:23
!
MVARouter(config)# dial-peer voice 2 voip
MVARouter(config-dial-peer)# destination-pattern 2...
MVARouter(config-dial-peer)# session target ipv4:10.1.1.1
MVARouter(config-dial-peer)# incoming called-number 9T
MVARouter(config-dial-peer)# codec g711ulaw
```

Example 10-1 shows a sample configuration of an H.323 gateway. Note that in the
example an incoming translation profile strips the called number to four digits and is
applied to the ISDN voice port. Therefore, all other dial peers that are applicable to calls
from the PSTN refer to four-digit called numbers only.

The following happens when a remote user dials the MVA number 1-511-555-2999: The call is routed to the router's voice port, and the PSTN delivers a ten-digit national number that is then stripped to four digits by the translation profile. The called number 2999 matches the incoming POTS dial peer 29991, which is configured with the call application **service mva** command. The service MVA is configured with the URL of the MVA VXML call application. (This is located at the CUCM server where the Cisco Unified MVA service has been activated.)

When the call is passed on to the MVA media resource, the number configured at the MVA media resource in Step 4 is used. In this case, it is 2999.

> **Note**   The number that is used to start the call application on incoming PSTN calls (1-511-555-2999) does not have to match (or partially match) the number used for the call leg from the H.323 gateway to the CUCM MVA media resource. However, it is recommended that you use the same number to avoid confusion.

The outgoing VoIP dial peer that is used for this call leg (dial peer 29992) has to be configured for DTMF relay, and voice activity detection (VAD) has to be disabled.

All other dial peers that are shown in the example apply to incoming PSTN calls to directory numbers other than 2999 (**dial-peer voice 1 pots** and **dial-peer voice 2 voip**) and outgoing PSTN calls. (All received VoIP calls use incoming **dial-peer voice 2 voip** and outgoing **dial-peer voice 1 pots**.) These outgoing PSTN calls include normal calls placed from internal devices and calls initiated from remote phones that are using MVA to place enterprise calls to the PSTN.

Summary

The following key points were discussed in this chapter:

- Mobile Connect enables users to receive calls placed to their enterprise number at the enterprise phone and remote phones such as cell phones. MVA extends the Mobile Connect functionality by allowing enterprise calls placed from a remote phone to first connect to the enterprise and then break back out to the called number using the user's enterprise number as the calling number.

- MVA requires an H.323 gateway providing an IVR application to MVA users.

- The Cisco Unified MVA service must be activated in the CUCM cluster for MVA.

- Implementation of Cisco Unified Mobility includes the configuration of access lists, RDP, and remote destinations.

- If an MGCP gateway is used for PSTN access, an additional H.323 or SIP gateway is required for the MVA feature.

- Proper CSS and access-list configuration is required for MVA and Mobile Connect.

References

For additional information, refer to the following:

Cisco Systems, Inc. Cisco Collaboration Systems 10.x Solution Reference Network Designs (SRND), May 2014. http://www.cisco.com/c/en/us/td/docs/voice_ip_comm/cucm/srnd/collab10/collab10.html

Cisco Systems, Inc. Cisco Collaboration Systems 11.x Solution Reference Network Designs (SRND), July 2015. http://www.cisco.com/c/en/us/td/docs/voice_ip_comm/cucm/srnd/collab11/collab11.html

Cisco Systems, Inc. Features and Services Guide for Cisco Unified Communications Manager, Release 10.0(1), May 2014. http://www.cisco.com/c/en/us/td/docs/voice_ip_comm/cucm/admin/10_0_1/ccmfeat/CUCM_BK_F3AC1C0F_00_cucm-features-services-guide-100/CUCM_BK_F3AC1C0F_00_cucm-features-services-guide-100_chapter_0110011.html

Review Questions

Use these questions to review what you have learned in this chapter. The answers appear in Appendix A, "Answers Appendix."

1. **Cisco Unified Mobility consists of what two features? (Choose two.)**

    a. Single-number connect

    b. Mobile Connect

    c. Mobile IVR

    d. Mobile voice access

    e. Mobile voice connect

2. **What is indicated as the calling number for a call placed from a remote destination to an internal directory number?**

    a. The MVA number

    b. The number of the remote destination

    c. The directory number of the office phone the remote destination is associated with

    d. The directory number of the called office phone, if it's associated with the calling number of the remote destination

3. **What is not a requirement for Cisco Unified Mobility?**

    a. Remote destinations must be external numbers.

    b. An H.323 or SIP gateway providing the MVA IVR application.

c. Out-of-band DTMF.

d. A transcoder running at the gateway providing an MVA IVR application.

4. **What configuration element is not used to implement Cisco Unified Mobility?**

a. Softkey templates

b. User accounts

c. Access lists

d. Remote destination profiles

e. Remote destinations

f. Enterprise parameters

5. **Which two of the following are requirements to implement Cisco Unified Mobility? (Choose two.)**

a. CUCM version 6.x or later

b. CUCM version 4.x or 5.x

c. An H.323 or SIP gateway providing MVA IVR application

d. Extension Mobility enabled for the cluster and users

6. **Which of the following statements about Cisco Unified Mobility is the most accurate?**

a. Cisco Unified Mobility is a replacement for CUCM extension mobility for traveling users.

b. Cisco Unified Mobility allows the user to no longer require a cellular phone.

c. Cisco Unified Mobility allows the user to have a single phone number contact, even if he or she requires several different phones.

d. Cisco Unified Mobility functions properly only if implemented with extension mobility.

7. **Which two of the following signaling protocols on the IOS gateway are supported to implement Cisco Unified Mobility? (Choose two.)**

a. SCCP

b. H.323

c. SIP

d. MGCP

Chapter 11

Cisco Video Communication Server and Expressway Deployment

The adoption of video and video technologies has increased multifold in recent years. Users have embraced the use of video-capable devices such as cell phones and even web technologies such as Skype, Microsoft Lync, and WebEx for video enablement of traditional communications. No longer is it "acceptable" to have a standard audio-only conversation; most users welcome the addition of video because it adds a more personable characteristic to a traditional conversation. Being able to see visual expressions and the person you are communicating with gives the caller a sense of connectivity not achieved with a standard audio-only call. Cisco has embraced video technologies, and now many of the endpoints sell support video capabilities such as the 8900, 9900, SX, DX, MX, or EX series endpoints as well as Cisco Jabber. It is estimated by Cisco and Gartner that by 2018, more than 80 percent of all calls will be video enabled in the enterprise.

The integration of enterprise video into the Unified Communications (UC) environment introduces new protocols, servers, and challenges. One of the most important design considerations for video enablement is call control and video processing. Cisco TelePresence Video Communication Server (VCS) provides call-processing services in Cisco video-enabled deployments. Traditionally, video signaling and call control is performed on the VCS, which is capable of terminating and processing signaling for legacy video endpoints. When the term *legacy* video is mentioned, it refers to the video technologies and endpoints that are non-Cisco-branded video endpoints (video endpoints such as Tandberg, Polycom, LifeSize, Sony, for example, existed prior to the acquisition of Tandberg by Cisco in early 2010). These endpoints often use the H.323 or Session Initiation Protocol (SIP) protocols to register as endpoints on the VCS. In modern video-enabled Cisco UC environments, it is recommended that any Cisco-branded endpoints (SX, DX, MX, EX, 9900, 8900, and so on) be registered with the Cisco Unified Communications Manager (CUCM) to leverage additional features and capabilities.

One of the challenges that arises with video enablement is remote or teleworkers who reside outside the traditional LAN or WAN of the enterprise. How do you video enable remote sites or workers? Cisco introduced the VCS Expressway and Cisco Expressway series to combat this issue. VCS can be paired with VCS Expressway Core (VCS Expressway C) and VCS Expressway Edge (VCS Expressway E) to provide edge services such as firewall traversal, business-to-business communications, and secure remote access that does not require virtual private network (VPN) technologies such as Cisco AnyConnect, Secure Sockets Layer (SSL), or WebVPN. VCS Expressway is paired with VCS for legacy video devices that need to secure connectivity back to the LAN/WAN in the enterprise. For Cisco UC endpoints and Cisco Jabber, Cisco Expressway series offers edge services for CUCM deployments via Expressway Core (Expressway C) and Expressway Edge (Expressway E). It is important to differentiate that these are two separate sets of services that are paired with either VCS or CUCM.

Upon completing this chapter, you will be able to meet these objectives:

- Describe the roles of Cisco VCS and Cisco Expressway series servers

- Describe Cisco VCS and Cisco Expressway series deployment options

- Describe Cisco VCS and Cisco Expressway series platforms, license options, and features

- Describe Cisco VCS and Cisco Expressway series clustering options

- Describe Cisco VCS and Cisco Expressway series initial configuration steps

Cisco VCS and Expressway Series Overview

Cisco UC and Collaboration systems support many deployment scenarios for video enabling environments. One important topic is video call control and signaling servers and their placement in the environment. This section describes the roles of the Cisco VCS, Cisco VCS Expressway, and Cisco Expressway series servers and how they fit into a modern UC environment, including the following:

- Cisco Unified Communications Manager (CUCM)

- CUCM with Cisco Expressway Series (Cisco Expressway Core and Cisco Expressway Edge)

- Cisco VCS Control (VCS-C)

- Cisco VCS-C with Cisco VCS Expressway

- CUCM and Cisco VCS-C (combined solution)

CUCM performs endpoint registration and call-processing capabilities for analog and IP-based phones such as the 6800, 6900, 7800, 7900, 8800, 8900, and 9900 series endpoints. In addition to traditional audio IP phones, CUCM provides call signaling and registration capabilities for various Cisco-branded video endpoints such as the SX, DX, MX, and EX series.

It is recommended that any modern Cisco-branded video endpoint be registered directly with CUCM and *not* registered with the VCS. This recommendation allows for more advanced capabilities and features for the Cisco-branded video endpoints, such as the enhanced locations call admission control (CAC), video regions capabilities, survivable remote site telephony (SRST), and advanced video-conferencing capabilities. CUCM now provides voice, video, TelePresence, instant messaging (IM) and Presence, messaging, mobility, web conferencing, and security services.

CUCM with Cisco Expressway Series

Many small to large enterprises and businesses are expanding their UC portfolio to include Cisco Jabber. Cisco Jabber is an IM platform that provides a tightly coupled integration with the Cisco UC suite. This integration allows advanced features such as the ability to check voicemail, initiate a WebEx, see Presence status information when a user is busy or in a meeting, as well as file transfer, picture transfer, and many additional features.

A common challenge is how organizations can offer these powerful IM and collaboration capabilities outside of the organization to mobile and remote workers. Cisco Expressway series can be added to a CUCM deployment to provide edge features and services to devices that are located outside the enterprise firewall in a secure way without the need for a virtual private network (VPN) solution.

The Cisco Expressway series solution consists of two servers; a Cisco Expressway Core and Cisco Expressway Edge server. Some UC engineers refer to these as the *Expressway C* and *Expressway E* servers. An important note is that Cisco Expressway C and E cannot register endpoints directly with them; think of them as a proxy service, which passes the registration request through to CUCM. In essence, this acts as a man in the middle brokering connectivity from external Jabber or video endpoints to the internal CUCM servers and users.

Cisco VCS Control

Many large enterprise organizations have been doing standards-based H.323 or SIP video conferencing for years. This video-conferencing technology is typically referred to as *legacy video* because it involves non-Cisco-related manufacturers such as Polycom, LifeSize, Tandberg, Sony, and so on. These manufacturer devices register on an IP network as either H.323 or SIP endpoints.

In 2010, Cisco acquired the Tandberg Company, which was a major player in the legacy video space. The acquisition of Tandberg allowed Cisco to add a core component directly to the UC suite. The Tandberg Video Communications Server Control (VCS Control) was rebranded as Cisco VCS Control (VCS-C).

The Cisco VCS-C performs endpoint registration and call processing for legacy video products or third-party video endpoints that cannot directly register with CUCM. It provides video services for SIP and H.323 endpoints. Cisco VCS-C acts as a SIP registrar, a SIP proxy server, an H.323 gatekeeper, and as a SIP-to-H.323 gateway server. This allows the VCS core to provide internetworking between SIP and H.323 devices. In addition, the VCS Core can provide complex translation between IPv4- and IPv6-enabled devices.

Cisco VCS-C with Cisco VCS Expressway

Enterprise organizations that want to expose legacy video capabilities to remote or offsite users or perform business-to-business (B2B) features outside of the traditional LAN and WAN can install Cisco VCS-C along with the Cisco VCS Expressway series servers. Cisco VCS Expressway can be added to a Cisco VCS-C deployment to provide edge features and services to devices that are located outside the enterprise firewall in a secure way without the need for VPN technologies.

A great example of this is the need to perform a legacy video conference with a third party or perhaps a B2B video call using legacy H.323 or SIP endpoints. If Company A has a conference room with a legacy Tandberg or Polycom endpoint and wants to dial Company B to perform a legacy video call, they can use the Cisco VCS-C and VCS Expressway to achieve this. It is important to note that this is an all-IP-based solution routing the video call across the Internet using IP.

In the late 1980s and 1990s, many organizations used Integrated Services Digital Network (ISDN) video dialup technology where the video endpoints communicated across ISDN basic rate interface (BRI) or primary rate interface (PRI) technologies or perhaps a T1-CAS circuit. This technology can still be used. However, it requires a special ISDN-to-IP gateway such as the Cisco TelePresence ISDN gateway.

CUCM and Cisco VCS-C (Combined Solution)

CUCM and Cisco VCS-C can be interconnected with each other to provide the best of both worlds: connectivity between legacy video devices and Cisco UC. CUCM and VCS-C can be interconnected via SIP trunks to provide connectivity between both solutions. Both solutions can also have the appropriate extension (Cisco Expressway series or Cisco VCS Expressway series) added for edge service. In complex enterprise environments, it is not uncommon to see a mixture of both Cisco UC and legacy video technology using all the aforementioned technologies (CUCM, Expressways, VCS-C, and VCS Expressway).

Common Terminology for Cisco Video and Legacy Video

This section describes some commonly used terminology for both Cisco TelePresence and legacy video deployments and includes the following:

- **Collaboration edge:** An umbrella term commonly used to describe the Cisco Expressway Core and Cisco Expressway Edge servers used in combination with CUCM to provide edge features such as firewall traversal, B2B communications, and protocol interworking of Cisco Jabber and video-enabled endpoints. Collaboration Edge is an architecture for exposing enterprise UC services to external users in a secure fashion without requiring a VPN connection.

- **Cisco VCS:** Provides advanced legacy video support for standards-based H.323 and SIP endpoints. Think of Cisco VCS as a "CUCM for legacy video," offering registration, signaling, bandwidth management, and internetworking of various protocols and features. Cisco VCS Expressway is an optional extension to Cisco VCS-C that provides edge features and services to Cisco VCS-C.

- **Cisco Expressway series:** Provides edge firewall traversal and video interworking to CUCM and always consists of two components: Cisco Expressway Core (Expressway C) and Cisco Expressway Edge (Expressway E).

- **Mobile Remote Access (MRA):** Is a special feature of the Cisco Expressway series that offers secure, VPN-less remote access for Cisco Jabber clients and specific video endpoints. MRA allows roaming users to stay connected to the enterprise voice and video environment without requiring a VPN connection. Cisco Jabber clients on Windows, OS X, smartphones, and tablets can use MRA to connect back to the enterprise. It is recommended to check the latest release notes for Cisco Expressway and IP phone and video endpoints when determining which fixed endpoints are supported with MRA. MRA features and its implementation are discussed in detail in Chapter 14, "Cisco Unified Communications Mobile and Remote Access."

Note   At the time of this writing, MRA is considered an "experimental" feature for endpoints such as 8900, 9900, and the MX/SX/DX/EX series phones and video endpoints. If the endpoint supports the feature, this allows endpoints to roam perhaps to an executive's home office and securely connect back to the enterprise without requiring a VPN connection.

- **Cisco Jabber Guest:** Cisco Jabber Guest is a special feature that can be provisioned in Cisco Expressway Core and Edge servers. It enables you to connect visitors to a corporate website or mobile application directly to one or more employees via instant real-time voice and video. Think of Jabber Guest as a "click to talk to an expert" button on a website. When visitors click the button, they launch a web browser-based video call through the Expressway Edge and Core servers to a subject matter expert or perhaps knowledge worker inside the enterprise with a video-enabled endpoint. This allows browser-based video calls using the latest technologies such as HTML 5 and WebRTC. Jabber Guest is a special provisioning mode of Cisco Expressway C and E.

Note   It is important to note that if an environment wishes to use both MRA and Jabber Guest that these services and features run independently of one another, thus requiring multiple sets of Expressway Core and Edge servers dedicated for either MRA or Jabber Guest. The two features cannot be run at the same time on the same hardware.

- **Cisco Jabber for TelePresence (Jabber Movi):** Cisco Jabber for TelePresence (Jabber Movi) is a specialized software soft client that supports video-only capabilities. Jabber Movi is a product that registers with Cisco VCS-C and enables users to make video calls to both internal and external users if the Cisco VCS

Expressway has been properly implemented. Jabber Movi is now end of sale by Cisco, but is still heavily used by many organizations.

Cisco Jabber for TelePresence was sold with a special web camera called the Cisco Movi cam. This was a special webcam that plugged into users' laptops or desktops and was capable of producing high-definition (HD) 1080p 30 frames per second video resolution during use.

The terms listed above relate to Cisco VCS-C, Cisco VCS Expressway series, and Cisco Expressway series. There are additional techniques to extend enterprise video and telephony services to external users. One such technique is the use of Cisco Unified Border Element (CUBE). CUBE allows external connectivity using SIP URI dialing between business entities or organizations, thus allowing B2B communication.

Cisco VCS and Cisco Expressway Series Deployment Options

This section examines the various Cisco VCS and Cisco Expressway series deployment options.

Cisco VCS Deployment

This section describes a typical Cisco VCS deployment scenario. Figure 11-1 shows a typical Cisco VCS-C and Cisco VCS Expressway implementation.

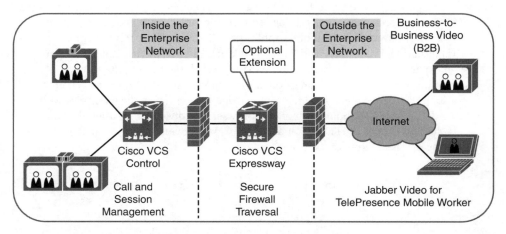

Figure 11-1   *Cisco VCS Deployment*

Figure 11-1 illustrates a Cisco VCS deployment for video-enabled collaboration using legacy video products that support SIP and H.323 industry standards. Cisco VCS-C provides video call and session control, registration, and security for legacy video endpoints and for certain Cisco TelePresence endpoints.

Cisco VCS Expressway allows video traffic to traverse the firewall securely, enabling rich video communications with partners, customers, suppliers, mobile users, and teleworkers. Cisco VCS Expressway is placed in a demilitarized zone (DMZ) and must use Network Address Translation (NAT) along with access lists to limit traffic to and from the device to specific TCP and UDP port ranges and protocols. In addition, X.509 certificates are used to securely trust devices along all segments of the video call.

In a Cisco VCS deployment, video-enabled legacy or TelePresence endpoints that are located inside the enterprise network are registered with Cisco VCS-C. Cisco VCS Expressway is not needed to establish calls between endpoints within the enterprise network. When making calls to endpoints that are outside the enterprise network, Cisco VCS Expressway uses secure firewall traversal functionality to allow B2B and remote endpoints to securely connect back to the enterprise.

Note   Cisco VCS Expressway is an advanced deployment for many organizations, so it requires coordination between the enterprise firewall team, UC team, and security team. It must have advanced knowledge of firewalls and X.509 certificates as well as advanced video telephony knowledge, including Cisco VCS-C, SIP, and H.323 protocols.

Cisco VCS Expressway offers the following features to externally connected endpoints:

- **B2B video communications:** By using URI dialing, you can collaborate with customers, partners, and suppliers as easily and securely as you do with e-mail messages.

- **Remote Cisco Jabber Video for TelePresence users:** Secure mobile access that is based on Transport Layer Security (TLS) allows you to make and receive calls on remote endpoints, such as Cisco Jabber Video for TelePresence (formerly known as Movi), without the need for a VPN connection.

- **Teleworkers and remote endpoints:** Teleworkers and remote video endpoints can use their personal Cisco TelePresence-enabled endpoints for video-enabled communications with colleagues, customers, partners, and suppliers from their home office or remote location.

Cisco Expressway Series Deployment

This section examines the various Cisco Expressway series deployment options. Figure 11-2 shows a typical CUCM deployment with Cisco Expressway solution.

The following figure illustrates a typical CUCM with Cisco Expressway series implementation. It includes voice, video, IM, and Presence collaboration features. With Cisco Expressway series, video endpoints and Cisco Jabber clients external to the enterprise are proxied through a secure firewall traversal using Cisco Expressway Core and Edge servers to the Cisco UC applications such as CUCM, Unity Connection, and IM and Presence.

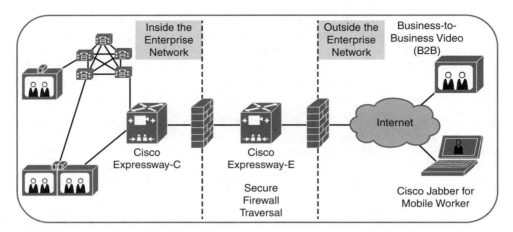

Figure 11-2 *Cisco Expressway Deployment*

Cisco Expressway series is an advanced Collaboration gateway that extends voice, video, IM, and Presence communications to an external network without the need for VPN. Cisco Expressway series plays a big part in the Collaboration edge architecture where CUCM is used for call processing.

Note   Cisco Expressway is an advanced deployment for many organizations that require coordination between the enterprise firewall team, UC team, and security team. It requires advanced knowledge of firewalls and X.509 certificates as well as advanced video telephony knowledge, including CUCM, Unity Connection, IM and Presence, SIP, and H.323 protocols.

A deployment with CUCM and Cisco Expressways offers the following features and communication modalities:

- **Business-to-consumer (B2C) video communications:** By using the Cisco Jabber Guest feature set, you can collaborate with customers, partners, and suppliers using secure, browser-based communication into the enterprise.

- **B2B video communications:** Expressways can be configured to route B2B calls using URI dialing, thus enabling B2B communications. It is important to note that any B2B communication requires a Rich Media license. Licensing of Expressway Core and Edge is discussed in a later section in this chapter.

- **Remote Cisco Jabber users:** By using the mobile and remote-access feature of the Cisco Expressway series, remote users can securely communicate with each other, customers, partners, and suppliers from their home office or remote location without the need of a VPN connection.

CUCM and Cisco VCS-C Interconnection

This section examines the deployment option where CUCM and Cisco VCS-C are deployed in parallel and interconnected with one another. Figure 11-3 shows a typical CUCM deployment with Cisco VCS-C.

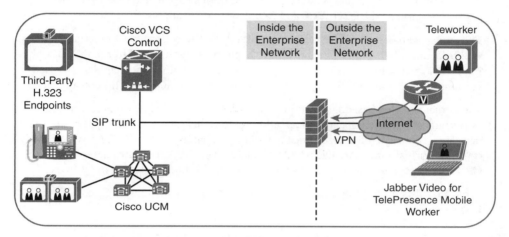

Figure 11-3 *CUCM and Cisco VCS-C Interconnection*

Figure 11-3 illustrates a Cisco VCS-C deployment that provides call processing for third-party H.323 and SIP video endpoints. Cisco-branded video endpoints such as SX, MX, DX, and EX series are registered against the Unified CM cluster. Both clusters are logically and physically separated; they both maintain their own device registrations and process signaling separately. The two independent systems can be integrated using a SIP trunk, thus enabling legacy video endpoints registered with Cisco VCS to dial phones and Cisco video endpoints registered with the CUCM servers.

This scenario is typically found in larger organizations with a substantial investment in legacy video technologies. Although this scenario is not ideal or the preferred architecture for Cisco Collaboration system deployments, it can be used when Cisco VCS-C is already present, or when you need Cisco VCS-C to support third-party video endpoints. CUCM provides registration and processing for all Cisco Unified IP phones and Cisco video endpoints. This deployment model adds to complexity because there are two separate call control servers (CUCM and Cisco VCS), each with its own device registrations and dial plans. Any time a new device is added, both servers potentially need to have their dial plan updated to reflect additions and changes.

In Figure 11-3, neither Cisco VCS Expressway nor Cisco Expressway series is deployed. Therefore, for secure integration of remote endpoints, a network-based firewall solution such as IPsec, SSL VPN, WebVPN, or Cisco AnyConnect is required. Notice in Figure 11-3 the Teleworker video screen pictured in the upper-right corner uses an IPsec-based VPN located on a small office/home office (SOHO) router or teleworker router such as a Cisco 800 series. In addition, Jabber Video for TelePresence mobile workers require a WebVPN or AnyConnect solution to be installed on their laptop to allow firewall traversal to the Cisco VCS-C for registration and signaling.

Cisco VCS and Cisco Expressway Series Platforms, Licenses, and Features

This section describes the Cisco VCS and Cisco Expressway series supported platforms, license options, and features. Cisco has released both an appliance-based and virtual machine version of the Cisco VCS and Expressway software.

The appliance-based hardware version is built on the Cisco Unified Computing System C220M3 and C220M4 platforms that have been specially designed to run the VCS or Expressway software as a bare-metal chassis-based installation. These appliance-based versions of VCS and Expressway are referred to as the CE 500 and CE 1000 appliances.

The virtual machine-based installation requires a VMware ESXi or vSphere hypervisor as the underlying operating system. In addition to requiring a VMware hypervisor, the installation comes in three sizes: small, medium, and large. These installations are based on Open Virtualization Archive (OVA) standards. The OVA templates determine the CPU resources, RAM, and hard drive specifications for the installation in VMware.

Table 11-1 lists both the appliance and VMware-based installations of Cisco VCS and Expressway.

Table 11-1 *Cisco VCS and Expressway Series Platform Installation Options*

Small and Medium Template, Appliance or CE 500 Installations	Large Template or CE 1000 Installations
The capacity of 1 Cisco VCS: ■ 2500 device registrations ■ 500 nontraversal calls ■ 100 traversal calls ■ 1000 subzones	The capacity of 1 Cisco VCS: ■ 2500 device registrations ■ 500 nontraversal calls ■ 500 traversal calls ■ 1000 subzones
The capacity of 1 Cisco Expressway: ■ 2500 proxy registrations to CUCM ■ 100 video calls ■ 200 audio calls	The capacity of 1 Cisco Expressway: ■ 2500 proxy registrations to CUCM ■ 500 video calls

Note   The capacities that are shown in Table 11-1 assume the use of a 10-Gbps network interface when choosing the large VM template or when deploying the CE 1000 appliance.

Table 11-1 makes reference to nontraversal and traversal calls. The definition of a nontraversal as well as a traversal call is outlined in the next section, which covers VCS and Expressway licensing. It is also important to study the maximum traversal and proxy connection limits for both types of installations.

Cisco VCS and Cisco Expressway Licensing

Cisco VCS and Cisco Expressway series have different licensing options due to the fact that each performs vastly different functions in the overall UC portfolio.

Cisco VCS has the following license options:

- **Nontraversal license:** These licenses are used when only signaling messages traverse Cisco VCS (for example, when calls between two locally registered SIP endpoints are placed).

- **Traversal license:** These licenses are used when signaling messages and media packets traverse Cisco VCS (for example, when B2B calls, or calls between SIP and H.323 endpoints, are placed).

Cisco Expressway series software licenses for virtualized installations are available at no additional charge to customers who have a license and a valued support contract for CUCM or CUCM Business Edition (BE6k or BE7k) Versions 9.1.2 or later.

Cisco Unified Workspace Licensing (CUWL) Standard or Professional Smartnet agreement or User Connect Licensing (UCL) Enhanced or Enhanced Plus Smartnet contracts are required to order the necessary licenses for Cisco Expressway. These Smartnet contract types include the necessary zero-dollar SKUs for both Cisco Mobile Remote Access (MRA), and Jabber Guest, or support for physical endpoints such as MX, EX, SX, and C series Cisco TelePresence endpoints to traverse and connect via the Expressways to CUCM.

Each call that terminates on an endpoint that is not registered with CUCM (for example, B2B calls and calls that involve Jabber Guest) requires a Cisco Expressway Rich Media Session license. Also, each call that requires interworking requires a Cisco Expressway Rich Media Session license. Examples of such calls include H.323 to SIP interworked calls or H.264 SVC to H.264 AVC calls. H.264 SVC to AVC interop is common if a B2B call is placed between Cisco Jabber and Microsoft Lync.

> **Note**   Calls that are placed between two endpoints that are registered to CUCM do not consume any license on the Cisco Expressway series, regardless of the other attributes of the call.

Cisco VCS and Cisco Expressway Feature Comparison

Table 11-2 lists the features of Cisco VCS and Cisco Expressway deployments. The following table can be used as a comparison of the various Cisco VCS and Expressway deployment options.

Table 11-2 *Cisco VCS and Cisco Expressway Series Feature Comparison*

Feature	Cisco Expressway Supported	Cisco VCS-C Supported	Cisco VCS-C and Cisco VCS Expressway Supported
Mobile and Remote Access	Y	N	N
Jabber Guest	Y	N	N
Business-to-Business Video	Y	N	Y
Video Interworking (IPv4 to IPv6, H.323 to SIP, MS H.264 SVC to AVC)	Y	Y	Y
Enhanced Security	N	Y	Y
Video/TelePresence Device Registration and Provisioning	N	Y	Y
Video Session Management and Call Control	N	Y	Y

Note Cisco Expressway series are always deployed as an extension to CUCM. The pair of Cisco Expressway Core and Cisco Expressway Edge servers provides edge features to endpoints that are registered with CUCM.

Cisco VCS and Cisco Expressway Clustering

Clustering is supported for Cisco VCS and Cisco Expressway series deployments. Clustering offers high availability and scalability for the Cisco VCS and Cisco Expressway solution. The following options are available when clustering the Cisco VCS and Cisco Expressway series servers:

- You can have up to six servers in a cluster. A cluster includes one master server and up to five additional servers or peers. Essentially, this creates a Publisher/Subscriber model similar to CUCM. Cisco VCS or Cisco Expressway would have one Publisher server and five Subscribers. This allows you to scale to 12 total servers (6 servers for Cisco VCS-C or Expressway Core, and 6 servers for Cisco VCS Expressway or Cisco Expressway Edge).

- You must use the same OVA template size for all servers within a cluster for VMware-based installations. An example is that deploying three large VMware templates and three small VMware templates is not supported; they all need to be the same size.

- Cisco VCS supports different numbers of peers in a Cisco VCS-C and Cisco VCS Expressway cluster. For example, a Cisco VCS-C cluster with two servers can be connected to a Cisco VCS Expressway cluster with fours servers.

- Cisco Expressway series remote access is limited to one domain per cluster. This regulation prevents Cisco Expressway from serving more than one cluster or organization. Expressways are tied to a single CUCM cluster.

- Cisco VCS and Cisco Expressway series support multiple clusters for the same domain. This allows you to scale to 24 servers or 2 full Expressway or VCS clusters servicing a single CUCM cluster.

- The master or publisher of each cluster replicates the configuration to all peers. All changes performed on the master server are replicated to all other servers that are in the same cluster. If any changes are done directly on a nonmaster or Subscriber server, the changes are removed during the next replication.

Clustering Considerations

Several design considerations must be taken into account while planning a Cisco VCS or Cisco Expressway cluster. Before you join servers into a cluster, you must consider the following prerequisites:

- All cluster members must run the same version of code. The only time members are allowed to run different versions of code is for the short period of time while a cluster is being upgraded from one software version to another software version. During the upgrade period, the cluster operates in a portioned fashion.

- A DNS SRV record must be available for the cluster, and the SRV record must have an address record (IPv4 or IPv6 A record) for each member of the cluster.

- The members must have different IP addresses. The IPv4 address and the IPv6 address (if enabled) must be unique on each member.

- Each member of the cluster must be able to reach all other cluster members within a 15-ms delay (30-ms round-trip delay). This can geographically limit deployment distance depending on WAN circuit speeds and propagation delays.

Note It is important to note this delay is different from CUCM round-trip time (RTT). CUCM servers use an 80-ms delay RTT or 40-ms one-way delay.

- Each member must have a directly routable path to all other members. Network Address Translation (NAT) must not be used between members of a cluster.

Note If there is a firewall between any two cluster members, make sure that you allow IP communications between the members.

- All members of a cluster must use a hardware platform (appliance or virtual machine) with equivalent capabilities. For example, you can cluster servers that are running on standard appliances with servers running on two medium-sized virtual machines, but you cannot cluster a server that runs on a standard appliance with servers running on eight large-sized virtual machines.

■ All members must have the same set of options keys installed.

■ Each member must have a unique system or hostname.

■ H.323 mode must be enabled on all members. Even if all endpoints in the cluster are using SIP, H.323 signaling is required because it is used for endpoint location searching and for sharing bandwidth usage information between members of a cluster.

Cluster Deployment Overview

This section provides examples and explains the clustering options for Cisco VCS and Cisco Expressway series servers. Figure 11-4 depicts example clustering options for both Cisco VCS and Expressway servers.

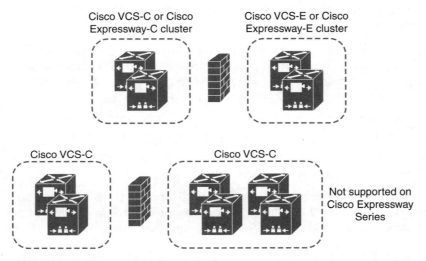

Figure 11-4   *Clustering Options for Cisco VCS and Cisco Expressway Series Servers*

Figure 11-4 shows two examples of Cisco VCS and Cisco Expressway clusters:

■ In the top portion of Figure 11-4, the same number of servers is deployed in a Cisco VCS-C or Cisco Expressway Core and the attached Cisco VCS Expressway or Cisco Expressway Edge cluster. This deployment option is supported on both Cisco VCS and Cisco Expressway series. Essentially, you are matching the same number of Cisco VCS or Cisco Expressway servers for both the Core and Edge servers. There can be up to six member servers in a cluster.

■ In the bottom portion of Figure 11-4, an unequal number of servers are used in a Cisco VCS-C cluster and the attached Cisco VCS Expressway cluster. This deployment option is only on Cisco VCS. It is not supported on Cisco Expressway series. When using Cisco Expressway series, you must deploy the same number of servers in the Cisco Expressway Control and the attached Cisco Expressway edge clusters.

Cisco VCS and Cisco Expressway Series Initial Configuration

This section discusses the initial configuration of the Cisco VCS and Cisco Expressway series servers. The initial configuration consists of setting high-level system settings and is typically performed only once during the system initialization and configuration.

Note The configuration of the IP address depends on the platform that is being used. There are three options to set the IP address for an appliance. For a virtual machine-based installation, the IP address is either configured as part of the VM deployment wizard in VMware vCenter or vSphere or via the VMware vSphere console. It is a requirement and best practice that the IP address for the Cisco VCS or Cisco Expressway servers be statically configured.

The procedure to configure a static address on the vSphere console is the same for Cisco VCS and Cisco Expressway:

Step 1.   Log in to the server via the console or command-line interface (CLI). The initial username is admin and the password is TANDBERG.

Note It is strongly recommended to change the admin password and the root password. You will receive a warning message when logging into the GUI until the default password has been changed.

Step 2.   Enter **Y** at the Run Initial Wizard prompt to start the Install Wizard.

Step 3.   Follow the Install Wizard instructions to enter the IP configuration of the server. You must set the IP protocol, IP address, subnet mask, default gateway, and Ethernet speed.

Step 4.   Reboot the server when prompted by the Installation Wizard, or use the **xcommand boot** command.

Step 5.   To access the administration GUI, use a web browser to navigate to https://<*Server IP Address*>, as shown in Figure 11-5. Expressway-C configuration has been shown as an example.

Note The first page that is shown after the login is the Overview page, which displays system uptime, software version, IPv4 address, license options, and resource usage. The Overview page of Cisco VCS and Cisco Expressway series differs slightly in terms of license and resource usage. The License and Resource Usage page on Cisco Expressway series contains the Rich Media licenses that are only available on Cisco Expressway. These licenses do not exist on Cisco VCS-C or VCS Expressway.

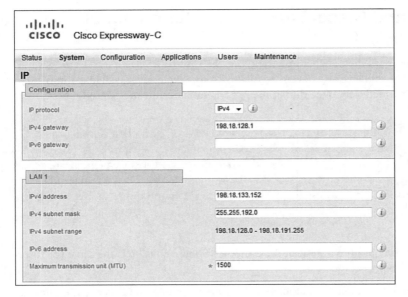

Figure 11-5 *IP Address Configuration on Cisco Expressway*

Step 6. The system name appears in various places in the web interface, and in the display on the front panel of the appliance-based installation, so that you can identify the appliance when it is in the rack with other servers. The system host name is also important in terms of zones as well as certificates. Choose **System > Administration**. Configure the system name, as shown in Figure 11-6. Click **Save**.

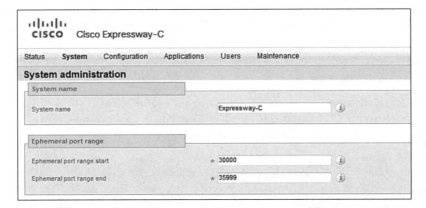

Figure 11-6 *System Host Name Configuration on Cisco Expressway*

Step 7. DNS is extremely important in both Cisco VCS and Cisco Expressway deployments because the majority of video devices, Cisco Jabber, and X.509

certificates require proper DNS resolution and DNS-related parameters such as a valid domain name. To configure DNS, choose **System > DNS**. Configure the system hostname, domain name, and default DNS servers address 1, as shown in Figure 11-7. Click **Save**.

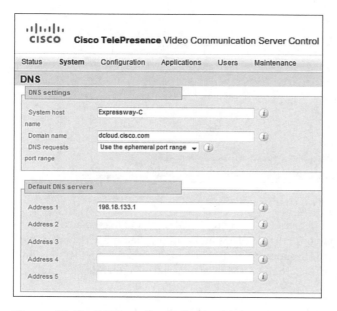

Figure 11-7   *DNS configuration on Cisco Expressway*

Note   The DNS server addresses are the IP addresses of up to five DNS servers. At least one DNS server must be set for address resolution. Cisco VCS and Cisco Expressway series servers send queries to one DNS server at a time. If that server does not respond, another backup server from the list is queried.

To configure Network Time Protocol (NTP) server address and time zone information, perform the following steps:

Step 8.   Both Cisco Expressway and Cisco VCS rely heavily on proper NTP server synchronization. This affects log files, system dates, and certificate validation. The NTP server address fields set the IP address or FQDNs of the NTP servers that are used to synchronize the system time. The time zone sets the local time zone of the Cisco VCS or Expressway servers. To configure NTP server choose **System > Time**.

Step 9.   Configure the NTP server 1 with the IP address of the first NTP server, as shown in Figure 11-8. For redundancy, you can configure additional NTP servers. Set the time zone. Click **Save**.

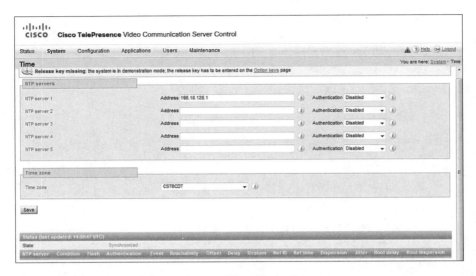

Figure 11-8   *NTP Configuration on Cisco Expressway*

Another common integration is to send SNMP traps and logs to a collector such as Cisco Prime or perhaps SolarWinds Orion collectors. Cisco VCS and Cisco Expressway both support SNMP Versions 1, 2, and 3. SNMP Version 3 adds the ability to secure the SNMP traffic using an authentication and encryption.

To enable and configure SNMP, perform the following steps:

Step 10.   Activating SNMP enables Cisco VCS and Cisco Expressway series to be integrated with an SNMP management system such as Cisco TelePresence Management Suite (Cisco TMS) in the case of Cisco VCS for managing Cisco Jabber Movi clients. To configure SNMP, go to **System > SNMP**.

Step 11.   Choose one of the following SNMP modes:

- Disabled

- v3 secure SNMP

- v3 plus TMS support (Choose this mode when integrating with Cisco TMS)

- v2c

Click **Save**. Figure 11-9 illustrates SNMP configuration with SNMP v3 plus TMS support.

Note   SNMP v3 plus TMS support is required to integrate Cisco VCS and Cisco TMS for Cisco Jabber TelePresence Movi provisioning capabilities.

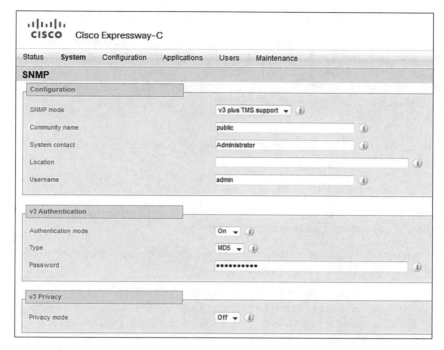

Figure 11-9 *SNMP Configuration on Cisco Expressway*

Step 12. Go to **Configuration > Domains**. Click **New**. Enter the domain name in the Name field, as shown in Figure 11-10. Click **Create Domain**.

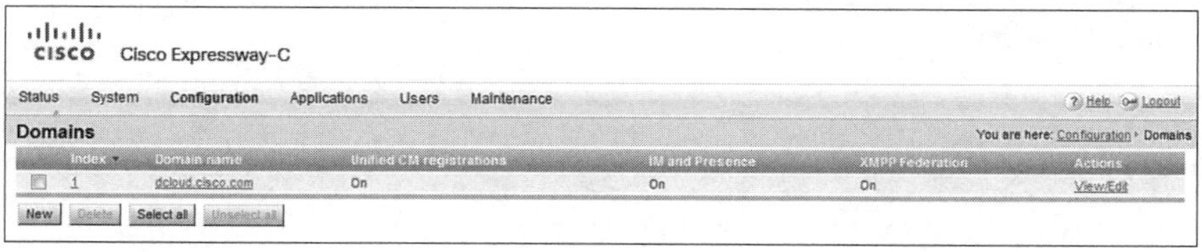

Figure 11-10 *SIP Domain Configuration on Cisco Expressway*

Note   Both Cisco VCS and Cisco Expressway series can act as a SIP registrar for configured SIP domains. After the SIP domains have been configured, the servers can accept registration requests from any SIP endpoint that attempts to register with an alias that matches a configured domain. In the case of Cisco Expressways, the SIP domain may be a hybrid of both the internal and external domain aliases. For example, if a company uses @company.local for an internal domain name and @company.com for an external domain name, both SIP domains must be configured on the Expressway.

After you create the domain, the domains page displays all configured SIP domain names.

> **Note**   Cisco Expressway series remote access allows you to enter multiple domain names. However, it is limited to servicing one enterprise. For example, Cisco Expressway cannot be multihomed and service two separate organizations; it must be homed to a single CUCM cluster and accompanying applications.

At this time, Cisco VCS and Cisco Expressway servers are configured with initial settings.

Summary

The following key points were discussed in this chapter:

- Collaboration Edge is an umbrella term describing the entire architecture for edge features in a Cisco Collaboration System solution.

- Cisco VCS-C can be deployed with or without Cisco VCS Expressway. The Cisco Expressway series consists of Cisco Expressway Core and Cisco Expressway Edge servers and is an optional extension to CUCM. Both provide firewall traversal capabilities without requiring VPN solutions.

- Cisco VCS-C and Cisco Expressway can be installed on specs-based virtual machines as well as an appliance or server. The appliance-based installation requires a CE 500 or CE 1000 server platform.

- Up to six peers can be a part of a Cisco VCS or Cisco Expressway cluster.

- Initial configuration of a Cisco VCS and Cisco Expressway series includes IP, DNS, and NTP settings.

References

For additional information, refer to the following:

Cisco Systems, Inc. Cisco Collaboration Systems 10.x Solution Reference Network Designs (SRND), May 2014. http://www.cisco.com/c/en/us/td/docs/voice_ip_comm/cucm/srnd/collab10/collab10.html

Cisco Systems, Inc. Cisco Collaboration Systems 11.x Solution Reference Network Designs (SRND), July 2015. http://www.cisco.com/c/en/us/td/docs/voice_ip_comm/cucm/srnd/collab11/collab11.html

Cisco Systems, Inc. Features and Services Guide for Cisco Unified Communications Manager, Release 10.0(1), May 2014. http://www.cisco.com/c/en/us/td/docs/voice_ip_comm/cucm/admin/10_0_1/ccmfeat/CUCM_BK_F3AC1C0F_00_cucm-features-services-guide-100/CUCM_BK_F3AC1C0F_00_cucm-features-services-guide-100_chapter_0110011.html

Cisco Systems, Inc. Video Communication Server (VCS) Configuration Guides, Oct 2015. http://www.cisco.com/c/en/us/support/unified-communications /telepresence-video-communication-server-vcs/products-installation-and- configuration-guides-list.html

Cisco Systems, Inc. Expressway Configuration Guides, Oct 2015. http://www. cisco.com/c/en/us/support/unified-communications/expressway-series/products- installation-and-configuration-guides-list.html

Review Questions

Use these questions to review what you have learned in this chapter. The answers appear in Appendix A, "Answers Appendix."

1. True or false: Collaboration Edge is a technology that refers to Cisco VCS-C and Expressway.

   a. True

   b. False

2. Cisco VCS-C paired with Cisco VCS Expressway provides which three core features to externally connected endpoints? (Choose three.)

   a. NAT translations

   b. Business-to-consumer (B2C) video communication

   c. Business-to-business (B2B) video communication

   d. Teleworkers

   e. Cisco IPMA support

   f. Remote Cisco Jabber Video for TelePresence (Movi)

3. Which Cisco Edge solution provides business-to-consumer capabilities?

   a. Cisco CUBE

   b. Cisco AnyConnect VPN

   c. Jabber Movi

   d. Jabber AnyConnect

   e. Jabber Guest

4. What is the maximum number of subzones that can be configured on a Cisco VCS-C?

   a. 5

   b. 25

    c. 100

    d. 500

    e. 1000

5. **Which two appliance models in which Cisco VCS and Expressway may be installed? (Choose two.)**

    a. ASA 5510

    b. CUCM BE7k

    c. CUCM BE6k

    d. CE 1000

    e. ISR 3945

    f. CE 500

6. **Which two licensing options are available for Cisco VCS? (Choose two.)**

    a. Traversal

    b. UCL Enhanced

    c. Nontraversal

    d. CUWP

    e. CUWL Pro

    f. UCL Enhanced Plus

7. **Please select the maximum number of servers that can belong to a VCS or Expressway Cluster.**

    a. 2

    b. 4

    c. 6

    d. 7

    e. 8

    f. 10

8. **To successfully cluster VCS or Cisco Expressway, which mode must be enabled?**

    a. SIP

    b. SIPS

    c. TLS

    d. H.264 SVC to AVC

    e. DNS

    f. H.323

9. True or false: An odd number of servers is supported in a Cisco Expressway cluster.

    a. True

    b. False

10. Please choose the default IP address for a Cisco VCS or Expressway installation.

    a. 10.252.20.1

    b. 192.168.0.1

    c. 172.16.32.1

    d. 192.5.41.41

    e. 192.168.255.1

    f. 10.255.255.1

    g. 172.16.255.254

    h. 192.168.0.100

Chapter 12

Deploying Users and Endpoints in Cisco VCS Control

Cisco TelePresence Video Communication Server Control (VCS-C) is a call control processing server for legacy video endpoints and non-Cisco branded video endpoints. Typically, legacy video endpoints like Tandberg, Sony, LifeSize, Polycom, and other manufacturers register with a VCS-C. Cisco VCS-C is often referred to as *Cisco VCS* if not deployed in tandem with Cisco VCS Expressway servers to extend legacy video outside of the enterprise. Cisco VCS-C supports different registration and authentication options for different types of endpoints. It is extremely important to think about security in terms of video deployments and Unified Communications (UC). It is easy to be hacked and incur hundreds or perhaps thousands of dollars in toll fraud charges before the hack is detected. Many organizations hire consultants who specialize in penetration testing to perform audits of their video deployments. SIPVicious suite is a set of tools that can be used to perform Session Initiation Protocol (SIP) audits on video and UC infrastructure. There are many security tools suites available. As you are reading this chapter, keep security in the forefront of your mind. SIP hacking and distributed denial-of-service (DDoS) attacks are on the rise and will only continue to increase throughout the first quarter of this century. For more information about security aspects with regard to Cisco UC, consider reading the Cisco Press publication *Securing Cisco IP Telephony Networks*.

This chapter describes user and device authentication options and how endpoints register with Cisco VCS-C. It also examines the configuration options for calls between locally registered endpoints. Lightweight Directory Access Protocol (LDAP) integration provides the ability to centralize user and device authentication management to a single user database. An overview of LDAP authentication options for endpoints and users as well as examples on how to provision the Cisco Jabber Video for TelePresence (previously named Movi) clients are also provided in this chapter.

Upon completing this chapter, you will be able to meet these objectives:

- Describe the user authentication options in a Cisco VCS deployment
- Describe the endpoint registration options in Cisco VCS deployments

- Describe the endpoint authentication options in Cisco VCS deployments

- Describe the role of Cisco TelePresence Management Suite for endpoint provisioning

- Describe the function of zones in Cisco VCS deployments

- Describe the functions of links in Cisco VCS deployments

- Describe the functions of pipes in Cisco VCS deployments

Cisco VCS User Authentication Options

This section describes how user authentication is achieved in Cisco VCS deployments. Figure 12-1 shows how end users are authenticated against an LDAP database. A Microsoft Active Directory LDAP database is shown running on a Microsoft server.

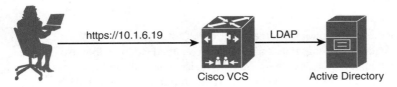

Figure 12-1 *Cisco VCS User Authentication Support*

Cisco VCS-C and VCS deployments support authentication of users either by locally defined user accounts on the VCS-C (manual creation on the VCS) or by using an LDAP server.

When no LDAP configuration exists on Cisco VCS, local users are employed for administrator and end-user authentication. In the case of no LDAP integration, administrative access to the admin web pages or graphical user interface (GUI) and the Cisco FindMe features for Cisco Jabber Movi are authenticated using locally defined user accounts on the VCS-C. Manually defining user accounts on the VCS-C works, but it requires duplication of user accounts on both LDAP and VCS. It also adds to the complexity because there are multiple authentication databases involved in the enterprise. If LDAP is not present in the enterprise, the only alternative is to use locally defined user accounts on the VCS.

Note   When only LDAP authentication is configured, no users are allowed to access the admin or user pages when there is an LDAP failure. This means if the connection or LDAP bind between the VCS and LDAP fails or cannot establish a connection, no users can authenticate into the VCS. Other logins, including the serial login and Secure Shell (SSH) login, will continue to use the local administrator account that is locally configured on the Cisco VCS to prevent a total and complete lockout scenario.

Note   User web login is only applicable when FindMe is used without Cisco TelePresence Management Suite (TMS). FindMe and TMS feature capabilities are examined later in this chapter.

LDAP Authentication Configuration Example

LDAP configuration on the Cisco VCS can seem tricky to a new engineer. Figure 12-2 shows the settings found on the Cisco VCS LDAP Configuration screen.

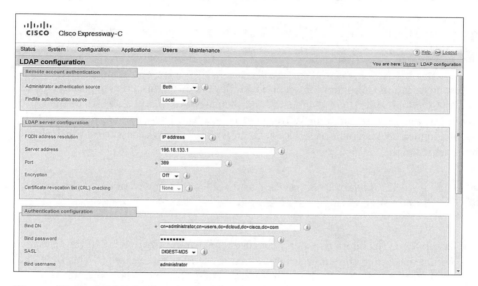

Figure 12-2   *LDAP Authentication Configuration*

To properly configure LDAP for Cisco VCS, the administrator must browse to the **Users > LDAP Configuration** web page inside the Administration GUI.

In Figure 12-2, the Administrator Authentication Source is set to Both, which allows Cisco VCS to use locally defined accounts and LDAP accounts for user authentication to the Administration GUI. Several options are available for authentication, including Local Only, Remote Only, or Both. With Local Only authentication, only locally defined administrator accounts are examined when a user tries to authenticate to Cisco VCS. With Remote Only authentication, only LDAP-defined administrator accounts are examined when a user tries to authenticate with Cisco VCS. Using both allows Cisco VCS to process authentication requests first against the locally defined user database, and if a match is not found, then and only then is the LDAP-defined database connection used to attempt to authenticate user credentials.

Note   You cannot log in to the Administration GUI using a locally configured administrator account, including the default admin account, if Remote Only authentication is configured.

The Fully Qualified Domain Name (FQDN) Address Resolution field allows you to set the connection method to Microsoft LDAP. The available options are via address record (A or AAAA record), service record (SRV), or IP address. In the preceding screenshot, the connection is made using an IP address of 10.1.5.14 for the server address. By default, the TCP port setting for a nonsecure LDAP connection is 389 and 636 for a secure connection.

Note   To use TCP port 636 for a secure LDAP bind or connection, set the encryption type to TLS.

The Authentication Configuration section includes the distinguished name (Bind DN), the sAMAccountName (Bind Username), and the password (Bind Password) of the LDAP account that is used to search the Active Directory server. These settings represent the LDAP administrator account or service account with read-only access to all other objects in the search path.

Note   The LDAP account must have at least read access to the relevant parts of the MS Active Directory where user accounts are stored.

The Base DN for Accounts parameter specifies the distinguished name of the search base within the Active Directory structure that is used for user searches. This field effectively sets where the user account lookups are performed. Many organizations have a folder in LDAP defined specifically for users. By default, all LDAP users are created in the Users folder. Nesting of folders is permitted, and the Base DN will traverse all subfolders under the Users folder when searching for a valid user account and credentials. The Base DN for Groups parameter specifies the distinguished name of the search base within the Active Directory for any group-based searches. Cisco VCS enables you to search for a group of users in addition to a single user account.

Note   If the Base DN for Groups is not configured, the Base DN for Accounts location is used for group searches.

Endpoint Registration

This section describes the endpoint registration options in Cisco VCS. As Figure 12-3 illustrates, several configuration settings are required prior to device registration inside Cisco VCS-C.

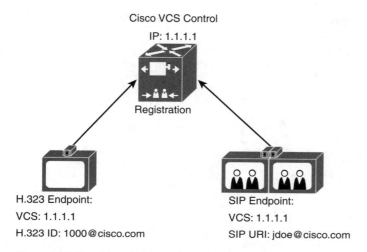

Cisco VCS Control

IP: 1.1.1.1

Registration

H.323 Endpoint:
VCS: 1.1.1.1
H.323 ID: 1000@cisco.com

SIP Endpoint:
VCS: 1.1.1.1
SIP URI: jdoe@cisco.com

Figure 12-3   *Cisco VCS Endpoint Registration*

Cisco VCS-C allows devices to register by H.323 or SIP protocols. Many legacy video endpoints will use these two protocols. To successfully register an endpoint with Cisco VCS, the following configuration is required on Cisco VCS:

■ The initial configuration includes the system name, Domain Name Service (DNS), Network Time Protocol (NTP), Simple Network Management Protocol (SNMP), and Session Initiation Protocol (SIP) domains, as discussed in Chapter 11, "Cisco Video Communication Server and Expressway Deployment."

■ H.323 or SIP or both protocols are activated on the VCS server. By default, these protocols are turned on, but may require protocol customization such as the port ranges used by Cisco VCS.

Once the initial configuration settings and protocol-specific settings are configured, devices such as legacy video endpoints must be properly configured to register with Cisco VCS. Many third-party and legacy video endpoints offer a graphical user interface (GUI) or telephony user interface (TUI) menu screen in which an administrator can configure the settings for the device to attempt registration with Cisco VCS. At the endpoints, you must configure the following settings:

■ IP address of Cisco VCS (registrar server)

■ The SIP URI of the endpoint when using SIP to register the endpoint with Cisco VCS

■ The H.323 ID of the endpoint when using H.323 to register the endpoint with Cisco VCS

Note   Third-party and legacy video products often require manual configuration by the administrators. The use of DNS SRV records may aid in the configuration process.

Figure 12-4 depicts the registration process for the Cisco Jabber Video for TelePresence (Movi) clients. Some endpoints, such as the Movi software-based endpoints, require Cisco TMS to register with Cisco VCS. These endpoints are not supported by Cisco VCS without the integration of Cisco TMS.

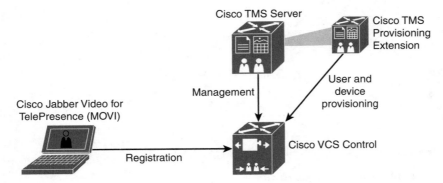

Figure 12-4   *Cisco VCS Endpoint Registration for Jabber Movi Clients*

To Support Cisco Jabber Video for TelePresence, Cisco VCS must be managed by Cisco TMS. Cisco TMS must control Cisco VCS to perform provisioning tasks on Cisco VCS. Cisco VCS must be configured with special SNMP v3 settings that allow TMS to control and provision VCS. In addition, a special provisioning extension must be installed on Cisco TMS. Finally, the provisioning license must be installed on Cisco VCS.

On Cisco TMS, the Cisco TMS Provisioning Extension must be installed. The Cisco TMS Provisioning Extension is a Java-based application that runs on the Cisco TMS server. This application offers user and software-based device provisioning on the Cisco VCS.

Endpoint Authentication

Cisco VCS can control which H.323 or SIP endpoints are allowed or disallowed to register using a registration policy. A registration policy can be thought of as an allow or deny list. Cisco VCS can also require endpoints to authenticate by username and password. With the rise in SIP hacking attempts in recent years, it is recommended to configure both an allow or deny list in combination with a username and password. Remember to use all available mechanisms to thwart hacking attempts and distributed denial-of-service (DDoS) attacks in the environment.

Cisco VCS attempts to verify the credentials that are presented near the endpoint by first checking them against its local database of usernames and passwords. The local database also includes checking against credentials that are supplied by Cisco TMS if the system is using device provisioning by Cisco TMS.

If the username is not found in the local database, Cisco VCS may check the credentials over a real-time LDAP connection to an external H.350 directory service. H.350 is the protocol used for LDAP connections to both Microsoft Active Directory and to

OpenLDAP servers. As mentioned earlier, the directory service, if configured, must have an H.350 directory schema for either a Microsoft Active Directory LDAP server or an OpenLDAP server.

Along with one of these methods, for those devices that support NT LAN Manager (NTLM) challenges (for example, Cisco Jabber Video for TelePresence software clients), Cisco VCS can alternatively check credentials directly against an Active Directory server using a Kerberos connection.

Cisco VCS Authentication Methods

Figure 12-5 depicts the various authentication methods on the Cisco VCS-C server.

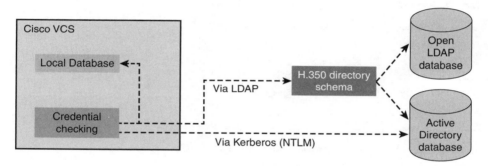

Figure 12-5 *Cisco VCS Authentication Methods*

When device authentication is enabled on Cisco VCS, any endpoint that attempts to communicate with Cisco VCS is challenged to present its credentials (typically based on a username and password). Cisco VCS then verifies those credentials, according to its authentication policy, and accepts or rejects the message accordingly.

Cisco VCS supports the following authentication methods:

- **Local database:** The local database can be used for SIP and H.323 device authentication. Usernames and passwords are stored in the local database.

- **H.350 directory service lookup via LDAP:** This method can be used for authenticating any SIP or H.323 endpoint. For H.350 authentication, the H.350 schema must be downloaded from Cisco VCS and installed on the LDAP server.

- **Active Directory database:** This method can be used for devices that support NTLM (for example, Cisco Jabber Video for TelePresence). The credentials are authenticated via direct access to an Active Directory server using a Kerberos connection. Active Directory database authentication can be enabled at the same time as either the local database or H.350 directory service authentication, because NTLM authentication is only supported by certain endpoints.

Note   When an endpoint that supports NTLM responds to the NTLM challenge, Cisco VCS uses NTLM in preference to the other authentication methods.

Registration Restriction Policy

This section describes how to use the registration Restriction Policy option to allow or deny endpoint registrations on the Cisco VCS-C. Figure 12-6 shows where the registration allow and deny lists are located on the Cisco VCS-C.

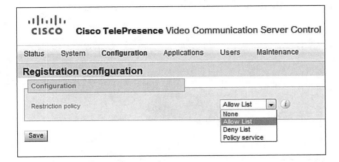

Figure 12-6   *Registration Restriction Policies on Cisco VCS*

Cisco VCS can control which devices are allowed or denied to register based on a restriction policy.

The registration restriction policy can be configured with one of the following options:

- **Allow List:** This option requires you to configure one or more allow lists. Endpoints that match an entry in one of the allow lists are allowed to register with Cisco VCS. Endpoints that do not match any entry are not allowed to register with Cisco VCS.

- **Deny List:** This option requires you to configure one or more deny lists. Endpoints that match an entry in one of the deny lists are not allowed to register with Cisco VCS. Endpoints that do not match any entry are allowed to register with Cisco VCS.

- **Policy Service:** This option can be used when an external policy server should be used to allow or deny endpoint registrations. An external policy server can be a centralized web server that is running a Call Processing Language (CPL) script; typically, this is housed on an Internet Information Services (IIS) or Tomcat web server. CPL Scripts are Extensible Markup Language (XML)-based scripts that provide allow and deny rules for Cisco VCS endpoints.

Allow or deny lists contain patterns by using the exact, prefix, suffix, and regex pattern types. You can use regular expressions to match a range of endpoints, match based on the suffix or prefix of the registration request or match on the exact registration request. An example of a registration request is a directory number or URI followed by the @ symbol and the domain name or IP address of the endpoint.

Note   Cisco VCS supports the configuration of only one registration policy at the same time.

Cisco TMS Provisioning

Cisco TelePresence Management Suite (TMS) offers control and management of multiparty conferencing, infrastructure, and endpoints to centralize management of all video collaboration and the Cisco TelePresence network. One of the more prevalent issues that arises with video technologies is the use and adoption of video-enabled conference calls and video-conference rooms. TelePresence conferences are most effective and attended when they can be set up easily. With TMS, you do not need to be concerned with the equipment being used or where people are located. To schedule a meeting, you tell Cisco TMS which rooms you want to use and how many people will be calling in. Cisco TMS then automatically books the rooms and reserves the conference bridges and resources and the conferencing ports needed for the conference.

To help include the people you need in the meeting, Cisco TMS integrates and searches directories and external information sources. It also integrates with Microsoft Outlook so that users can book Cisco TelePresence meetings directly from an Outlook client or Outlook Web Access. Because this is done automatically, the video-conferencing costs are lower and user satisfaction increases. Cisco TMS also has extensive reporting and analytics capabilities, so you can see resource utilization, conference capacities, and historical trends for your conferencing resources.

Cisco TMS configuration can be quite complex. The topics and techniques in TMS are quite extensive and beyond the scope of this book. Only the main features and characteristics of the solution are covered in this chapter.

The major benefits of Cisco TMS include the following:

- **Scalable provisioning:** Cisco TMS can support rapid, large-scale deployments of up to 100,000 Cisco TelePresence users, endpoints, and soft clients across disparate customer locations, including up to 5000 direct-managed devices.

- **Centralized administration:** Cisco TMS automates and simplifies the management of Cisco TelePresence meetings and Cisco TelePresence infrastructure resources, reducing complexity and the total cost of ownership of a Cisco TelePresence solution. TMS allows you to create templates for the provisioning of additional video endpoints. The templates can be quite detailed and list video calling characteristics, conferencing capabilities, and Movi configuration characteristics.

- **Flexible scheduling:** Cisco TMS makes scheduling Cisco TelePresence meetings more accessible with a range of tools, including a simple and intuitive Smart Scheduler option, extensions for Microsoft Exchange integration, and advanced booking capabilities for experienced administrators.

■ **Natural user experience:** Cisco TMS reduces complexity and makes it easy for users to start and join meetings on time with One Button to Push (OBTP) capability for select Cisco TelePresence systems and intuitive instructions for other participants, including "one click to join" for people joining with WebEx and in-meeting message notifications.

■ **Easy contacts management:** Cisco TMS allows the use of integrated phonebooks that offers fast, accurate dialing and consistency with company directories, and enhanced reachability with FindMe.

To enable many of the provisioning features inside Cisco VCS, Cisco TMS must have a special extension installed called the Cisco TelePresence Management Suite Provisioning Extension (TMSPE). Cisco TMSPE is an extension for Cisco TMS that enables rapid provisioning of Cisco TelePresence users and endpoints, such as Cisco Jabber Video for TelePresence. Cisco TMSPE also enables users to set optional Cisco VCS FindMe preferences so that they can be reached based on availability, location, and preferred devices. Cisco VCS FindMe is similar to single number reach (SNR) on CUCM.

Deploying Cisco Jabber Video for TelePresence

Deploying Cisco Jabber Video for TelePresence is an advanced topic and requires multiple configuration settings across various servers in the Cisco TelePresence suite. The entire configuration is beyond the scope of this book, but the main configuration steps required are as follows:

Step 1.   Perform the initial configuration of Cisco VCS-C (IP, system name, DNS, NTP, SNMP, SIP domain, and so on).

Step 2.   Configure authentication policies in Cisco VCS.

Step 3.   Enable NTLM authentication in Cisco VCS.

Step 4.   Enable provisioning on Cisco TMS.

  1. Verify that a valid provisioning license is installed on Cisco TMS.

  2. Install the TMS Provisioning Extension on TMS.

  3. Add the Cisco VCS-C server to the Cisco TMS server.

  4. Configure Cisco VCS provisioning.

Step 5.   Configure users and groups in Cisco TMS.

Step 6.   Configure Cisco Jabber Video for TelePresence (Movi) clients.

Cisco VCS Zones

Zones are one of the most important concepts inside of Cisco VCS-C. Cisco VCS uses a zone concept to enforce call admission control (CAC) and registration policies. The collection of all endpoints, gateways, multipoint control units (MCUs), and content servers registered with the Cisco VCS make up what is known as the VCS's *local zone*.

The local zone is made up of subzones. These include an automatically created default subzone and up to 100 manually configurable subzones. Each manually configured subzone specifies a range of IP addresses. When an endpoint registers with the VCS, it is allocated to the appropriate subzone based on its IP address. If the endpoint's IP address does not match any of the subzones, it is assigned to the default subzone. Subzones can be thought of as device pools in CUCM, because they logically group similar video endpoints.

Subzones are used for the purposes of bandwidth management. After you have set up your subzones, apply bandwidth limits for the following:

- Individual calls between two endpoints within the subzone

- Individual calls between an endpoint within the subzone and another endpoint outside of the subzone

- The total of calls to or from endpoints within the subzone.

The VCS also has a special type of subzone known as the *traversal subzone*. This is a conceptual subzone; no endpoints can be registered with it, but all traversal calls (that is, calls for which the VCS is taking the media in addition to the signaling) must pass through it. The traversal subzone exists to enable you to control the amount of bandwidth used by traversal calls, because these can be particularly resource intensive.

Traversal subzones are typically used for calls and media that traverse a firewall or demilitarized zone (DMZ) in order to be publically accessible. An example is a Cisco VCS-C paired with a Cisco VCS Expressway in the DMZ for firewall traversal and external connectivity. The traversal zone concept is also used in dealing with remote teleworkers on the Cisco Expressway series with Cisco Jabber clients located outside of the enterprise.

Local Zone

By default, a local zone contains one default subzone and one traversal subzone, and it may have user-defined subzones located within. No devices can register directly to a local zone; the local zone represents a collection of all devices registered with the Cisco VCS-C. Video endpoints must register with the subordinate default subzone or to a user-defined subzone. The local zone is considered independent of network topology, and may consist of multiple network segments. Figure 12-7 is an example of a local zone inside a Cisco VCS-C server.

In Figure 12-7, the local zone represents the superior zone that by default contains one default subzone and one traversal subzone. In addition, the local zone includes user-defined subzones that represent sites in New York, San Jose, and St. Louis (abbreviated in the diagram as NY, SJ, and SL). Endpoints register with the subzones based on their suffix, prefix, or IP address. The figure also shows an endpoint that does not match any configured subzone matching criteria. This endpoint is associated with the default subzone.

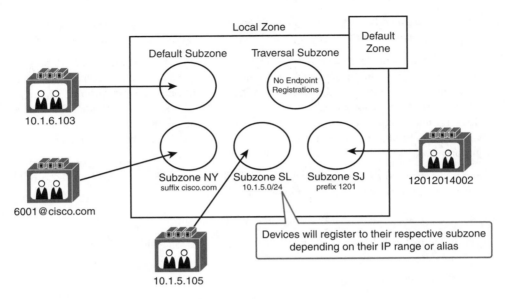

Figure 12-7   *Local Zone Inside Cisco VCS*

A video endpoint can be bound to a subzone by matching criteria. Matching criteria determine how the endpoint is placed into a particular subzone and can be configured to match based on IP addresses, suffix patterns, prefix patterns, and subnet masks.

Default Subzone

Endpoints that do not match the criteria of a user-defined subzone register in the default subzone when they are not denied by a registration restriction policy. The default sub-zone is used as a "catchall" zone in which devices that are not defined in a specific zone are still allowed to register and can have bandwidth policies and CAC applied from the default subzone.

> **Note**   As a leading practice, known and planned endpoints should register with a zone created by the administrator. The default zone is provided specifically for unrecognized endpoints.

Any incoming calls from endpoints that are not recognized as belonging to any of the existing configured zones are deemed to be coming from the default zone. The VCS comes preconfigured with the default zone and default links between it and both the default subzone and the traversal subzone. The purpose of the default zone is to allow you to manage incoming calls from unrecognized endpoints with the VCS.

You can manage calls from unrecognized endpoints by performing several actions in the VCS-C, such as the following:

- Deleting the default links. This prevents any incoming calls from unrecognized endpoints.

- Applying pipes to the default links. This enables you to control the bandwidth consumed by incoming calls from unrecognized endpoints. Links and pipes are covered later in the chapter.

- Using Call Processing Language (CPL) scripts. These scripts can be used to prevent unknown endpoints from registering and placing calls.

- Using device-level authentication and registration policies discussed in Chapter 11.

Subzone

A subzone is a logical collection of devices, such as endpoints, MCUs, and gateways. When a device registers with Cisco VCS, it is assigned to a specific subzone based on its IP address or pattern match. If there is no match, the endpoints register with the default subzone. Each subzone can have a bandwidth restriction for calls within the subzone.

Traversal Subzone

A traversal subzone is used for firewall traversal, or when Network Address Translation (NAT) or Port Address Translation (PAT) is used. To create a traversal zone on the Cisco VCS-C, it must be paired with a Cisco VCS Expressway in the DMZ located on a supporting firewall. A special zone relationship is set up on the Cisco VCS-C and VCS Expressway to allow a secure channel for communications from outside the enterprise into the LAN. The concept of the traversal subzone is also prevalent in the Cisco Expressway series for remote Cisco Jabber clients.

Each subzone can have CAC limits configured. These limits apply to calls within the zone, and they are configured as follows:

- **Within:** This parameter defines the maximum allowed bandwidth of all calls within the subzone. If two devices are registered with the same subzone, the call between these two devices is classified as a call within the subzone.

- **In&Out:** This parameter defines the maximum allowed bandwidth of all calls that are established out of and into a subzone. If one device of a call is registered in the subzone and the other device is not registered in the same subzone, the In&Out limit of the subzone applies to the call.

- **Total:** This parameter defines the maximum allowed bandwidth of all calls that involve a subzone: calls within the subzone and calls that involve one device within the subzone and one device that is located outside the subzone.

Links

Links are logical connections between subzones or zones. Links are created automatically among the default subzone, traversal subzones, and default zone, but they can be deleted or modified if desired. Figure 12-8 depicts several links between the default subzone, and several other subzones representing sites in New York, San Jose, and St. Louis in the United States.

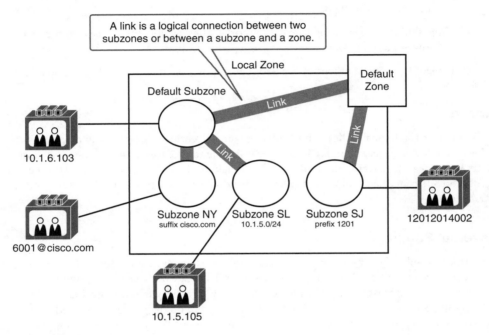

Figure 12-8 *Links in Cisco VCS-C*

You can freely configure links between any subzones and zones. Calls between endpoints that are located in two different zones or subzones are only possible if there is a path between the two zones or subzones. The path does not have to be a direct link between the two zones or subzones. It can involve multiple zones or subzones and links. Configure subzones and links so that they match the physical layout of your network. The subzones and links are similar to locations and links between locations in a Cisco UC system. In Cisco VCS, however, they are not only used for CAC but also for call routing, because calls are only possible if the two endpoints have a path configured between their zones or subzones.

Before pipes are discussed, let's go over the zone bandwidth restrictions Within, In&Out, and Local. The next section addresses these.

Zone Bandwidth Restrictions: Within

This section describes the zone bandwidth restrictions within a zone. Figure 12-9 is an example of a call in which the Within bandwidth restriction parameter applies.

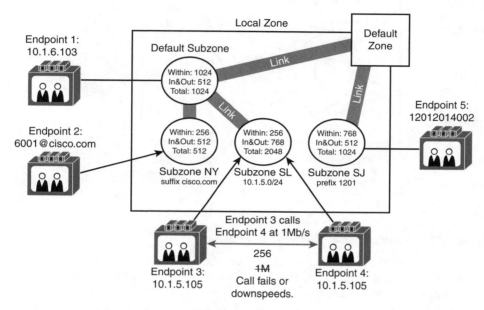

Figure 12-9 *Zone Bandwidth Restrictions: Within*

Endpoint 3 at IP address 10.1.5.105, which is connected to the St. Louis subzone, attempts to call endpoint 4 at IP address 10.1.5.105. The video call setup and the endpoints attempt to negotiate video codecs and bandwidth at 1 Mbps, but the bandwidth is downsized to 256 kbps because of the bandwidth restriction within the zone. The Within bandwidth is set to 256 kbps on the subzone for St. Louis, and because both endpoints are registered with this subzone, the Within limit is considered for the call.

Within bandwidth limitations can be beneficial in sites with multiple video endpoints because they help to prevent the LAN from becoming saturated with video calls. Even though you have 100 MB or 1000 MB links from a LAN perspective, video is greedy by nature and will try to use the most bandwidth possible for the best quality call. It is possible in a LAN to have several executives attempting a conference call between one another. When this occurs, video saturation is possible on the LAN. The Within zone-based bandwidth restriction is set in place to prevent this scenario from occurring.

Zone Bandwidth Restrictions: In&Out

This section describes the zone bandwidth restrictions in and out of a zone. Figure 12-10 is an example of a call where the In&Out bandwidth restriction parameter applies.

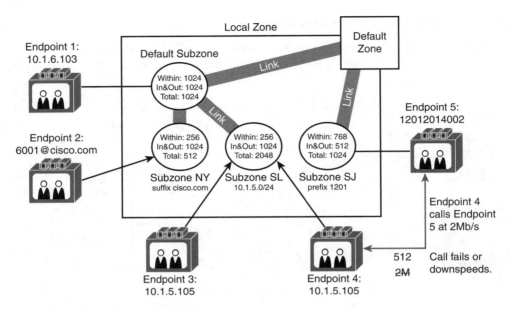

Figure 12-10   *Zone Bandwidth Restrictions: In&Out*

Endpoint 4 at IP address 10.1.5.105, which is connected to the St. Louis subzone, attempts to call endpoint 5 in the default subzone. The video call setup and the endpoints attempt to negotiate video codecs and bandwidth at 2 Mbps, but the bandwidth is downsized to 512 kbps because of the bandwidth restrictions that apply to calls that go beyond a single subzone. The In&Out bandwidth is set to 768 kbps on the St. Louis (SL) subzone and to 512 Kbps on the San Jose (SJ) subzone. Both limits are applied because the two endpoints that are involved in the call are located in these two zones. The lower limit is enforced, and therefore the call results in a bandwidth of 512 kbps.

Think of in and out zone bandwidth restrictions as analogous to regions in CUCUM. In and out zone bandwidths affect or indicate the maximum amount of bandwidth allowed to be consumed between video endpoints located in different zones inside Cisco VCS-C.

Zone Bandwidth Restrictions: Total

This section describes the zone bandwidth restrictions as a total (within + in and out) for a zone. Figure 12-11 is an example of a call where the Total bandwidth restriction parameter applies.

Endpoint 2, which is connected to the New York subzone, attempts to call endpoint 3 at 10.1.5.105 in the St. Louis subzone. The video call setup and endpoints attempt to negotiate video codecs and bandwidth at 2 Mbps, but the bandwidth is downsized to 512 kbps because of the bandwidth restrictions that apply to calls that go beyond subzone NY. While the call between endpoint 2 and endpoint 3 is active, endpoint 4 calls endpoint 5 at 2 Mbps. The bandwidth of the call between endpoint 4 and endpoint 5 is downsized to 256 kbps because the total limit of subzone SL would be

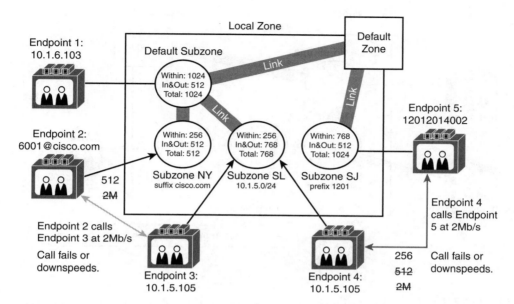

Figure 12-11 *Zone Bandwidth Restrictions: Total*

exceeded if more bandwidth were used. The first call consumed 512 kbps, and the remaining total bandwidth left in subzone SL is only 256 kbps.

Think of the Total zone bandwidth restriction as the total of all bandwidth allowed for all calls in a particular subzone. This can be used to place a bandwidth cap on conference calls in a particular subzone or to ensure bandwidth maximums are not exceeded for all video calls in a particular subzone.

Pipes

This section describes the concept of pipes. Figure 12-12 shows the concept of a pipe in Cisco VCS-C and VCS Expressway.

A pipe defines bandwidth limitations between two subzones or between a subzone and a zone. Pipes are analogous to CAC in CUCM. Pipes in Cisco VCS-C are very powerful and more closely resemble the enhanced locations CAC concept, where you can build a full topology and bandwidth restrictions between sites and internal to sites.

A pipe allows the application of bandwidth limits to links. By configuring subzones, links, and applying pipes to links, you can create a model of the physical network and its bandwidth limits on network connections such as WAN links. You can apply bandwidth restrictions per call and for the total bandwidth that goes over a link. When calls are placed between endpoints in different subzones, you can control the bandwidth that is used on each link that is along the path between the two subzones.

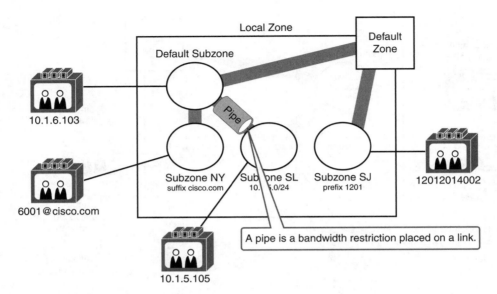

Figure 12-12   *Cisco VCS-C Pipes*

First you must create one or more pipes and configure them with the desired bandwidth limits. Then you must assign the pipes to the links to apply the desired limits. All calls that traverse one or more links that have a pipe applied must find enough available bandwidth on all links that are involved in the call. This implementation of CAC is similar to an enhanced locations CAC in CUCM.

Pipe Bandwidth Restrictions

This section shows an example of bandwidth restrictions that are applied to links via pipes. Again, think of pipes as a CAC mechanism in Cisco VCS-C allowing a maximum amount of bandwidth to be used between zones or subzones. Figure 12-13 shows an example of a video call where pipe bandwidth restrictions are applied.

Endpoint 1 at IP address 10.1.6.103 located in the upper-left corner of the diagram, calls endpoint 3 at IP address 10.1.5.103. The bandwidth requested for the call is 2 Mbps, but the bandwidth is downsized to 256 kbps because of the per-call pipe limitation that is applied to the link between the subzones that are involved in the call. If there were no pipe applied to the link, 512 kbps of bandwidth would be used for the call because this bandwidth is the maximum bandwidth that is permitted based on the bandwidth limits configured at the two subzones.

If another call were to be made between the default subzone and subzone SL, 256 kbps of bandwidth would be used for the second call. A third call between the two subzones would fail, but not because of the total bandwidth limit that is configured at the pipe (2048 kbps), but because of the In&Out bandwidth limit that is configured at the default subzone (512 kbps). In the example, the maximum bandwidth allowed between zones or

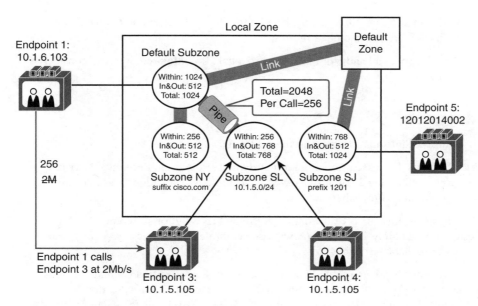

Figure 12-13 *Pipe Bandwidth Restriction*

subzones is the lowest In&Out value configured on both zones (in this case, 512 kbps). The default subzone allows 512 kbps of bandwidth in and out of the zone, whereas the subzone for SL allows 768 kbps of bandwidth in and out of the zone, thus the lowest value is the maximum allowed for calls between both zones. Think of the pipe as a governor or CAC mechanism indicating the maximum amount of bandwidth allowed between zones.

Summary

The following key points were discussed in this chapter:

- Cisco VCS administrators and end users can be authenticated locally or via an LDAP server.

- You must activate SIP and H.323 protocols if you want to allow endpoints to register with both protocols with the VCS-C. VCS-C can provide interworking capabilities between H.323 and SIP devices.

- Endpoint registration can be controlled by device authentication and by registration restriction policies.

- Cisco TMSPE enables provisioning of Cisco Jabber Video for TelePresence.

- Cisco VCS uses a zone concept for CAC and registration policies.

- Links are logical connections between two zones or subzones. Calls between two zones or subzones are only possible if a path between the two zones or subzones exists. In addition, the call must be admitted by CAC, based on the bandwidth limits that are configured at the zones.

■ A pipe applies a bandwidth limitation to a link. If configured, it is also considered by CAC.

References

For additional information, refer to the following:

Cisco Systems, Inc. Cisco Collaboration Systems 10.x Solution Reference Network Designs (SRND), May 2014. http://www.cisco.com/c/en/us/td/docs/voice_ip_comm/cucm/srnd/collab10/collab10.html

Cisco Systems, Inc. Cisco Collaboration Systems 11.x Solution Reference Network Designs (SRND), July 2015. http://www.cisco.com/c/en/us/td/docs/voice_ip_comm/cucm/srnd/collab11/collab11.html

Cisco Systems, Inc. Features and Services Guide for Cisco Unified Communications Manager, Release 10.0(1), May 2014. http://www.cisco.com/c/en/us/td/docs/voice_ip_comm/cucm/admin/10_0_1/ccmfeat/CUCM_BK_F3AC1C0F_00_cucm-features-services-guide-100/CUCM_BK_F3AC1C0F_00_cucm-features-services-guide-100_chapter_0110011.html

Cisco Systems, Inc. Video Communication Server (VCS) Configuration Guides, Oct 2015. http://www.cisco.com/c/en/us/support/unified-communications/telepresence-video-communication-server-vcs/products-installation-and-configuration-guides-list.html

Review Questions

Use these questions to review what you have learned in this chapter. The answers appear in Appendix A, "Answers Appendix."

1. Which three ways can a user be authenticated inside Cisco VCS-C? (Choose three.)

   a. Locally defined accounts

   b. ACS server

   c. TACACS

   d. Microsoft Active Directory

   e. OpenLDAP

2. Which two protocols must be configured and activated on the Cisco VCS-C to allow device registrations? (Choose two.)

   a. MGCP

   b. H.323

c. SIP TLS

d. SIP

e. SCCP

f. TACACS

3. Which extension must be configured on the Cisco TMS Server to enable device provisioning of Cisco Jabber Video for TelePresence (Movi) clients?

a. ARP

b. DHCP

c. Prime Provisioning Extension

d. CDP

e. NTP

f. TMSPE

g. TMSXP

4. Which three mechanisms can limit device registrations on a Cisco VCS-C server? (Choose three.)

a. Allow lists

b. Block lists

c. Deny lists

d. Append lists

e. Policy maps

f. Service policies

g. Policy service

h. Policing

5. Registration restriction policies can contain patterns that use which four of the following pattern types or matches? (Choose four.)

a. Exact

b. Prefix

c. Access list

d. Priority list

e. Suffix

f. Regex

g. Regular extensions

h. Dynamic extensions

6. Which four of the following are valid Cisco VCS zone types? (Choose four.)

   a. Traversal subzone

   b. Transformation subzone

   c. Local zone

   d. Default subzone

   e. Transformation subzone

   f. Subzone

7. Which concept provides a logical connection between subzones or zones?

   a. Transformation

   b. Translation

   c. Region

   d. Location

   e. Link

   f. Pipe

   g. Coupler

8. What are three ways that bandwidth restrictions can be enforced and measured when using links? (Choose three.)

   a. In and out

   b. Total

   c. Single

   d. Exclude

   e. Within

9. Which concept provides bandwidth limitations between two subzones or zones?

   a. Transformation

   b. Translation

   c. Region

   d. Location

   e. Link

   f. Pipe

   g. Coupler

Chapter 13

Interconnecting Cisco Unified Communications Manager and Cisco Video Control Server

Cisco Unified Communications Manager (CUCM) and Cisco Video Communications Server (VCS) can be interconnected to allow endpoints that are registered on the two separate systems to communicate with each other. This can bring a tremendous amount of flexibility and interoperability to an existing Unified Communications (UC) and legacy video environment. The configuration or interworking of the two separate systems is fairly simple to achieve using the Session Initiation Protocol (SIP). A SIP trunk must be built between the servers with a few special settings. There are dial plan components that must be mastered on both the CUCM and VCS control servers for calls to properly route between the two systems. This chapter focuses on the dial plan components on both CUCM and VCS as they relate to interworking the separate systems together. It examines CUCM components such as route patterns, route lists, route groups, and SIP trunks. The VCS control call transforms, search rules, admin policies, and neighbor zones are covered.

This chapter describes the integration methods and components for both CUCM and VCS control.

Upon completing this chapter, you will be able to meet these objectives:

- Provide an overview of how to interconnect CUCM and Cisco VCS

- Describe the call flow between CUCM and Cisco VCS

- Describe the dial plan components of Cisco VCS

- List the configuration steps to interconnect CUCM and Cisco VCS

- List the configuration steps to implement FindMe

Cisco Unified Communications Manager and Cisco VCS Interconnection Overview

Figure 13-1 provides an overview on how the interconnection between CUCM and Cisco VCS is achieved.

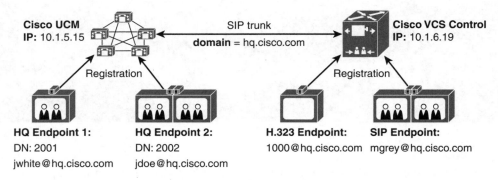

Figure 13-1 *Cisco Unified Communications Manager and Cisco VCS Deployment Overview*

CUCM and Cisco VCS control are connected together by a SIP trunk. In CUCM, there is a SIP trunk with a special SIP profile and SIP trunk security profile that is applied to the trunk, which points to one or more IP addresses or fully qualified domain name (FQDN) addresses of a VCS control or VCS control cluster.

In CUCM, proper dial plan components are configured to route calls to the VCS control. These dial plan components include route patterns or SIP route patterns, route lists, route groups, and the SIP trunk that contains the VCS control members.

The Cisco VCS control also has additional dial plan components and a connection to CUCM. This connection is a special zone called a *neighbor zone*. The neighbor zone is configured to point to the CUCM cluster. All IP addresses or FQDNs are added to the neighbor zone to route calls to the necessary CUCM Subscriber servers in the cluster. Once the neighbor zone is built, several dial plan components route calls to CUCM. The dial plan components include transforms, admin policies, FindMe, and search rules. All these components are addressed in this chapter.

In Figure 13-1, endpoints in CUCM and in Cisco VCS use the same domain (hq.cisco.com). The CUCM server has extensions 2001 and 2002 in its routing table as well as Uniform Resource Identifiers (URIs) for jwhite@hq.cisco.com and jdoe@ hq.cisco.com. The Cisco VCS has URIs 1000@hq.cisco.com and mgrey@hq.cisco.com in its routing table. If the destination extension or URI of a call that is placed to an endpoint on both systems with the hq.cisco.com suffix is not found in the local routing table, the call is routed to the other system over the SIP trunk and neighbor zone.

Call Flow Between CUCM and Cisco VCS

This section describes the call flow for calls between CUCM and Cisco VCS. The call flow process is described in Figure 13-2.

Call flow from Cisco Unified Communications Manager

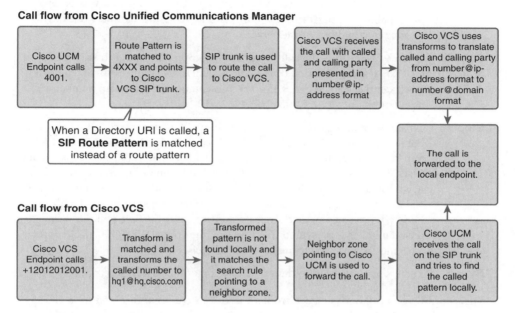

Figure 13-2 *Call Flow Between CUCM and Cisco VCS*

The call flow for a call that is placed from a CUCM registered endpoint is as follows:

Step 1.   An endpoint that is registered with CUCM dials extension 4001. The endpoint that is placing the call is registered to CUCM using either Skinny Client Control Protocol (SCCP) or SIP; the call routing is agnostic of the underlying protocol of the IP phone or video endpoint.

Step 2.   Inside Cisco CUCM, a route pattern defined as 4XXX is matched and refers to a route list, route group, and SIP trunk that points to Cisco VCS.

Note   If the endpoint dialed a URI such as mgrey@hq.cisco.com, a SIP route pattern that matches the domain or host portion of the called URI must be matched. The SIP route pattern then refers to a route list > route group > SIP trunk that points to Cisco VCS.

Note   Route patterns and SIP route patterns can refer directly to a specific SIP trunk or they can use a route list and route group construct to refer to one or more SIP trunks. It is best practice to point the route pattern or SIP route pattern to a proper route list > route group > and SIP trunk to VCS control.

Step 3. The SIP trunk that points to Cisco VCS sends the call setup message (SIP INVITE) to Cisco VCS. Examine the SIP INVITE messages by using Cisco Real Time Monitoring Tool (RTMT) or performing a Wireshark Packet Capture (PCAP) trace either by sniffing the wire or performing this capture directly inside Cisco VCS control or CUCM. Both servers provide the capability of performing a Wireshark PCAP capture of network traffic flowing into and out of the server.

Step 4. Cisco VCS receives the call through a neighbor zone connecting to Cisco CUCM. The presentation of the calling number is number@ip_of_cucm (for example, 2001@10.1.5.15). The presentation of the called number is number@ip_of_vcs (for example, 4001@10.1.6.19).

Step 5. To allow return of the call from the Cisco VCS endpoint to CUCM in the instance of a missed call or redial situation, a transform in Cisco VCS translates the calling number to a pattern with a domain (2001@hq.cisco.com). Cisco VCS will route calls to any unknown domain of hq.cisco.com to the neighbor zone connecting to CUCM. A second transform is also used for the called pattern. The called pattern must be transformed to the URI address (4001@hq.cisco.com) that is registered with Cisco VCS. If you recall the dialed number is presented to VCS using the format 4001@10.1.6.19 format, VCS must change the format to 4001@hq.cisco.com.

Step 6. The call is forwarded to the called endpoint (4001@hq.cisco.com).

The call flow for a call that is placed from a Cisco VCS endpoint destined for a CUCM registered endpoint is as follows:

Step 1. An endpoint registered with Cisco VCS makes a call to +12012012001. The user can dial the number or the user can dial the number@domain format.

Step 2. The called number matches a transform that translates this number to the format number@domain (+12012012001@hq.cisco.com).

Step 3. Because the translated pattern is not found on a locally registered endpoint, Cisco VCS tries to find a search rule. A search rule is matched and it points to a neighbor zone.

Step 4. The neighbor zone points to CUCM and a SIP INVITE message is sent to CUCM.

Step 5. CUCM receives the call over the SIP trunk that points to Cisco VCS. CUCM tries to find the called number on a registered endpoint. It is important to note that on the SIP trunk in CUCM, the incoming calling search space (CSS) must allow the call to reach the endpoint. Significant digits or translation patterns can be used to manipulate the digits in CUCM as well.

Step 6. The call is forwarded and delivered to the called endpoint that is registered to CUCM.

Cisco VCS Dial Plan Components

This section describes dial plan components of Cisco VCS. Just like CUCM, there are several components involved with a dial plan on the Cisco VCS control. This section examines each VCS call routing component and explains how it is used.

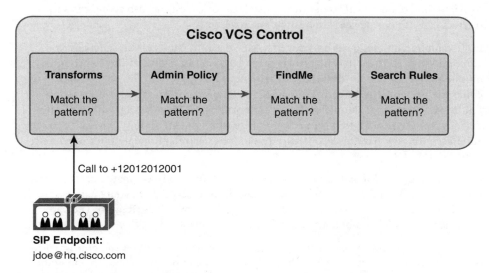

Figure 13-3 *Cisco VCS Dial Plan Overview*

Figure 13-3 illustrates the search flow and call routing path when a call is placed from an endpoint that is registered to Cisco VCS control (VCS-C).

When a call is placed from the endpoint that is registered to Cisco VCS, the VCS call routing engine first tries to match the pattern against any matching transforms. A transform in Cisco VCS is analogous to a translation pattern inside Cisco CUCM. Think of a transform as a black box with an input and output. It takes a number or URI pattern in as input and changes it into a completely separate number or URI as output. If the matching dialed pattern is found inside a configured transform, it is translated based on the rules that are configured in the transform.

If no matching transform pattern or URI exists, Cisco VCS searches for a match in the admin policies. When an admin policy is found, the call can be rejected or allowed, based on the admin policy configurations. An admin policy can be thought of as call policy or call policy script. In complex installations of VCS control or multisite deployments of VCS control, complex call policies and call policy scripts (CPL) may need to exist. These rules can be centralized using CPL scripts. The CPL scripts can exist as Extensible Markup Language (XML) files and be served from any web server. On a simpler scale, an admin policy can be used to admit or deny a specific call in VCS. Think of this as a white list or black list for calls.

If the call is not rejected by an admin policy, Cisco VCS tries to find a match in FindMe. FindMe can be thought of as single number reach or unified mobility for video endpoints registered to VCS control. FindMe lets you control how you are contacted: on any device, at any location, through a single FindMe ID. You can set up a list of locations such as "at home" or "in the office" and associate your devices (endpoints, Movi, mobile phones, and so on) with those locations. You can also set up rules to redirect calls if your devices are busy or unanswered.

For example, you could set up your FindMe so that it calls you on your desktop videophone first. If there is no answer after 10 seconds, it diverts the call to your mobile phone, or if your desktop videophone is busy, it could divert the call to your colleague's telephone instead. Your system administrator can also set up a group FindMe for a team of people such as a support desk. This works by calling all the devices associated with that group when the single FindMe group ID is called. Your administrator may have also configured your external telephone number to route to your FindMe ID.

If no matching FindMe pattern is found, Cisco VCS tries to find a match in the search rules. Search rules are perhaps the closest concept to Cisco CUCM route patterns. Search rules can match extensions, URIs, and regular expressions (RegEx). They are extremely powerful component inside Cisco VCS control.

If a match is found in a search rule, the call is forwarded based on the configuration of the search rule. A call that matches a search rule can be forwarded to a neighbor zone, zone, subzone, or traversal zone. If no match is found in a search rule, the call fails.

Transforms

This section describes how transforms work inside Cisco VCS control. A called number transform inside Cisco VCS control changes a dialed URI or number into another format. Think of a transform as a translation pattern in Cisco CUCM. A VCS transform changes an alias that matches certain criteria into another alias. Presearch rule transforms are applied to an alias before a search is conducted. Transforms can be compared to translation patterns in CUCM. It is important to remember that transforms can be configured as an exact match, prefix match, suffix match, or can match a regular expression. Transforms are configured with a unique priority (range is 1–65534) and are searched in order. Once a match is made, no further transforms are searched.

Admin Policy

Admin policies are also called *call policies*, and are a set of rules that can be entered into Cisco VCS to allow or deny calls from certain endpoints to others. Call policy rules can be created by configuring allow or deny lists on Cisco VCS, or they can be applied from an external policy server. Call policies are applied after transforms but before search rules are applied. Complex call policies can be uploaded directly on the Cisco VCS if they are written in a native CPL script language that resembles XML. Only one

CPL script can be applied directly to a Cisco VCS control server at one time. In complex multiserver environments and deployments, an external call policy script can be used. An example of an external call policy script is a web server that pushes an XML call policy script to a series of VCS servers.

FindMe Feature

This section describes how FindMe works. Figure 13-4 shows an example of the Cisco VCS FindMe feature.

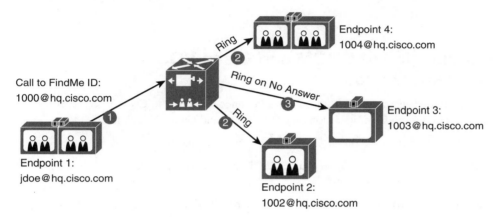

Figure 13-4 *Cisco VCS FindMe*

The FindMe feature provides the ability to specify which endpoints (video and audio-only) should ring when someone calls the FindMe ID of a user. FindMe also allows a user to specify fallback devices that are called if any of the primary devices are busy, and to specify fallback devices that are called if none of the primary devices are answered.

An important feature of FindMe is that the administrator can configure the caller ID that is displayed on the endpoint of a called party as the FindMe ID of the caller, rather than the ID of the endpoint of the caller. This approach allows a user to appear with the same ID, regardless of which device was used to place the call.

Always using the same ID as the caller ID ensures that when a call is returned, the return call is placed to the FindMe ID of the user. As a result, all active FindMe destinations of the called user ring rather than ringing only the endpoint that was used for the original call.

FindMe is similar to Cisco Unified Mobility in CUCM.

The example in Figure 13-4 shows a FindMe ID of 1000@hq.cisco.com. Calls that are placed to this FindMe ID are first sent to endpoints 2 and 4. If the call is not answered on one of these two endpoints, the call is sent to endpoint 3.

Search Rules

Search rules are searched at the end of the call routing process in the Cisco VCS control. A search rule defines how Cisco VCS routes calls (to destination zones) in specific call scenarios. When a search rule is matched, the destination can be modified according to the conditions defined in the search rule. Search rules in Cisco VCS are applied after any FindMe IDs have been discovered. A search rule is analogous to a route pattern in CUCM.

Note   Search rules apply even when the FindME ID has not been matched.

Configuration of CUCM and Cisco VCS Interconnections

This section lists the steps required to interconnect CUCM and Cisco VCS control. The procedure requires configuration on both CUCM and Cisco VCS control.

Configuration required on CUCM includes the following steps:

Step 1.   Go to **System > Security > SIP Trunk Security Profile** to configure a SIP trunk security profile.

Step 2.   Go to **Device > Device Settings > SIP Profile** to configure a SIP profile.

Step 3.   Configure a SIP trunk on CUCM to connect to Cisco VCS control. This can be done by browsing to **Device > Trunk** and selecting trunk type as **SIP Trunk**. This SIP trunk should be configured as any other normal SIP trunk.

Step 4.   Configure the cluster fully qualified domain name by browsing to **System > Enterprise Parameters**.

Step 5.   Create a route group by browsing to **Call Routing > Route/Hunt > Route Group**. Subsequently, create a route list and route pattern. This route pattern will be used to route calls to Cisco VCS.

Step 6.   Assign the previously created SIP trunk to the route group. Subsequently, assign this route group to a route list and to a route pattern that points the calls to Cisco VCS.

Step 7.   Go to **Call Routing > SIP Route Pattern** and create a SIP route pattern on CUCM to route calls to Cisco VCS.

Configuration required on the Cisco VCS control includes the following steps:

Step 1.   Go to **Configuration > Zones** and create a neighbor zone on Cisco VCS to connect to Cisco CUCM, as shown in Figure 13-5.

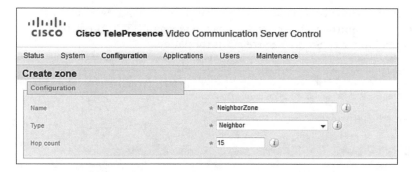

Figure 13-5 *VCS Neighbor Zone Configuration*

Step 2. Go to **Configuration > Dial Plan > Search Rules** and create a search rule on Cisco VCS for calls to CUCM, as shown in Figure 13-6. By default, the LocalZoneMatch rule accepts any calls.

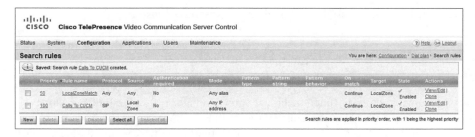

Figure 13-6 *VCS Search Rules Configuration*

Step 3. Configure any required transforms by browsing to **Configuration > Dial Plan > Transformations**.

FindMe Configuration Procedure

This section lists the steps required to implement FindMe. The configuration listed includes the steps to enable the FindMe feature on Cisco VCS.

Note FindMe can be configured on Cisco VCS only or VCS with Cisco TelePresence Management Suite (Cisco TMS). The latter gives the option of using mass provisioning of FindMe accounts. If a VCS only deployment is subsequently migrated to a system that does use Cisco TMS, any FindMe accounts that were configured on the Cisco VCS are deleted and replaced by account data provided by Cisco TMS.

The configuration steps for Cisco TMS and TMS Provisioning Extensions are beyond the scope of this text. Configuration required on Cisco VCS control to enable FindMe is as follows:

Step 1.   Verify the option keys on Cisco VCS control for the FindMe feature by browsing to **Status > Overview**, as depicted in Figure 13-7. Alternatively, you can browse to **Status > System > Information**.

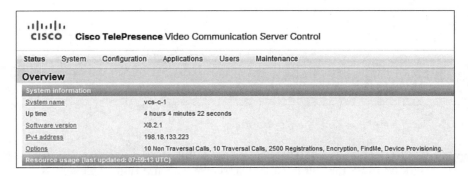

Figure 13-7   *VCS Options Overview*

Step 2.   Configure a cluster name by going to **System > Clustering**.

Note   This name usually is the same as the FQDN name of the VCS.

Step 3.   Enable FindMe functionality, as shown in Figure 13-8. Set **FindMe Mode** to On and define the **Caller ID** (default is FindMe Caller ID). You can optionally restrict user access to create their own devices, and define a device creation message that users will receive upon creating their devices.

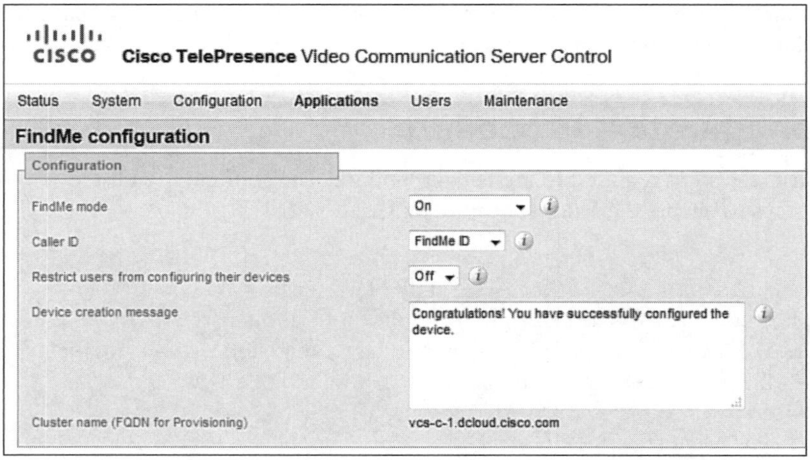

Figure 13-8   *VCS FindMe Configuration*

Step 4.   Go to **Users > FindMe Accounts** and create a new user account. Define the username (used to login to account), display name (used for endpoint phonebook), phone number (E.164 number), FindMe ID (unique alias for the user), principal device address (initial device ID of FindMe), password, and FindMe type (user or group), as shown in Figure 13-9. Repeat this step for as many user accounts as you want to create.

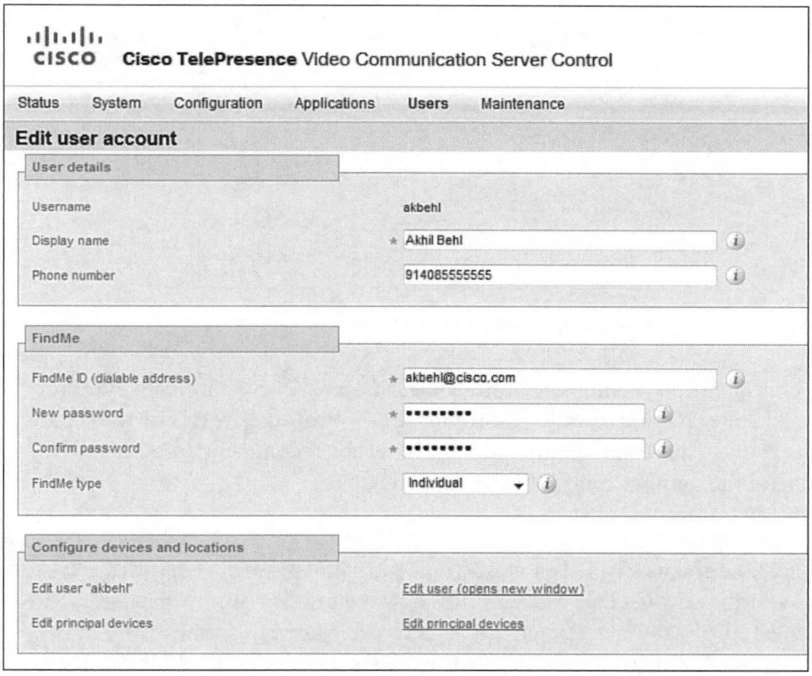

Figure 13-9   *FindMe User Account Creation*

Note   Additional FindMe devices can be added by clicking the Edit User link.

Step 5.   Set up the user authentication for FindMe to Local (on Cisco VCS) or Remote (via LDAP) by browsing to **Users > LDAP Configuration**, as shown in Figure 13-10.

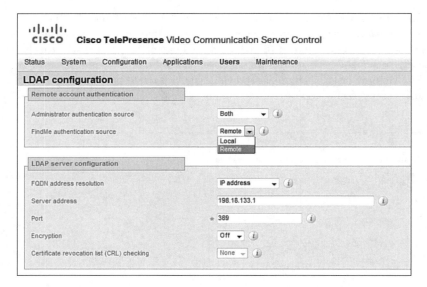

Figure 13-10 *FindMe User Authentication Method*

The FindMe configuration is complete at this time on Cisco VCS. Optionally, you can set up the H.323 caller ID prefix under **Configuration > Protocols > H.323** to include or exclude the gateway prefix. Excluding the prefix is not recommended because the user has to amend the number every time while calling back.

Note If Cisco VCS interworks an E.164 H.323 call, it will create a caller ID with a domain set to IP address of VCS that carried out the interworking. Appropriate search rules should be created to handle routing of such calls. Alternatively, a transformation can be implemented that converts number@IP-of-VCS into number@Local-Sip-Domain.

Summary

The following key points were discussed in this chapter:

- CUCM and Cisco VCS control are connected together using a SIP trunk across an IP network.

- The Cisco VCS dial plan consists of transforms, admin policies, FindMe, and search rules.

- For CUCM to accept calls with the domain in the directory URI, the cluster fully qualified domain name must be configured.

- The configuration of the caller ID pattern is mandatory for normal operation of the FindMe feature.

References

For additional information, refer to the following:

Cisco Systems, Inc. Cisco Collaboration Systems 10.x Solution Reference Network Designs (SRND), May 2014. http://www.cisco.com/c/en/us/td/docs/voice_ip_comm/ cucm/srnd/collab10/collab10.html

Cisco Systems, Inc. Cisco Collaboration Systems 11.x Solution Reference Network Designs (SRND), July 2015. http://www.cisco.com/c/en/us/td/docs/voice_ip_comm/ cucm/srnd/collab11/collab11.html

Cisco Systems, Inc. Features and Services Guide for Cisco Unified Communications Manager, Release 10.0(1), May 2014. http://www.cisco.com/c/en/us/td/docs/ voice_ip_comm/cucm/admin/10_0_1/ccmfeat/CUCM_BK_F3AC1C0F_00_cucm-features-services-guide-100/CUCM_BK_F3AC1C0F_00_cucm-features-services-guide-100_chapter_0110011.html

Cisco Systems, Inc. Video Communication Server (VCS) Configuration Guides, Oct 2015. http://www.cisco.com/c/en/us/support/unified-communications/telepresence-video-communication-server-vcs/products-installation-and-configuration-guides-list.html

Cisco Systems, Inc. Cisco TelePresence Management Suite Provisioning Extension, October 2014. http://www.cisco.com/c/dam/en/us/td/docs/telepresence/infrastructure/ tmspe/install_guide/Cisco_TMSPE_Deployment_Guide_1-0.pdf

Review Questions

Use these questions to review what you have learned in this chapter. The answers appear in Appendix A, "Answers Appendix."

1. **Which mechanism interconnects a Cisco VCS control and Cisco CUCM?**

   a. PIMG

   b. ACS server

   c. MGCP gateway

   d. H.323 gatekeeper controlled trunk

   e. H.323 non-gatekeeper-controlled trunk

   f. SIP trunk

2. **Which component inside Cisco VCS control most closely resembles a translation pattern inside Cisco CUCM?**

   a. Admin policy

   b. Transform

    c. Transformation

    d. Translation

    e. Neighbor zone

3. **Which type of zone is configured on the Cisco VCS to connect and communicate with Cisco CUCM?**

    a. Traversal zone

    b. Subzone

    c. Default zone

    d. Zone

    e. Neighbor zone

    f. No-fly zone

4. **Which ways can a transform match a dialed extension or URI? (Choose four.)**

    a. Non match

    b. Exact match

    c. Prefix

    d. Translation profile

    e. Regular expression

    f. Transformation

    g. Suffix

5. **Transforms are configured with a priority value. What is the maximum priority value allowed inside Cisco VCS control?**

    a. 64535

    b. 65534

    c. 65531

    d. 65533

    e. 65536

    f. 32768

    g. 65532

6. True or false: More than one call policy script can be uploaded to a Cisco VCS control at one time?

   a. True

   b. False

7. Which type of SIP message is exchanged for a new call arriving from Cisco VCS to Cisco CUCM?

   a. SIP REFER

   b. SIP OOB

   c. SIP ACK

   d. SIP BYE

   e. SIP TRYING

   f. SIP INVITE

Chapter 14

Cisco Unified Communications Mobile and Remote Access

Today with an "on-the-go workforce," remote users and mobile workers are increasing many fold. Moreover, organizations are encouraging on-the-go relationships both internally and with customers, which leads to the notion of collaboration everywhere and anywhere. Cisco Mobile Remote Access (MRA) is a feature of Collaboration Edge that enables remote/mobile workers access to Cisco Jabber without the use of a virtual private network (VPN) connection. This in turn enables the remote (hereafter also referred to as mobile) user to leverage enterprise-grade voice, video, instant messaging (IM), and voice messaging functions.

Upon completing this chapter, you will be able to meet the following objectives:

- Define Cisco Unified Communications mobile and remote access

- Define Cisco Unified Communications mobile and remote access components

- Define Cisco Unified Communications mobile and remote access operations

- Define Cisco Unified Communications mobile and remote access configuration and troubleshooting

Cisco Mobile Remote Access Overview

As discussed earlier, Mobile and Remote Access (MRA) is a feature of Collaboration Edge that enables remote individuals access to Enterprise Collaboration services via Cisco Jabber without the use of a VPN connection. This is often referred to as VPN-less Jabber leveraging Transport Layer Security (TLS). This feature is available on both the Video Communications Server (VCS) and Expressway series servers. In MRA, the client is able to build a secure connection back to the enterprise (corporate) network via Domain Name System service records (DNS SRV), as shown in Figure 14-1.

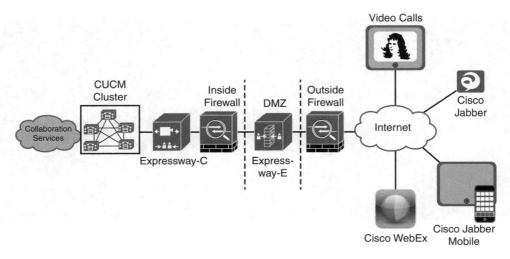

Figure 14-1   *Cisco MRA Solution Overview*

Note   Business applications within the enterprise such as corporate e-mail and intranet are not accessible, and connections to the Internet are expected to be made directly from the device than through the enterprise. Cisco MRA solution gives access to remote clients to enterprise wide Cisco Collaboration services, as shown in Figure 14-1.

As shown in Figure 14-1, Jabber clients and video endpoints from various locations (public Internet) connect via the Expressway-E (Edge). When the client reaches the Expressway-E, the traffic is sent to the Expressway-C via a secure connection and then over to the various collaboration services, whereby the client registers and is able to get services such as the following:

- IM
- Voice messaging
- Video
- Voice calls

Cisco Expressway solution delivers the following key capabilities enabling the Expressway Mobile and MRA feature:

- Extensible Communications Platform (XCP) router for Extensible Messaging and Presence Protocol (XMPP) traffic
- HTTPS reverse proxy
- Proxy Session Initiation Protocol (SIP) registrations to CUCM

Figure 14-2 depicts the traffic flows pertinent to the said protocols in Cisco Expressway solution.

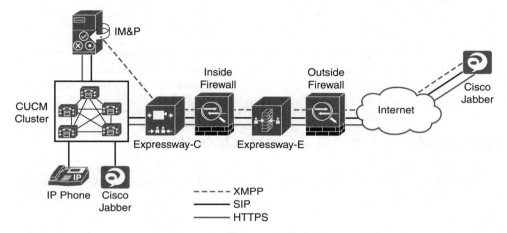

Figure 14-2 *Cisco Expressway Solution: Protocol Flows*

Cisco MRA solution consists of various components that are described in the following section.

Cisco Mobile Remote Access Components

The following are functions of various elements in a Cisco MRA solution, from external/outside to internal/inside (in that order):

- **Remote/external clients:** Jabber being the external client, and connection origins can be anywhere from a mobile device on the go to a home user connected to Wi-Fi to a user in airport transit in a cafe.

- **Outside firewall:** This is a typical perimeter firewall such as a Cisco ASA55XX series, which demarcates Outside (unsecure zone) from DMZ (partially secure zone). A number of ports are required from the external network into the DMZ (described in the next section).

- **Expressway-E (Expressway-Edge):** The Expressway-E is a SIP proxy for devices that are located outside the internal network (for example, Jabber clients across the Internet and [business] partners making or receiving calls from the enterprise network).

- **Inside firewall:** This is a typical corporate (internal) firewall and can be yet again a Cisco ASA55XX series that demarcates Inside (secure zone) from DMZ (partially secure zone). A number of ports are required from the inside network into the DMZ, as described in the next section.

- **Expressway-C (Expressway-Core):** Expressway-C connects through internal firewall with Expressway-E and acts as a proxy for the remote client such that remote clients communicate with Expressway-C and on their behalf it communicates with CUCM and other Collaboration services (for example, voicemail and Instant Messaging & Presence (IM&P).

- **Cisco Unified Communications Manager (and other Collaboration services):** CUCM allows the registration of Cisco Jabber client as Cisco Unified Client Services Framework (CSF). This gives the remote clients an ability to leverage CUCM as call control and other services of IM&P and voice messaging.

Cisco Mobile Remote Access Operation

The Cisco MRA solution has various components that work in synergy to deliver enriched user experience. This section examines Cisco MRA solution operation and how the Expressway solution works.

Cisco Mobile Remote Access Firewall Traversal

This section addresses the Expressway firewall traversal concepts that explain the traffic flow from outside of enterprise to the core systems. Figure 14-3 illustrates the call flow across the firewalls, Expressway-E, and Expressway-C.

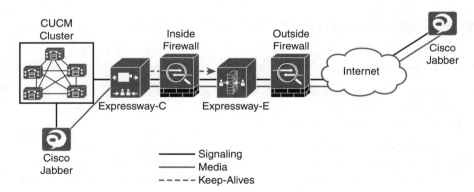

Figure 14-3 *Cisco MRA Solution: Firewall Traversal*

> **Note**   Expressway-E is the traversal server installed in DMZ. Expressway-C is the traversal client installed inside the enterprise network.

These steps are common for inbound or outbound calls:

Step 1.   Expressway-C initiates traversal connections outbound through the firewall to specific ports on Expressway-E with secure login credentials.

Step 2.   Once the connection has been established, Expressway-C sends keep-alive packets to Expressway-E to maintain the connection.

These steps are applicable to inbound calls:

Step 3a.   When Expressway-E receives an incoming call (from outside of enterprise [for example, a Jabber client on the go]), it sends an incoming call request to Expressway-C.

Step 4a.   Expressway-C then routes the call to CUCM to reach the called user or endpoint (video endpoint or phone). The call is established and media traverses the firewall securely over an existing traversal connection.

These steps are applicable to outbound calls:

Step 3b.   For outbound calls (from inside corporate network), CUCM sends a SIP invite to Jabber with the Expressway-C IP address. (CUCM knows that the Jabber client is registered through Expressway-C as proxy server.)

Step 4b.   Expressway-C forwards SIP invite across the Secure Shell (SSH) tunnel (Unified Communications Traversal Zone) to Expressway-E.

Step 5.   Call is connected with the remote Jabber client.

The following firewall ports are required to be opened on the DMZ and internal firewalls:

- For internal firewall (facing Expressway-C on inside and Expressway-E on outside), no inbound ports are required to be opened/allowed.

- Internal firewall requires the following outbound connections from Expressway-C to Expressway-E:

  - SIP: TCP 7001

  - Traversal media: UDP 36000 to 36011

  - XMPP: TCP 7400

  - HTTPS (tunneled over SSH between C and E): TCP 2222

- External firewall requires the following inbound connections to Expressway:

  - SIP: TCP 5061

  - HTTPS: TCP 8443

  - XMPP: TCP 5222

  - Media: UDP 36012 to 59999

- Cisco MRA firewall traversal

HTTPS Reverse Proxy

The Expressway-E server provides reverse proxy services for clients requesting access to services inside the enterprise. The reverse proxy function acts as an intermediary for servers inside the enterprise. This allows the clients outside the enterprise to be directed to the correct server inside the enterprise based on port number. Reverse proxy also serves to hide the internal IP addresses of the enterprise servers, and functions as a load balancer for redundant servers inside the enterprise.

Expressway-E server listens on TCP 8443 for HTTPS traffic (reverse proxy traffic). This allows inbound authenticated HTTPS requests to the following destinations on the enterprise network:

- All discovered CUCM nodes (for TFTP file download and device registration) TCP 6970 (TFTP file requests) and TCP 8443 (UDS API)

- All discovered IM&P nodes (for presence information) TCP 7400 (XCP router) and TCP 8443

- HTTPS traffic to any additional hosts (must be administratively added to the Expressway-C allow list) such as voicemail servers

Note   The allow list provides a mechanism to support Visual Voicemail access, contact photo retrieval, Jabber custom tabs, and so on.

DNS SRV Setup

Cisco Expressway solution relies heavily on Domain Naming Service (DNS) service records (SRV). This implies that both for internal and external name resolution, Expressway-C and Expressway-E, respectively, have to be set up for DNS records. This section describes the DNS SRV requirements from the Cisco Expressway solution point of view.

Note   In the following examples, the domain name used is fictitious, used purely to explain the DNS SRV concept.

For Expressway-E, external DNS SRV should be set up as shown in Table 14-1. Note that this applies to Public DNS SRV.

Table 14-1   *Cisco Expressway-E DNS SRV Requirements*

Expressway-E

Domain	Service	Protocol	Priority	Weight	Port	Target Host
dcloud.cisco.com	collab-edge	TLS	10	10	8443	expressway-e. dcloud.cisco.com

For Expressway-C, internal DNS should be configured as shown in Table 14-2 pertinent to DNS SRV.

Table 14-2 *Cisco Expressway-C DNS SRV Requirements*

Expressway-C

Domain	Service	Protocol	Priority	Weight	Port	Target Host
dcloud.cisco.com	cisco-uds	TCP	10	10	8443	cucm10.dcloud.cisco.com
dcloud.cisco.com	cuplogin	TCP	10	10	8443	cup1.dcloud.cisco.com

Keep in mind that the SRV records return a fully qualified domain name (FQDN) and not an IP address. To return an IP address upon querying an FQDN, an A record needs to be provisioned. In other words, upon querying an SRV record (for example, _collab-edge._tls.domain), the DNS SRV returns an FQDN, and upon querying that FQDN (with help of A record), an IP address is returned. Table 14-3 illustrates DNS SRV and DNS A record relation for Cisco Expressway servers.

Table 14-3 *Cisco Expressway-E DNS Resolution*

Type	Entry	Resolves To
SRV record	_collab-edge._tls.dcloud.cisco.com	Expressway-E.dcloud.cisco.com
A record	Expressway-E.dcloud.cisco.com	IP address of Expressway-E

Table 14-4 illustrates the DNS resolution for Expressway-C.

Table 14-4 *Cisco Expressway-C DNS Resolution*

Type	Entry	Resolves To
SRV record	_cisco-uds._tcp.dcloud.cisco.com	cucm10.dcloud.cisco.com
A record	cucm10.dcloud.cisco.com	IP address of CUCM
SRV record	_cuplogin._tcp.dcloud.cisco.com	cup1.dcloud.cisco.com
A record	cup1.dcloud.cisco.com	IP address of IM and Presence

Registering Remote Jabber Client with CUCM

This section describes the registration process for remote Jabber client (outside of enterprise perimeter) with the corporate CUCM cluster. Figure 14-4 illustrates the registration process for Jabber clients outside of corporate perimeter.

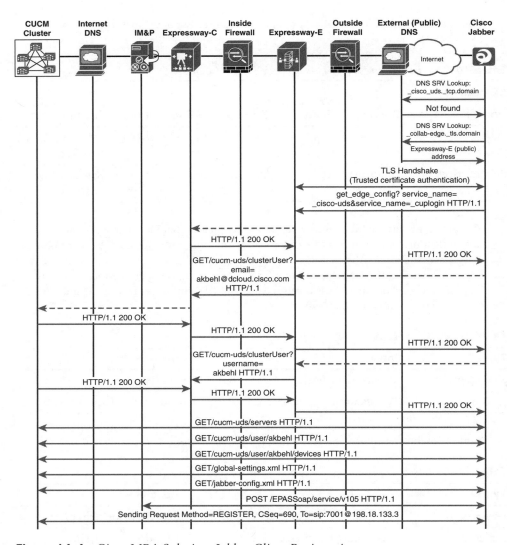

Figure 14-4   *Cisco MRA Solution: Jabber Client Registration*

The following steps cover the registration process:

Step 1.   Jabber client initiates connection to the corporate network when the user signs in. Initially, the connection is attempted as TCP to the external address; however, upon DNS SRV lookup, it is redirected as a Transport Layer Security (TLS) connection to Expressway-E's public-facing name.

Note   Expressway-E server keeps listening on TCP 8443 for HTTPS traffic.

Step 2. Once the trusted (TLS) handshake process is complete, the client requests cuplogin service and registration with CUCM. This request is intercepted by Expressway-E and sent over to Expressway-C, which acts as a proxy for the client (CUCM and IM&P for external clients) and the client for collaboration services.

Step 3. Expressway-C confirms the connection by sending 200 OK back to Expressway-E; the latter in turn returns 200 OK to the client.

Step 4. Expressway-C reaches out to CUCM using the information (user-email) shared by Expressway-E, and upon successful connection and verifying the credentials of the user, gets 200 OK from CUCM and sends 200 OK to Expressway-E, which in turn sends 200 OK to the Jabber client.

> **Note** Basic MRA configuration allows inbound authenticated Hypertext Transfer Protocol Secure (HTTPS) requests to the CUCM and IM&P nodes discovered in corporate network. For all discovered CUCM nodes TCP 6970 (for Trivial File Transfer Protocol [TFTP] file requests) and TCP 8443 (for User Date Service [UDS] application programming interface [API]) are used. For all discovered IM&P nodes, TCP 7400 (for XCP router) and TCP 8443 are used. For any other Collaboration services, HTTPS traffic to any additional hosts needs to be administratively added to the Expressway-C allow list that provides a mechanism to support visual voicemail access, contact photo retrieval, Jabber custom tabs, and so on.

Step 5. The same process as Step 4 is repeated for username authentication between Expressway-C and CUCM and Expressway-E, and the client is sent 200 OK.

Step 6. Upon getting 200 OK, the client communicates with CUCM and gets the client-specific settings. Also, the client interacts with IM&P to get access to Presence and IM services (SOAP API).

Step 7. The client sends out the SIP register request to CUCM, and then the usual SIP UA registration process continues. The client is registered with CUCM for call control and associated with Extensible Messaging and Presence Protocol (XMPP) router (IM&P) for IM and Presence services.

Cisco Unified Communications Mobile and Remote Access Configuration

As discussed in the earlier section about Cisco Unified Communications MRA functionality, additional elements must be configured:

- Cisco Unified Communications Manager

- Cisco Unified IM and Presence

- Expressway series

The sections that follow discuss the respective component configuration procedure.

CUCM Configuration for Cisco Unified Communications MRA

To configure CUCM to support Cisco Unified Communications MRA, follow these steps:

Note   If CUCM and Cisco Unified IM, Presence integration, and Jabber clients are configured, skip the CUCM and IM and Presence configuration sections, and proceed to the Expressway configuration section.

Step 1.   Go to **Device > Trunk** and click **Add New**. Add a SIP trunk. Configure the SIP trunk to Cisco Unified IM and Presence server. Configure the SIP trunk. Provide a name (this is used to publish the SIP trunk to IM and Presence server [in this case, IMPTrunk]). Enter the IP address of IM and Presence server, select the SIP trunk security profile as Non Secure, and set the SIP profile to Standard SIP Profile, as shown in Figure 14-5.

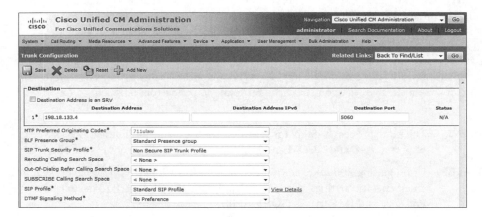

Figure 14-5   *IM and Presence SIP Trunk Configuration*

Step 2.   Go to **System > Enterprise Parameters**. Browse to Clusterwide Domain Configuration and set the organization top-level domain (OTLD) and cluster fully qualified domain name to their respective values (in this case, dcloud. cisco.com and CUCM10.dcloud.cisco.com, respectively), as depicted in Figure 14-6.

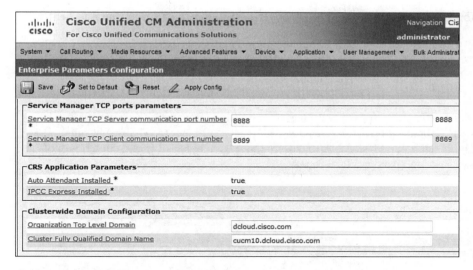

Figure 14-6 *CUCM Enterprise Parameters: Clusterwide in figure Domain Configuration*

Step 3. Go to **System > Service Parameters** and select **CUCM > CallManager Service**. Browse to IM and Presence Publish Trunk and, as shown in Figure 14-7, select the IM and Presence trunk (in this case IMPTrunk) that was previously created. CUCM will use this trunk to send publish messages that pertain to presence activities.

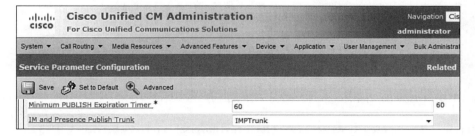

Figure 14-7 *Configuration of Publish IM and Presence Trunk*

Step 4. Go to **Device > Phone** and add a new phone type **Cisco Unified Client Services Framework**. Give the CSF a name and configure a (existing) user as owner as shown in Figure 14-8. Make sure the **Device Security Profile** is set to **Standard Non-Secure Profile** and **SIP Profile** is set **to Standard SIP Profile**. Configure Jabber DN.

> **Note** To enable CSF for video calling, set the **Video Calling** option under Product Specific Configuration Layout to **Enabled** and make sure that the **Override Common Settings** check box is checked. Also, ensure that region assigned to the device pool of the CSF has video bandwidth allocated.

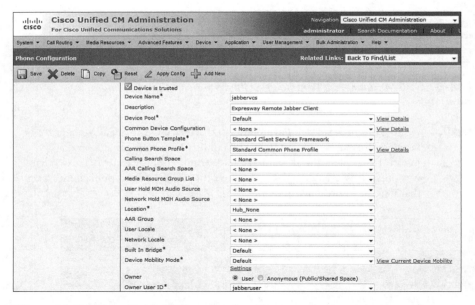

Figure 14-8 *Cisco Unified Client Services Framework*

Step 5. Configure UC services for Directory (Universal Directory Services [UDS]) and IM and Presence. Figure 14-9 illustrates UDS configuration. This allows the Jabber users to access the directory contacts and use the IM and Presence service via the mobile/PC client. It is important to note that UDS is the contact source used for Expressway MRA.

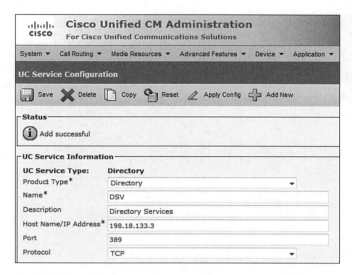

Figure 14-9 *Unified Directory Service*

Step 6.   Configure the service profile in Cisco Unified Communications Manager
for Instant Messaging and Presence (IM&P) and UDS, as shown in
Figure 14-10.

Note   When using Jabber as an IM&P client, an administrator can configure
Enhanced Directory Integration (EDI), Basic Directory Integration (BDI), and Cisco
Unified Communications Manager User Data Service (UDS) as contact sources.
Now, whatever the CUCM integration is with Lightweight Directory Access
Protocol (LDAP), the user Presence and contact information will follow the type of
service used.

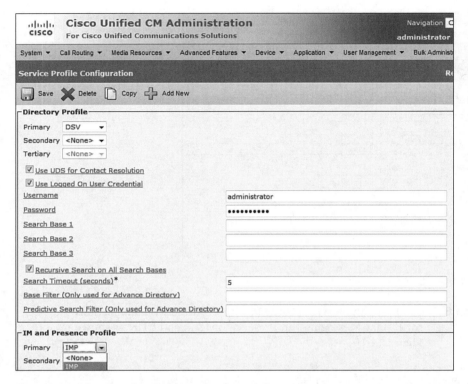

Figure 14-10   *Service Profile*

Step 7.   Enable user for Unified CM IM and Presence. Go to **User Management > End User** and ensure that the Cisco Jabber client configured is associated with the user. Enable the user for IM&P, as shown in Figure 14-11.

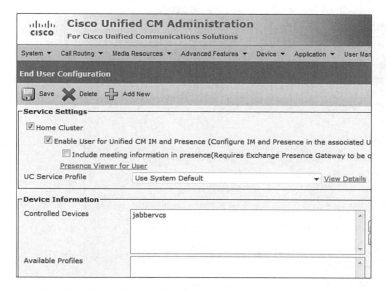

Figure 14-11 *End User Configuration*

> **Note**   Enable the user to allow control of CTI from CTI on the primary extension. Ensure that the user gets the necessary permissions from groups and roles (as standard CTI end user). For CTI, the extension can be controlled via the soft client (Jabber in this case).

At this time, the CUCM configuration is complete, and the next section covers the IM&P configuration for Jabber for Cisco MRA.

IM&P Configuration for Cisco Unified Communications MRA

The following steps explain the IM&P configuration required to support Cisco Unified Communications MRA:

Step 1.   Go to **System > Service Parameters** and configure the CUCM IM and Presence - Cisco SIP proxy service, as illustrated in Figure 14-12.

> **Note**   Cisco SIP proxy or Cisco Unified SIP Proxy (CUSP) simplifies call routing within multi-entity SIP networks using call routing rules to improve control and flexibility of the overall network. While Cisco IM&P server provides basic SIP proxy features, CUSP is a full-featured SIP Proxy solution for enterprise-grade networks.

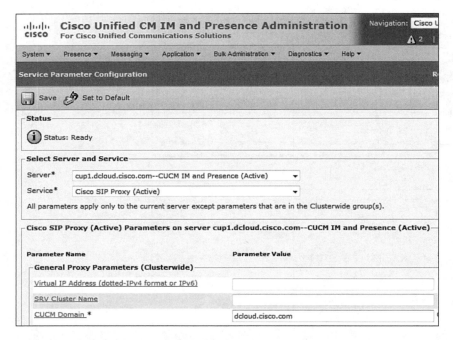

Figure 14-12 *Cisco IM&P Service Parameter Configuration*

> **Step 2.**   Go to **Presence > Settings > Standard Configuration** and select the IMP
> trunk created earlier, as shown in Figure 14-13.

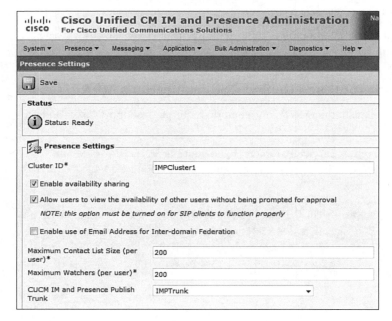

Figure 14-13 *Cisco IM&P Trunk Configuration*

Step 3.   Go to **Presence > Gateway** and configure the Presence gateway (CUCM), as shown in Figure 14-14.

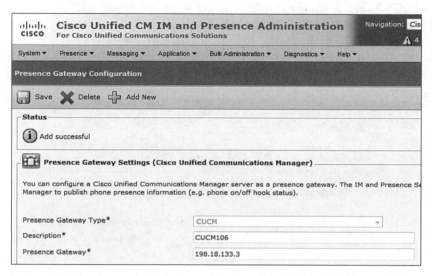

Figure 14-14 *Cisco IM&P Presence Gateway Configuration*

Step 4.   Go to **Application > Client Settings** and set up the TFTP server (CUCM), as shown in Figure 14-15. This TFTP server is used for hosting the client configuration settings (XML file), similar to a Cisco Unified IP phone's SEPCnfXML file. When users sign in, the client retrieves the XML file from the TFTP server and applies the configuration.

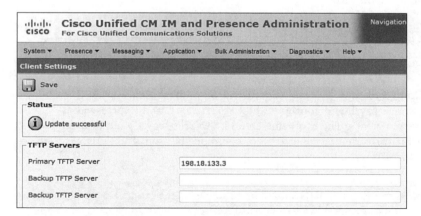

Figure 14-15 *Cisco IM&P Client Settings*

Check the IM&P status from diagnostics for any issues or configuration errors from the diagnostic dashboard, as shown in Figure 14-16.

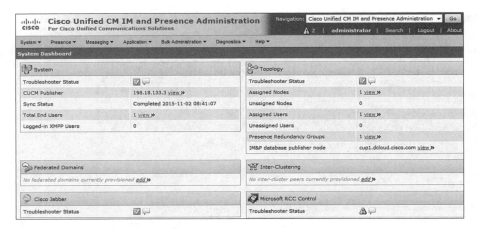

Figure 14-16 *Cisco IM&P Diagnostic Dashboard*

At this time, the IM&P server is set up for IM and Presence functions and integrated with CUCM to support Jabber clients both locally and Expressway-based Jabber connections.

The next section covers the Expressway (Expressway-C and Expressway-E) configuration for Cisco MRA.

Cisco Expressway (Expressway-C and Expressway-E) Configuration for Cisco Unified Communications MRA

The following steps walk you through the configuration for Cisco Expressway-C and Expressway-E to support MRA.

> **Note**   Because most of the configuration steps are similar for Expressway-C and Expressway-E, Expressway-C has been used as an example where applicable.

Step 1.   Go to **System > System Administration** on both Expressway-C and Expressway-E and configure the Expressway-C and Expressway-E system names. The hostname could be Expressway-C, and domain name (provided in the next step) could be abc.com. The FQDN is then Expressway-C.abc.com.

> **Note**   For Expressway-E, the FQDN should be routable via the public DNS and appears in the Alternate Name for Expressway-E certificate. Also, the domain name is reachable from outside of the corporate network.

Step 2.   Go to **System > DNS** and configure the DNS for Expressway-C and Expressway-E. Ensure that for Expressway-C corporate DNS is configured (because it is accessible from inside of the corporate network for locally present devices), whereas for Expressway-E Public DNS is configured (because clients from the public network or the Internet and the other entities of solution will be accessing it using the FQDN). Also confirm that the domain name (for example, abc.local) is accessible via both internal (corporate) and external (public) DNS resolution. Figure 14-17 illustrates the DNS configuration for Expressway-C.

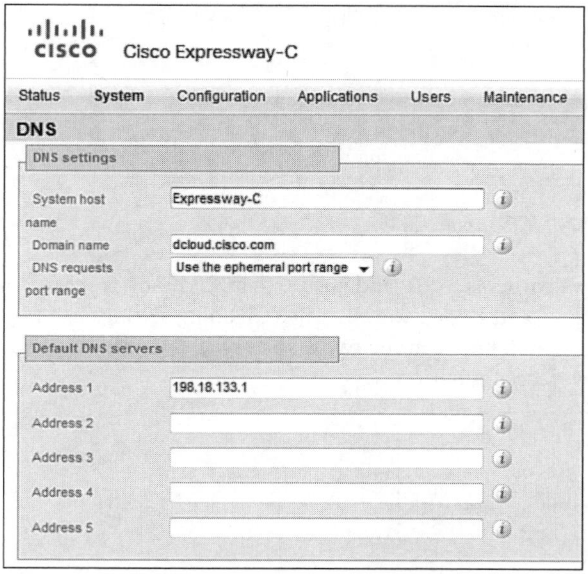

Figure 14-17   *Cisco Expressway-C DNS Configuration*

Step 3.   Go to **System > Time** and configure NTP servers on both Expressway-C and Expressway-E. This ensures that the servers are in sync with other collaboration elements. This is required for TLS handshake and certificate authentication with CUCM and IM and Presence. Figure 14-18 shows the NTP configuration for Expressway-C.

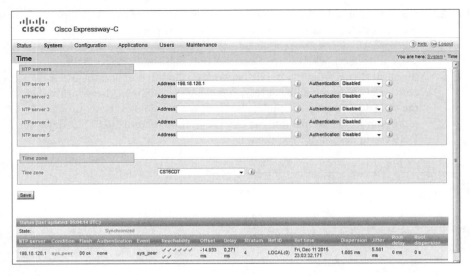

Figure 14-18   *Cisco Expressway-C NTP Configuration*

> **Step 4.**   Go to **Configuration > Protocols > SIP** and ensure that SIP mode is set to
> **On** on both Expressway-C and Expressway-E. The remaining settings can be
> left at default but must match between the two Expressways. Figure 14-19
> depicts the Expressway-C configuration.

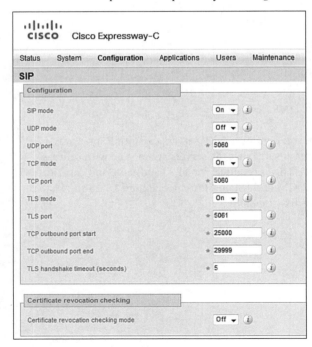

Figure 14-19   *Cisco Expressway-C Protocol Configuration*

Step 5.   On Expressway-C, go to **Configuration > Unified CM servers** and configure the CUCM Publisher address with CUCM administrator credentials. Figure 14-20 illustrates Expressway-C CUCM configuration.

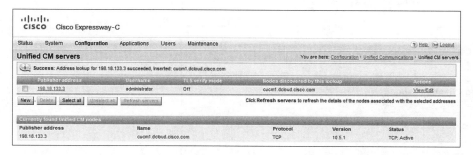

Figure 14-20   *Cisco Expressway-C CUCM Configuration*

Step 6.   On Expressway-C, go to **Configuration > IM and Presence Service nodes** and configure the IM&P Publisher address with IM&P administrator credentials (similar to Step 5).

Step 7.   On Expressway-C go to **Configuration > Domains** and configure the domain (same as configured in Step 2) and set (under supported services) **SIP Registrations and Provisioning on Unified CM** and **IM and Presence Service** to **On** so that the respective services are available to the clients connecting from the Internet. Figure 14-21 depicts the domain configuration for Expressway-C.

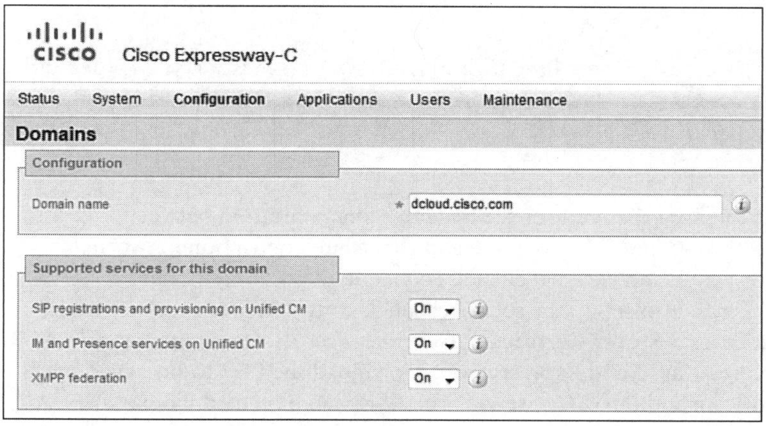

Figure 14-21   *Cisco Expressway-C Domain Configuration*

Step 8.   Enable MRA by browsing to **Configuration > Unified Communications** and setting Unified Communications mode to **Mobile and Remote Access** on both Expressway-C and Expressway-E. Figure 14-22 shows the configuration for Expressway-C.

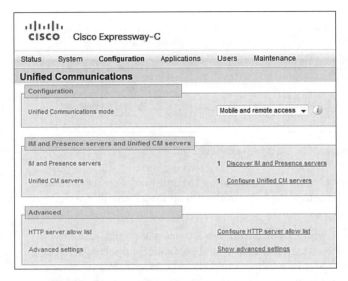

Figure 14-22   *Cisco Expressway-C MRA Configuration*

Step 9.   Go to **Configuration > Dial Plan > Search Rules**, and you should be able to see the search rules built automatically based on the configuration thus far.

Note   Search rules are created with CEtcp if the TLS verify option is set to Off (default) and with CEtls if the option is set to On during configuration for CUCM and IM&P nodes in Steps 5 and 6.

Step 10.   Certificates are required for secure (TLS) communication between Expressway-C and Expressway-E and for secure connection between Expressway-E and external clients. To enable Expressway-C and Expressway-E to set up the secure channel, go to **Maintenance > Security Certificates > Server Certificate**. Generate a certificate signing request (CSR) on Expressway-C and Expressway-E. Download the CSR to be signed by a certificate authority (CA) server. The CSRs can be signed by external CA (such as Verisign or GeoTrust) or an internal CA (for example, corporate MS CA). This would enable the certificates to be signed and authenticated by a single source of truth.

> **Note** For Expressway-E, it is recommended that the certificates be signed by a public CA because every device (for example, mobile phone, laptop) has the public CA provider's root CA certificate in its certificate trust store. For Expressway-C, it can be internal or external CA. However, maintaining public CA signed certificates throughout an organization is a leading practice.

Step 11. Upload the respective signed identity certificates (signed CSRs) to Expressway-C and Expressway-E. Go to **Maintenance > Server Certificate** and upload the signed certificates. Once the identity certificates are uploaded, go to **Maintenance > Trusted-CA Certificate** and upload the root CA certificate on both Expressway-C and Expressway-E. Figure 14-23 illustrates Expressway-C with valid identity certificate.

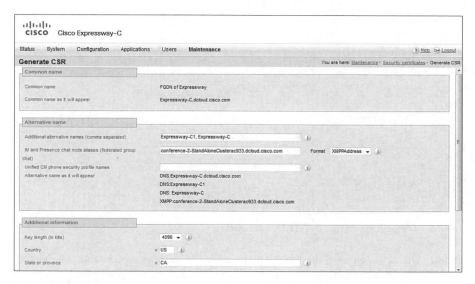

Figure 14-23 *Cisco Expressway-C Identity Certificate*

Step 12. Go to **Configuration > Zones** and configure Traversal Client on Expressway-C. This is required for Expressway-C to connect via secure tunnel to Expressway-E. Provide the name, connection credentials, port (TCP 7001) and peer address (DNS name of Expressway-E), as shown in Figure 14-24.

Step 13. Similar to Step 12, configure Expressway-E and provide TLS verify subject name (same as peer address of Expressway-C). Go to **Configuration > Zones** and check traversal zone status to Expressway-E on Expressway-C.

> **Note**   You can verify the SSH (secure) tunnel status. Go to **Status > Unified Communications > View SSH Tunnel Status** and ensure that the tunnel status shows as Active.

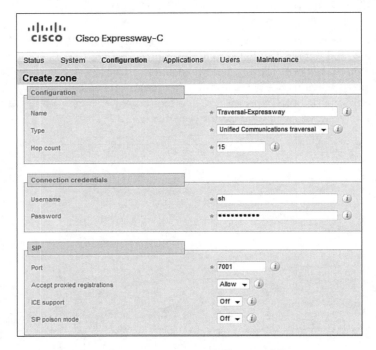

Figure 14-24   *Cisco Expressway-C Traversal Zone*

Step 14.   On a PC, log in to the Jabber client using the username@domain.com (where domain.com is the domain name set on Expressway series servers, IM&P, and CUCM server [that is, the enterprise domain]). Check the status of Jabber on **Status > Unified Communications** and ensure that the client shows up on Expressway-C. The client should show as registered on CUCM under **Device > Phone**.

> **Note**   Upon placing or receiving a call, go to Expressway-C/Expressway-E and browse to **Status > Call Status**. The call should appear as Type Traversal.

Troubleshooting Cisco MRA

The following tools can be used for troubleshooting call issues and system issues for Cisco MRA solution.

CUCM Real Time Monitoring Tool can be used for troubleshooting call and signaling related issues by using the following:

- Call activity

- Session trace log view

- SDL trace

- Called-party tracing

On Expressway series, the network log can be used to isolate issues. The network log is available under **Status > Logs > Network Log**. For call issues, refer to search history. This is accessible via **Status > Search History**.

Cisco Jabber offers the network log available under %user_profile%\AppData\Local\ Cisco\Unified Communications\Jabber\CSF\Logs to troubleshoot issues with Jabber on a PC. Combined with network log on Expressway series, issues can be correlated.

For further information on various tools used with MRA solution (and otherwise in a Cisco Collaboration setup) for troubleshooting, refer to the *Troubleshooting Cisco IP Telephony and Video (CTCOLLAB)* book.

Summary

The following key points were discussed in this chapter:

- The Cisco Unified Communications MRA service offers a flexible solution for remote/mobile workers for accessing on the go corporate IM, voice, video, and voice messaging services.

- Cisco Expressway series offers a sophisticated way to ensure that the clients from outside corporate network are allowed access to only collaboration services yet can leverage and harness the power of real-time interaction anywhere, anytime.

- Cisco Expressway-C and Expressway-E servers help connect Jabber users on the go and remote workers without the use of a dedicated VPN connection, on any device. Instead, the solution uses TLS and proxies the connection from users to CUCM, voicemail, IM&P, and other collaboration services.

References

For additional information, refer to these resources:

Cisco Systems, Inc. Cisco Collaboration Systems 10.x Solution Reference Network Designs (SRND), May 2014. http://www.cisco.com/c/en/us/td/docs/voice_ip_comm/ cucm/srnd/collab10/collab10.html

Cisco Systems, Inc. Cisco Collaboration Systems 11.x Solution Reference Network Designs (SRND), July 2015. http://www.cisco.com/c/en/us/td/docs/voice_ip_comm/ cucm/srnd/collab11/collab11.html

Review Questions

Use these questions to review what you have learned in this chapter. The answers appear in Appendix A, "Answers Appendix."

1. **CUCM and IM&P servers communicate with which Expressway server?**

   a. Expressway-E

   b. VCS Edge

   c. Internal Expressway firewall

   d. Expressway-C

2. **What protocol is supported by Cisco Expressway solution for integration of MRA?**

   a. SIP

   b. H.323

   c. XML

   d. SCCP

3. **Which protocols traverse through Expressway solution? (Choose three.)**

   a. MTP

   b. SIP

   c. XMPP

   d. RTP

4. **How does Expressway-E communicate with Expressway-C for call media and signaling?**

   a. Through the external firewall

   b. Secure SSH tunnel

   c. Through internal DNS

   d. Through CUCM proxy

5. **What Collaboration services are supported via a Cisco MRA solution? (Choose three.)**

   a. Voice

   b. Video

   c. Visual voicemail

   d. Intranet server access

6. **The Cisco MRA solution requires how many Expressway servers and firewalls?**

   a. A typical Cisco MRA solution requires two Expressway (Expressway-E and Expressway-C) servers and two firewalls.

   b. A typical Cisco MRA solution requires at least one Expressway server and two firewalls.

   c. A typical Cisco MRA solution requires two Expressway (Expressway-C and Expressway-D) servers and two firewalls.

   d. A typical Cisco MRA solution requires at least one Expressway server and three firewalls.

7. **The Cisco Expressway solution offers which of the following Collaboration options to remote users? (Choose four.)**

   a. Jabber voice calls

   b. Jabber video calls

   c. WebEx

   d. Jabber for mobile

   e. Native Lync client for mobile

8. Expressway-E and Expressway-C require certificates for secure tunnel creation through the internal firewall and for intercepting secure communication from the remote client. Which of following statements is true?

   a. Expressway-E and Expressway-C can be used with default certificates.

   b. Expressway-C and Expressway-E need to have a corporate (local) certificate issued and uploaded with a root certificate.

   c. Only Expressway-E needs a certificate.

   d. Both Expressway-C and Expressway-E need a certificate. Expressway-E needs a certificate that is issued by a public certificate authority or can be trusted by external devices. Expressway-C can also be assigned a certificate from a public CA, but it can be issued a certificate from a corporate (nonexternal) CA.

9. True or false: Expressway-C can only integrate with CUCM and IMP&P publisher servers.

   a. True

   b. False

Chapter 15

Cisco Inter-Cluster Lookup Service (ILS) and Global Dial Plan Replication (GDPR)

Cisco Collaboration solutions offer state-of-the-art facilities to reach people within and outside of organizations. As the communications networks evolve and expand, it becomes implicit (and sometimes imperative) that device-independent user addressing is available in the Collaboration solutions. As discussed in Chapter 4, "URI-Based Dial Plan for Multisite Deployments," user addressing via Universal Resource Indicator (URI) is becoming the means for identifying everything in a collaboration network. In fact, when a user can reach another user or device by name, there is no need to remember the directory number (DN) of the called entity.

Cisco expects the use of URIs for communication purposes to grow as communication continues to be integrated with data applications. URI dialing and URI click to call are increasingly becoming the norm in almost any large setup. This translates to a user being reachable via a URI/DN/E.164 number (that is, device-independent user addressing).

Upon completing this chapter, you will be able to meet these objectives:

- Describe Inter-Cluster Lookup Service (ILS)
- Describe ILS networking
- Describe ILS networking configuration
- Describe directory URI, enterprise alternate, and +E.164 alternate number exchange
- Describe Global Dial Plan Replication (GDPR)
- Describe GDPR configuration
- Describe global dial plan catalogs

Inter-Cluster Lookup Service Overview

Traditionally, Cisco Unified Communications Manager (CUCM) and other Cisco Collaboration applications have supported DNs to support user dialing within and outside of an enterprise environment. As discussed earlier in Chapter 4, Session Initiation Protocol (SIP) URIs are progressively being used as the enterprise Collaboration solutions continue to evolve and integrate with other applications and video solutions, where URI dialing is the default behavior.

However, the obvious issue with URIs is that they cannot be summarized in the same way as DN ranges. This is primarily because URIs are fully qualified and cannot be summarized (for example, bob@us.cisco.com, jim@eu.cisco.com) in a simple manner. Although it is possible to summarize them to, say, *.cisco.com, the granularity of information is lost. Inter-cluster Lookup Service (ILS) acts as a dynamic URI discovery mechanism. The following are some salient features of ILS:

- ILS runs as a CUCM service and helps distribute locally registered URIs to other clusters, dynamically.

- Cluster to cluster ILS connections can be hub and spoke or full mesh.

- URIs are associated with a cluster SIP route-string (label) (for example, cipt1.cisco.com, cipt2.cisco.com). (SIP route strings identify and distinguish clusters participating in ILS networking.)

- Eases adoption of powerful dial plan concepts such as tail-end hop-off (TEHO).

The next section discusses ILS networking in detail.

ILS Networking Overview

ILS networking is a foundation for exchange of information for SIP URIs. For ILS networking and URI propagation, it is important to consider the following:

- Call controls participating in ILS network form a hub and spoke topology.

- Each call control is hub or spoke. All hubs need to be full mesh.

- Each call control keeps a local copy of all URIs advertised by all other nodes in the ILS network.

- URIs and their route string are stored locally; partial ILS file updates are supported.

- Each call control periodically pulls in all changes in all URI catalogs advertised into ILS from directly connected call controls (interval 1–1440 minutes).

- URI catalog updates propagate through the ILS network hop by hop (maximum diameter is three hops).

Figure 15-1 illustrates ILS networking with call controls acting as hubs or spokes.

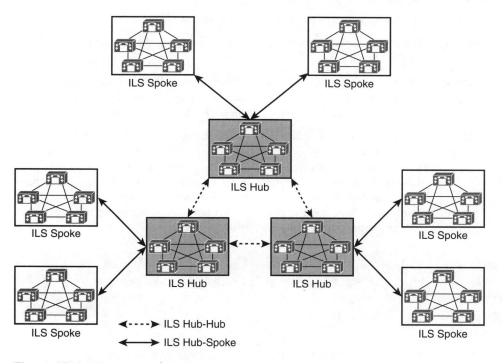

Figure 15-1 *ILS Networking Overview*

As shown in Figure 15-1, ILS networking establishes a replication relationship between various clusters such that the spokes in a logical/physical region can replicate data to the local hub and the hub communicates further with other hubs/spokes. The following rules apply in an ILS networking setup:

■ Maximum distance in an ILS network is three hops (that is, spoke<>hub<>hub<>spoke).

■ Hubs fetch (pull) information from all other hubs, whereas spokes fetch information from their directly associated hub.

■ Sync intervals can be configured independently per cluster (default 10 minutes). Consequently, maximum convergence time is the total of sync intervals along the synchronization path (default 30 minutes).

■ Per synchronization interval, cluster pulls in information from every other cluster that is directly associated.

ILS Networking Configuration

To configure ILS on clusters to participate as hub or spoke, follow these steps:

Step 1.   Browse to **Tools > Service Activation** and activate the ILS service from Cisco Unified Serviceability.

Step 2.   Ensure that a unique cluster ID is defined for each cluster that is going to participate in ILS networking. Cluster ID can be changed from default by browsing to **CUCM Administration GUI > System > Enterprise Parameters**.

Step 3.   ILS networking can be based on shared secret (password) or certificates. If using the certificates, the Publisher server in each cluster needs to exchange Cisco Tomcat certificates with other Publishers. Using Bulk Certificate Management option in **CUCM OS GUI > Security > Certificates**, exchange the Cisco Tomcat certificates of a Publisher server in every cluster with other Publisher servers in other clusters. Alternatively, configure a password that will be common among all clusters in the ILS network.

> **Note**   As with any cryptographic configuration, use of certificates over passwords is recommended for higher security. However, the passwords are easier to manage than certificates (in terms of managing certificate lifecycle [expiry, renewal, revocation, and so on]).

Step 4.   Go to **CUCM Administration GUI > Advanced Features > ILS Configuration**. Set the role for first (full-mesh connected) cluster as **Hub Cluster**. Others can be set as hub or spoke as per the topology of the network, as shown in Figure 15-2. Hub cluster can be the Session Manager Edition (SME) cluster.

Step 5.   On other clusters, to join an ILS network, set role as **Spoke Cluster** and configure **Registration Server** as the hub/SME IP address/hostname.

> **Note**   In case there is no SME, the server acting as an aggregation point should be termed as hub.

Step 6.   Configure SIP route patterns on leaf nodes and SME. SME needs specific SIP route patterns for each SIP route string pointing to the respective leaf cluster (for example, singapore.SIN.AS or sanjose.SJC.US), whereas leaf clusters only need a "catchall" SIP route pattern that matches on all SIP route strings and points up to SME such as *.*AS. See Figure 15-3 for more information on SIP route patterns and SIP route strings.

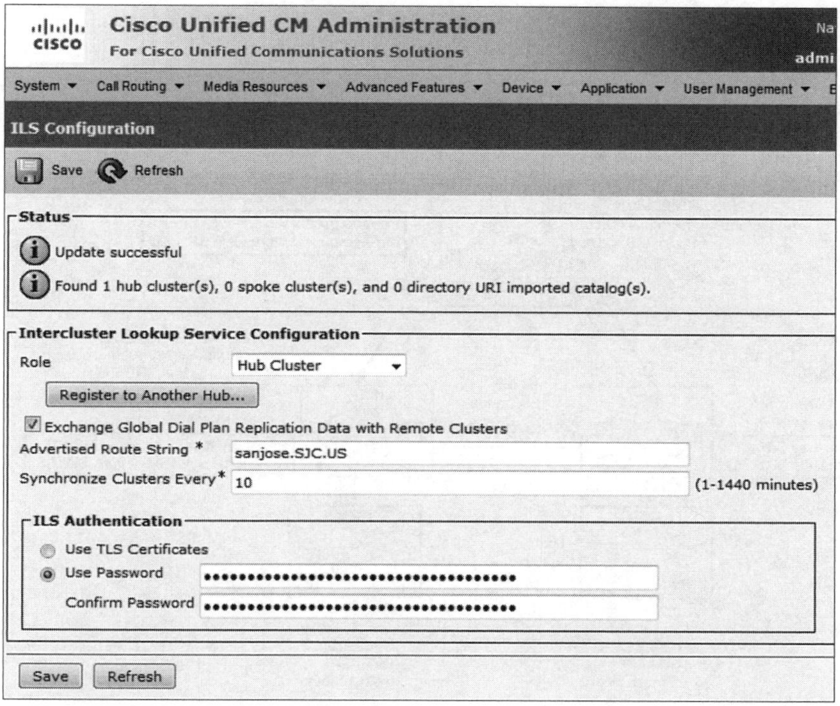

Figure 15-2 *ILS Configuration*

At this time, ILS configuration is complete. Navigate to **Advanced Features > Cluster View** for remote cluster information and available services. Also, the ILS Configuration menu allows monitoring sync state of URI data. Endpoint advertisement of enterprise alternate number/URI/+E.164 alternate number is covered later in the chapter.

ILS-Based SIP URI Dialing/Routing

As discussed in the previous section, the following are the components of end-to-end URI dialing/routing:

- ILS networking
- URI propagation
- SIP trunk
- SIP route pattern

Clusters can exchange URIs and create a URI to route string mapping table, as shown in Figure 15-3.

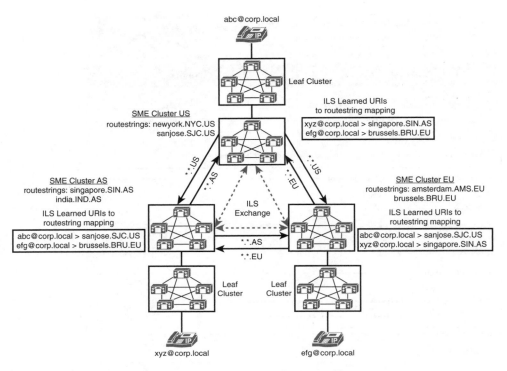

Figure 15-3   *ILS URI to Route String Mapping Between Clusters*

In Figure 15-3, ILS networked SME clusters in the United States (US), Asia (AS), and Europe (EU) are configured with route strings. As an analogy, think about ILS network as static routing protocol and the SME clusters as routers, where each router has static next-hop routes to another router in the network. For each packet (URI) destined from US > AS, the router (US SME cluster) knows exactly where to send the request/information as it has a route for anything *.*.AS in its routing table (ILS database). The SIP URI is mapped to respective route strings. (For example, abc@corp.local is mapped to sanjose.SJC.US, and xyz@corp.local is mapped to singapore.SIN.AS.) This helps the SME clusters to pinpoint the exact destination and route the calls to remote SME clusters. Similarly, the EU SME cluster knows that it can route anything with *.*.US to US SME cluster and *.*.AS to Asia SME cluster.

Note   It is important to note that ILS route strings allow summarization like IOS route summarization for routing. ILS route string summarization helps build a tightly held database.

Behind the scenes, ILS service allows propagation of individual alpha URIs between call controls (clusters) and helps bind alpha URIs with attributes that allow routing to the URI's home cluster. Each call control (represented by SME cluster in this case) replicates its alpha URIs to its neighbors and also announces SIP route string together with the alpha URIs. SIP route string can be routed based on SIP route patterns for intercluster call routing of alpha URIs, based not on URIs' host part, but on the SIP route string.

The following are some characteristics of ILS based call routing:

■ Internal numbers, E.164 numbers, and public switched telephone network (PSTN) failover numbers are advertised in ILS.

■ When a number is called, CUCM performs closest match routing. If the match from ILS is the best match, CUCM looks up the route string and routes based on route string.

■ If an exact match is not found in ILS or if ILS does not have the closest match, route patterns are used to route the call.

■ If the call fails, it is rerouted via the PSTN.

Important points to consider are that SIP connectivity is the foundation for call routing based on SIP route patterns, ILS networking is the foundation for exchange or URI reachability information, and URI propagation is enabled independent of ILS networking.

ILS Calls Via SIP Trunk and Cisco Unified Border Element

ILS-based calls can be dialed via SIP trunks (that is, SME to SME) or via Cisco Unified Border Element (CUBE) routers. ILS networking in SME clusters was discussed earlier. Figure 15-4 gives an insight to the ILS dialing process (using the URI or a number pattern using the SME cluster).

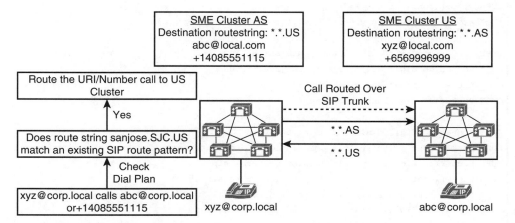

Figure 15-4 *ILS Call Process in SME Clusters*

CUBE can be used as an intermediary device in the case where business-to-business (B2B) communications are required or a single point of convergence is required where SMEs can interact with the IT service provider (ITSP). In such cases, calls are routed to CUBE, with the route string sent in SIP invite as a separate header and CUBE routes the call based on the dial peers without any ILS lookup. In other words, CUBE does not directly participate in ILS networking and can be leveraged as a hop to aggregate calls out to another cluster participating in ILS. Figure 15-5 illustrates the ILS-based call flow via CUBE.

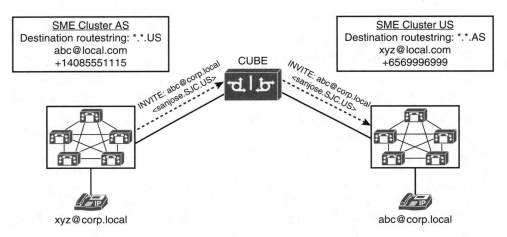

Figure 15-5 *ILS Call Process with CUBE*

Example 15-1 outlines the configuration of CUBE for the call flow shown in Figure 15-5.

Example 15-1 *CUBE Configuration for ILS Calls*

```
CUBE-Router(config)# voice service voip
CUBE-Router(conf-voi-serv)# sip
CUBE-Router(conf-voi-serv)# call-route dest-route-string
!
CUBE-Router(config)# voice class route-string 10
CUBE-Router(config-class)# pattern sanjose.sjc.us
CUBE-Router(config-class)# pattern *.sjc.us
CUBE-Router(config-class)# pattern *.us
!
CUBE-Router(config)# dial-peer voice 501 voip
CUBE-Router(config-dial-peer)# description ILS supporting dial-peer with higher pref
CUBE-Router(config-dial-peer)# session protocol sipv2
CUBE-Router(config-dial-peer)# destination route-string 10
CUBE-Router(config-dial-peer)# voice-class sip call-route dest-route-string
CUBE-Router(config-dial-peer)# session target ipv4:198.18.133.3
CUBE-Router(config-dial-peer)# preference 1
!
CUBE-Router(config-dial-peer)# dial peer voice 601 voip
CUBE-Router(config-dial-peer)# description Secondary (backup) dial-peer
CUBE-Router(config-dial-peer)# destination-pattern +14085551...
CUBE-Router(config-dial-peer)# session target ipv4:198.18.133.3
CUBE-Router(config-dial-peer)# preference 2
```

The next section describes the directory URI, enterprise alternate number (EAN), and +E.164 alternate number exchange and configuration.

Directory URI, Enterprise Alternate, and +E.164 Alternate Number Exchange

Previous sections covered the ILS networking and ILS routing concepts. You also read about ILS networking configuration. For a URI/DN/E.164 to be advertised and exchanged in an ILS network, follow these steps:

Step 1. Go to **CUCM Administration GUI > Device**, and on the endpoints that are going to participate in ILS networking, configure the enterprise alternate number and +E.164 alternate number. Consequently, enable advertisement of these via ILS via the Advertise Globally via ILS check boxes, as shown in Figure 15-6.

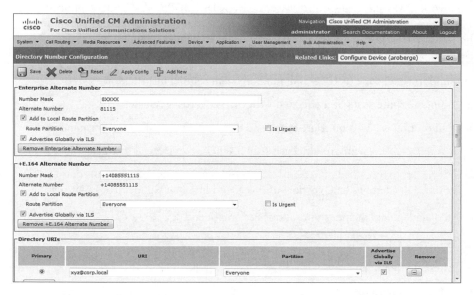

Figure 15-6 *Enterprise Alternate, +E.164 Alternate Number, and URI Advertisement*

Step 2. Ensure the user's (directory) URI is assigned/created (from Lightweight Directory Access Protocol [LDAP] import or manually) and the **Advertise Globally via ILS** check box is checked.

Step 3. Go to **CUCM Administration GUI > User Management > End User.** Check the box for users with **Home Cluster** (for Jabber Home Cluster Discovery), as shown in Figure 15-7.

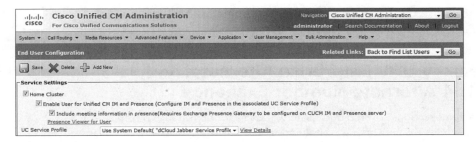

Figure 15-7 *End-User Home Cluster Discovery*

Global Dial Plan Replication Overview

Global Dial Plan Replication (GDPR) is a feature based on ILS and Call Control Discovery (CCD). CCD is discussed in Chapter 16, "Cisco Service Advertisement Framework (SAF) and Call Control Discovery (CCD)." GDPR uses the ILS application to exchange numbers and associated route strings for URI and numbered call routing. The topic of URI/DN (enterprise alternate and +E.164 alternate number) routing was discussed in previous sections. The following are characteristics of GDPR:

- Supports exchange of URIs and numbered call routing information.

- A unique route string is associated with each advertising cluster.

- Call routing is based on route string and not on actually called URI or number.

- Routing can be on any-to-any basis (full mesh) or hierarchical (for example, SME).

- Works best with a dial plan based on +E.164 numbers (globalized call routing, covered in Chapter 3, "Overview of PSTN and Intersite Connectivity Options").

Global dial plan data includes the following components:

- Directory URIs

- Alternate numbers

- Advertised patterns

- PSTN failover number

- Route strings

- Learned global dial plan data

- Imported global dial plan data

Directory URI, alternate numbers, and route strings were discussed earlier.

Alternate patterns allow creating summarized routing instructions for a range of enterprise/+E.164 alternate numbers. A contiguous range of numbers can be summarized (much like route summarization) and replicated throughout an ILS network. This permits all clusters in the ILS network to learn the summarized route to the destination cluster when URI dialing is not available or if a call is routed via PSTN.

CUCM uses a PSTN failover number to reroute only the calls placed to patterns, alternate numbers, or directory URIs learned through ILS (not used for local patterns, alternate numbers, or directory URIs).

Learned global dial plan data is the GDPR data learned through ILS networking. Every CUCM cluster keeps a local database of the learned (and local) data.

Imported global dial plan data is the GDPR data imported into the hub cluster through comma-separated value (CSV) files manually (exported from other clusters manually). This is used for Cisco TelePresence Video Control Server (VCS) and for any third-party IP private branch exchange (PBX) that cannot participate in an ILS network.

GDPR supports the exchange of numbered call routing information via ILS. GDPR supports both:

- Configured individually per directory number:
  - Enterprise number
  - +E.164 number

with PSTN failover number (either enterprise number or +E.164 number).

- Configured as summary patterns:
  - Enterprise patterns
  - +E.164 pattern

each with PSTN failover rule.

For PSTN failover number, enterprise alternate number +E.164 can be used, whereas for URI dialing, enterprise alternate number or +E.164 number can be used. The enterprise alternate number acts as an alias of a DN so that when you dial the enterprise alternate number, the phone that is registered to the associated DN rings. A DN can be associated to both an enterprise alternate number and a +E.164 alternate number at the same time.

Note   You may also choose one of the alternate numbers as the PSTN failover number for all alternate numbers and directory URIs that are associated to that directory number. This implies that a single DN can be associated with an enterprise alternate number, an +E.164 number, a PSTN failover number, and a SIP URI.

For GDPR, numbers and patterns can be advertised and assigned to partitions for class of service (CoS). Default partitions are assigned to the following:

- Enterprise alternate numbers (for example, 81115)

- +E.164 alternate numbers (for example, +14085551115)

- Enterprise patterns (for example, 8XXXX) and E.164 patterns (for example, +14085551XXX)

GDPR Configuration

To configure GDPR, follow these steps:

Step 1.   Configure ILS to exchange global dial plan data with other clusters in the ILS network. As shown in Figure 15-8, go to **Advanced Features > ILS Configuration** and check the check box **Exchange Global Dial Plan Data with Remote Clusters** and enter a route string in the Advertised Route String field (which identifies the local cluster).

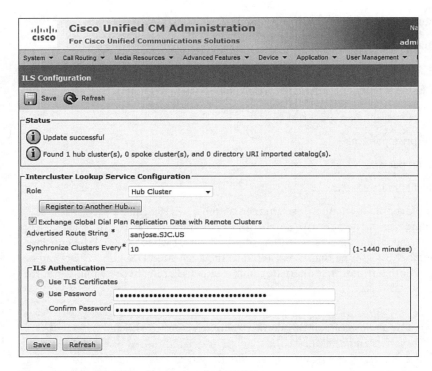

Figure 15-8   *ILS Configuration for GDPR*

Step 2. Set up the enterprise alternate number and +E.164 alternate number for the DNs that will participate in GDPR, as described in the earlier section "Directory URI, Enterprise Alternate, and +E.164 Alternate Number Exchange."

Step 3. Go to **Call Routing > Global Dial Plan Replication > Advertised Patterns** and select an existing pattern or add a new pattern, as shown in Figure 15-9.

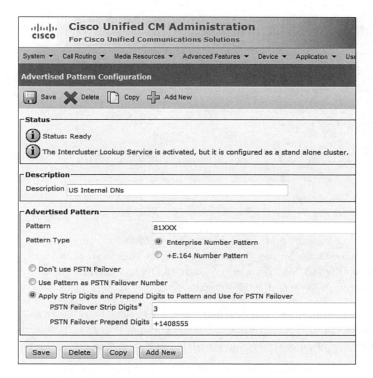

Figure 15-9 *Advertised Pattern Configuration*

Step 4. Go to **Call Routing > Global Dial Plan Replication > Partitions for Learned Numbers and Patterns** and set up partitions for ILS-learned numbers and patterns, as shown in Figure 15-10. Set up any existing partitions or use the default CUCM partitions for each learned number or pattern type.

Note CUCM has predefined partitions for learned alternate numbers and number patterns such as global learned enterprise numbers, global learned E.164 numbers, global learned enterprise patterns, and global learned E.164 patterns.

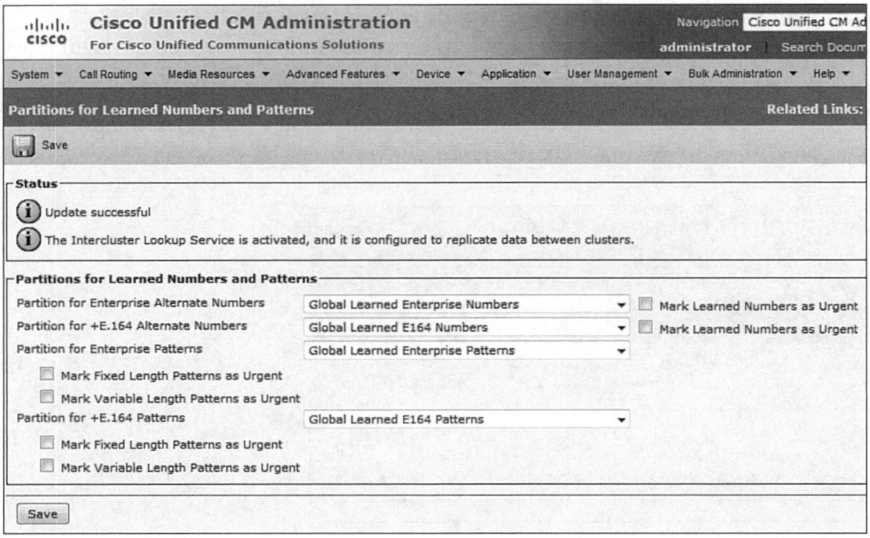

Figure 15-10   *Partitions Configuration*

Step 5.   If you need to change the number of learned entries/objects from other
clusters, go to **System > Service Parameters > Cisco Intercluster Lookup
Service** and set the **ILS Max Number of Learned Objects in Database** to a
number between 100,000 and 1000,000.

At this time, the GDPR configuration is complete and the ILS service can replicate numbers and patterns across the clusters participating in ILS networking. There are options to
further fine-tune the patterns or numbers learned via GDPR, as depicted in Figure 15-11.

Learning of information can be prevented (as required) by blocked learned patterns
based on the following:

■ Pattern

■ Prefix

■ Cluster ID

■ Pattern type

The learned patterns, numbers, and directory URIs show the ILS learned objects. The
imported global dial plan catalogs, directory URIs, and patterns show the imported
information from a third-party or a non-ILS-compatible server (where information was
manually imported into the local cluster and propagated using ILS to other ILS networked
clusters).

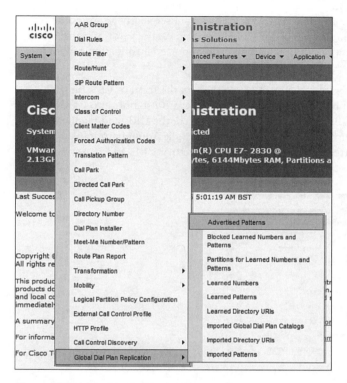

Figure 15-11 *GDPR Configuration Options*

Global Dial Plan Catalogs

The global dial plan catalogs (URIs and numbers) can be imported from third-party systems or a Cisco TelePresence VCS, as discussed briefly earlier in the chapter. This allows the IP PBXs or servers that cannot natively participate in ILS networking to exchange URI and number information with ILS networked clusters. Subsequently, global dial catalogs can be exported from ILS networked clusters and imported into non-ILS-compatible servers.

This occurs when the URIs and numbers are imported from another server or third-party PBX:

- Each import is stored in an imported global dial plan catalog.
- The CUCM administrator needs to assign a route string to each imported global dial plan catalog.

Information from imported global dial plan catalogs is treated as if it were learned via ILS global dial plan replication. Imported global dial plan catalogs are also propagated to other clusters on next pull (ILS replication update). The default limit for learned objects in an ILS network is 100,000 (can be changed from CUCM service parameters).

However, local URIs and numbers can be exported and subsequently imported on third-party systems that support import of global dial plan catalogs (for example, Cisco TelePresence VCS). This implies that ILS learned and imported information cannot be exported.

Both import and export of global dial plan catalogs are in the form of CSV files. The import of the CSV file can be done similarly to any other CSV import into CUCM using Bulk Administration Tool (BAT). Once the file is imported, the URIs and numbers can be inserted into the CUCM global dial plan catalog. Go to **Bulk Administration > Directory URIs and Patterns > Insert Imported Directory URIs and Patterns,** as shown in Figure 15-12.

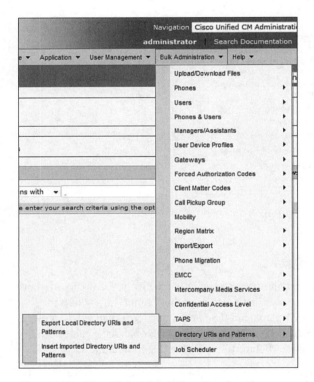

Figure 15-12   *Global Dial Plan Catalog Export and Import Options*

To export local directory URIs and patterns, select that option. The options are illustrated in Figure 15-13.

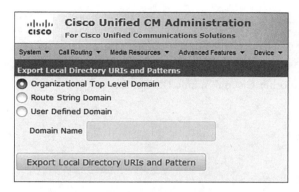

Figure 15-13 *Global Dial Plan Export Options*

Summary

The following key points were discussed in this chapter:

- URI-based dialing has been a norm in the video world for a long time. With the evolution of organizations and collaboration environments, URI-based dialing is becoming more popular. Users prefer dialing via name than remembering DNs or PSTN numbers.

- ILS and GDPR help larger enterprises by enabling them to leverage URI dialing while at the same time reducing the effort of N-N meshed networking.

- ILS helps dynamically populate remote clusters with information to reach a local cluster and vice versa.

- GDPR helps replicate number and pattern information so that URI dialing is not impacted by unavailability of network (IP layer). Users can reach others within and outside of an enterprise seamlessly.

References

For additional information, refer to these resources:

Cisco Systems, Inc. Cisco Collaboration Systems 10.x Solution Reference Network Designs (SRND), May 2014. http://www.cisco.com/c/en/us/td/docs/voice_ip_comm/cucm/srnd/collab10/collab10.html

Cisco Systems, Inc. Cisco Collaboration Systems 11.x Solution Reference Network Designs (SRND), July 2015. http://www.cisco.com/c/en/us/td/docs/voice_ip_comm/cucm/srnd/collab11/collab11.html

Review Questions

Use these questions to review what you have learned in this chapter. The answers appear in Appendix A, "Answers Appendix."

1. GDPR leverages ILS and which other Cisco dynamic DN/number pattern discovery service?

   a. LBM

   b. SRST

   c. CCD

   d. CMR

2. ILS allows dynamic building of which database? (Choose two options.)

   a. SIP URI

   b. Number pattern

   c. Route strings

   d. Route patterns

3. True or false: ILS allows propagation of only SIP DN associated with SIP URIs.

   a. True

   b. False

4. GDPR databases can be exported and imported. What information can be extracted from exported catalogs? (Choose three options.)

   a. Organization top-level domain (OTLD)

   b. User-defined domain

   c. CCD domain

   d. Route string domain

5. There can be multiple ILS hubs in an ILS network. How many hops can an ILS update traverse?

   a. 1

   b. 2

   c. 3

   d. 4

   e. N hops

6. **What is the default limit of learned objects in an ILS network?**

   a. 1000

   b. 10,000

   c. 100,000

   d. 1000,000

7. **True or false: Third-party IP PBXs can participate in ILS networking natively.**

   a. True

   b. False

8. **Data learned from GDPR catalog import will be replicated to how many nodes?**

   a. The next-hop hub node if the data was imported to a spoke node.

   b. To next two hops till next hub.

   c. Throughout the ILS network.

   d. The imported catalog information is not replicated across an ILS network.

9. **When CUBE is used within an ILS network, how is data replicated between ILS nodes and CUBE?**

   a. Data is replicated between the CCD and SAF service and the ILS networked CUCM clusters.

   b. Data is replicated in the flash of the CUBE router which is then passed to the next ILS networked node.

   c. Data is replicated directly between the two ILS nodes with CUBE as proxy agent.

   d. CUBE does not participate directly in an ILS network and can forward the calls based on SIP header information.

10. **In an ILS network, how is the hub-hub or hub-spoke communication trust transited? (Choose two options.)**

    a. The trust can be built using CUCM Tomcat certificates from Publisher nodes.

    b. The trust can be built using a password.

    c. The trust can be built using Tomcat certificates from all nodes.

    d. The trust can be built by setting a service parameter.

Chapter 16

Cisco Service Advertisement Framework (SAF) and Call Control Discovery (CCD)

More often than not, during the operate and run phase (post deployment) most Cisco Collaboration network administrators end up performing dial plan edits and changes to accommodate a growing network. However, when the scale of a Collaboration network is in hundreds of thousands of endpoints, including gateways and phones, making such changes can be very time-consuming and error prone. It can be both administratively difficult and time-consuming to configure trunks in point-to-point (for example, Cisco Unified Communications Manager [CUCM] <> Cisco Unified CME Express [CME]) or point-to-multipoint (for example, CUCM <> CME <> CUCM Session Management Edition [SME] <> Cisco Unified Border Element [CUBE]) networks for the dial plan. Cisco Service Advertisement Framework (SAF) and Call Control Discovery (CCD) services simplify the dial plan in large/multisite networks by virtue of auto-propagation and -proliferation of the dial plan through various network elements.

Upon completing this chapter, you will be able to meet these objectives:

- Describe complex dial plan implementation challenges
- Describe Cisco SAF
- Describe SAF characteristics and operation
- Describe CCD
- Describe CCD characteristics and operation
- Describe SAF and CCD configuration

Complex Dial Plan Implementation Challenges

Before getting into the nitty-gritty of SAF and how it is a value add in a Cisco Collaboration network, let's review some of the challenges that persist since the conception of IP-based communications. Traditionally, the IP networks have been used by the Collaboration elements as a purely transportation medium (without leveraging the real power of connectivity). Figure 16-1 compares a traditional point-to-point versus a centralized-based approach (CUCM, SME, CUBE) to dial plan planning.

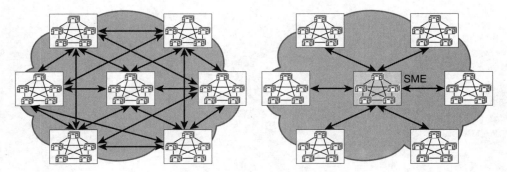

Figure 16-1   *Traditional Versus Modern Dial Plan Approach*

As shown in Figure 16-1, in the traditional approach every cluster in a Cisco Collaboration network has dial plan elements such as route patterns and dial peers pointing to the other clusters. This implies that there is n-1 relationship between the clusters (where n is the number of clusters in a Collaboration network). However, using SME, the dial plan can be centralized because all clusters home into SME and end up sharing their respective dial plans with a single source of truth for the Collaboration network.

The following are some of the issues with the traditional approach of deploying a dial plan:

■ A point-to-point dial plan involves configuration complexity.

■ Managing a complex dial plan on an ongoing basis leads to high operational cost and increased total cost of ownership (TCO).

■ Single point of failure if the trunks to other call controls fail.

As seen in the modern approach, CUCM SME clusters can ease the dial plan configuration by being the single point of contact (as described in Chapter 15, "Cisco Inter-Cluster Lookup Service (ILS) and Global Dial Plan Replication (GDPR)) for not just directory numbers (DNs) but also for Session Initiation Protocol (SIP) Uniform Resource Identifiers (URIs), route strings, and number patterns. In the context of SAF and CCD, the following are the benefits of using dynamic discovery and advertisement of numbers:

■ Allows call control agents to advertise/learn DN reachability.

■ Allows dynamic dial plan deployment.

■ Dynamic SAF-based learning and advertisement solves the complexity of managing a full mesh of static trunks without a single point of failure.

■ Automatic public switched telephone network (PSTN) rerouting if the IP route fails to route the call.

■ Dynamic reachability of information with readily available network capabilities.

Finally, with an SAF-enabled network where a dial plan is dynamically exchanged between the SAF network members (SAF-enabled routers and SAF clients—CUCM clusters), the network state looks like Figure 16-2.

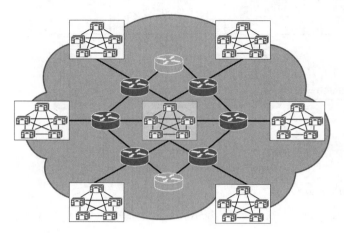

Figure 16-2   *SAF Network Layout*

The next section describes SAF and its architecture.

Cisco Service Advertisement Framework Overview

SAF is a network-based, scalable, bandwidth-efficient, and real-time framework that was developed by Cisco to enable the use of underlying network elements so that it can be used as a medium for automatically propagating service advertisements such as dial plan updates. It allows for automatic discovery of participating entities (call controls, gateways) in context of the service it is used for, which in this case is Call Control Discovery (dial plan).

SAF is based on Enhanced Interior Gateway Routing Protocol (EIGRP) as its underlying technology. However, SAF does not rely on EIGRP for routing (only for configuration) and thus is independent of IP routing protocol implemented in a network (for example static routing, Open Shortest Path First [OSPF], or Border Gateway Protocol [BGP]—any routing protocol can be used as long as the underlying routing protocol propagates SAF advertisements). SAF agents (SAF forwarder and SAF client, explained later in this section) form together what is known as an SAF network.

SAF allows administrators to control the scope of each supported service (for example, CCD) through domains, filtering, and virtual routing and forwarding (VRF), and it works with non-SAF-enabled nodes (also known as dark nets) for heterogeneous deployments. This implies that SAF can be deployed in a heterogeneous network where there is Cisco and non-Cisco networking equipment.

SAF Architecture

A Cisco SAF network consists of multiple entities and protocols. Figure 16-3 depicts the Cisco SAF architecture.

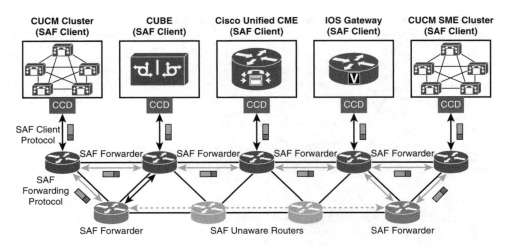

Figure 16-3   *Cisco SAF Architecture*

The following are the SAF terms used to describe the role of each SAF entity in a Cisco SAF network:

- **SAF client:** The SAF client is an application capable of advertising a service to the network or requesting a service from the network or both. Call controls such as CUCM, CUBE, and CME are examples of an SAF client.

- **SAF Client Protocol (SAF-CP):** The SAF network interface layer inside a call control for SAF applications. This is used between a call control agent/SAF client (for example, CUCM cluster and the SAF forwarder [IOS router]).

- **SAF forwarder:** The SAF forwarder is a Cisco IOS router feature that provides a relationship between the SAF client and the SAF framework and stores service information and consequently propagates it to other forwarders.

- **SAF Forwarding Protocol (SAF-FP):** The SAF forwarding protocol is used by an SAF forwarder to communicate (share/receive) SAF updates with other SAF forwarders in the SAF network.

- **CCD:** The CCD is one of the many potential services that SAF can support. An SAF client in this case would use CCD to advertise and consume dial plan components.

- **SAF (message) advertisement:** An SAF advertisement carries service information and consists of an SAF header and service data. Figure 16-4 depicts the SAF advertisement packet structure.

Figure 16-4   *SAF Advertisement Packet Structure*

- **SAF unaware (node) routers:** An SAF unaware node is any (non-SAF) router that does not run SAF protocols or is not configured to participate in an SAF network.

An SAF forwarder learns dial plan information from an SAF client (leveraging CCD service, described in the next section) using SAF-CP. Thereafter, an SAF forwarder exchanges learned call routing information with other SAF forwarders, using SAF-FP in the SAF network and other SAF clients so that the SAF enabled network is aware of all advertised and learned call routes. This helps overcome the full-mesh (point-to-point) manual dial plan creation and solves the complexity of managing a full mesh of static trunks without a single point of failure. By using the SAF-CCD framework, you can build networks that are not plagued by traditional dial plan configuration issues.

An SAF message consists of two parts: SAF header and SAF service data. The SAF header is significant to SAF forwarders and identifies the service type (for example, CCD) and includes information relevant for dynamic distribution of SAF services. SAF service data is significant to the SAF client and includes the IP address and port of the advertising SAF client. It also incorporates detailed client data describing the advertised service. (For example, CCD client data includes call routing information such as DNs, signaling protocol used by call control agent, public switched telephone network [PSTN] prefixes, and so on.)

An SAF learned dial plan and the SAF forwarder information can be monitored via Real-Time Monitoring Tool (RTMT). To view the learned routes, go to **CallManager > Report > Learned Pattern**. To view the SAF forwarders, go to **CallManager > Report > SAF Forwarders**, as shown in Figure 16-5.

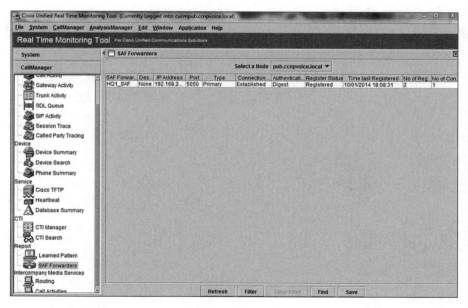

Figure 16-5   *RTMT-Based SAF Forwarder Monitoring*

The next section discusses the various SAF network entities.

SAF Characteristics and Operation

There are various entities in an SAF network. This section addresses these.

SAF Clients

SAF clients are call controls in the Cisco world. These are the central intelligence for call routing from where (leveraging the CCD service) the SAF protocol can dynamically advertise and learn DNs in an SAF network. The following is a list of currently supported SAF clients:

- CUCM
- SME
- CME
- CUBE
- Cisco IOS routers

SAF clients perform three major functions:

- Register to the network
- Publish services
- Subscribe to services

As shown in Figure 16-6, external clients such as CUCM communicate to an SAF forwarder via the SAF Client Protocol (SAF-CP), whereas Cisco IOS SAF clients communicate to a co-located SAF forwarder via an internal application programming interface (API).

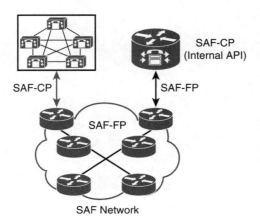

Figure 16-6 *SAF Client – SAF Client Protocol (SAF-CP) Relationship*

SAF Client Protocol

SAF-CP binds an SAF client to the SAF forwarder. In other words, SAF-CP is the bridge between an SAF client and an SAF network so that the SAF client can publish and consume information from the underlying routing infrastructure. The following are some key characteristics of the SAF-CP:

- SAF-CP assumes that the SAF client knows the IP address of the SAF forwarder.

- SAF-CP is a simple transport control protocol (TCP)-based binary type-length-value (TLV) protocol based on Session Traversal Utilities for NAT (STUN, RFC 3489).

- SAF-CP offers security by means of digest authentication based on a shared secret (user ID and password).

- The SAF client can publish and subscribe to multiple services, and the SAF forwarder notifies the client of services matching subscriptions.

- The SAF client sends a periodic register as a keepalive.

SAF Forwarders (SAF Forwarding Nodes)

SAF has various forwarder types within the SAF network, such as the following:

- **Forwarder:** A forwarder node learns services from other forwarders and locally attached clients as well as advertising services to other forwarders, as shown in Figure 16-7.

- **Stub forwarder:** A stub forwarder learns services from other forwarders and locally attached clients, but advertises only locally published services, or no services at all, as shown in Figure 16-7. This behavior is inherited from EIGRP routing, which is the underlying service for SAF and helps improve scalability and network stability.

- **Transitive forwarder:** A transitive forwarder node learns services from other forwarders only. As the name suggests, being transitive in nature, it does not have locally attached clients, as shown in Figure 16-7. It can advertise services to other forwarders and discards service data after forwarding, which improves resource scalability.

When an SAF forwarder receives an advertisement, the following process takes place:

Step 1.   It stores it in memory.

Step 2.   Then it sends it out through all SAF-enabled interfaces (except the interface where the update was received from, to avoid loops).

Note   EIGRP-style metrics and the DUAL algorithm are used to avoid loops and to provide fast convergence. The SAF forwarder sends updates only when there are service changes.

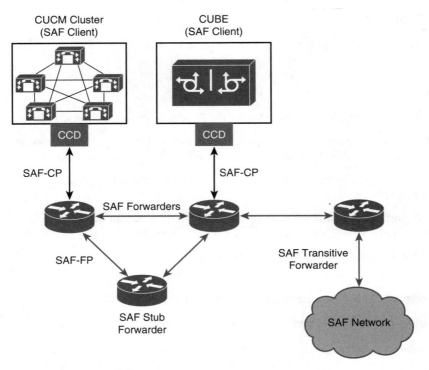

Figure 16-7 *SAF Forwarder Types in an SAF Network*

An important point to consider is the way the SAF forwarders can discover each other in an SAF network. The discovery can be classified into two categories:

- SAF-FP neighbor establishment - L2 adjacent

- SAF-FP neighbor establishment - Nonadjacent

Both adjacent and nonadjacent discovery mechanisms can be based on dynamic (multicast routing-based discovery) discovery and static (unicast with manual configuration) discovery.

Example 16-1 illustrates the SAF forwarder L2 adjacent configuration.

Example 16-1 *SAF Forwarder L2 Adjacent Configuration*

```
SAFRouter(config)# interface FastEthernet 0/0
SAFRouter(config-if)# ip address 1.1.1.1 255.255.255.0
!
SAFRouter(config)# router eigrp SAF-FDR
SAFRouter(config-router)# service-family ipv4 autonomous-system 1
SAFRouter(config-router-sf)# topology-base
```

Example 16-2 shows the SAF forwarder configuration for a nonadjacent peer.

Example 16-2 *SAF Forwarder Nonadjacent Peer Configuration*

```
SAFRouter(config)# interface Loopback 0
SAFRouter(config-if)# ip address 2.2.2.2 255.255.255.0
!
SAFRouter(config)# router eigrp SAF-FDR
SAFRouter(config-router)# service-family ipv4 autonomous-system 1
SAFRouter(config-router-sf)# neighbor 3.3.3.3 loopback0 remote 16
SAFRouter(config-router-sf)# topology-base
```

The next section describes the SAF Forwarder Protocol.

SAF Forwarder Protocol

SAF-FP works within the SAF network to advertise and learn DNs from one SAF
forwarder to another. As illustrated in Figure 16-3, SAF-FP enables SAF forwarders
(forwarder, stub, or transitive node) to communicate with each other in an SAF network.
The following are key properties of SAF-FP:

- Supports IPv4 and IPv6.

- Supports various topology scenarios such as local-area network (single Layer 2
  domain), multihop networks, campus networks (single Layer 3 administrative
  domain), multicampus networks, and service provider multihop bypass.

- Offers resource control by implying "service limits" and also enables controlled
  memory usage on SAF forwarders.

- Provides neighbor authentication using MD5 or HMAC-SHA2 algorithms.

- Thwarts malicious service injection or manipulation of services.

Example 16-3 illustrates Cisco IOS router's support for SAF.

Example 16-3 *SAF Enabled Router*

```
SAFRouter# show eigrp plugins detail
<output omitted for brevity>
ipv4-af : 2.01.01 : Routing Protocol Support
ipv4-sf : 1.00.00 : Service Distribution Support
external-client : 1.02.00 : Service Distribution Client Support
ipv6-af : 2.01.01 : Routing Protocol Support
ipv6-sf : 1.00.00 : Service Distribution Support
<output omitted for brevity>
```

For SAF (and CCD) configuration, see Example 16-4.

SAF Message

As described earlier in this chapter, an SAF header is used for propagating service advertisements between forwarders and holds the service ID, whereas the service data is provided by SAF clients and is consumed by the clients that are interested in the advertisement. Figure 16-8 shows the SAF message structure.

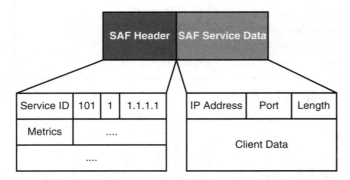

Figure 16-8   *SAF Message Structure*

An SAF header has the following characteristics:

- Identifies service type and unique instance
- Used by forwarders to propagate advertisements
- Metrics used to avoid loops and for future client use

SAF service data has the following characteristics:

- Service-specific information
- Meaningful only to clients of the given service
- Transparent to forwarders

Call Control Discovery Service Overview

Call Control Discovery (CCD) is an SAF service that enables call control agents to discover each other through the SAF network by advertising their reachability information along with the DN ranges they own. Subsequently, CCD requests data to learn about other call control agents in the network, as shown in Figure 16-9. Call control agents can then dynamically route calls to remote destinations based on received advertisements.

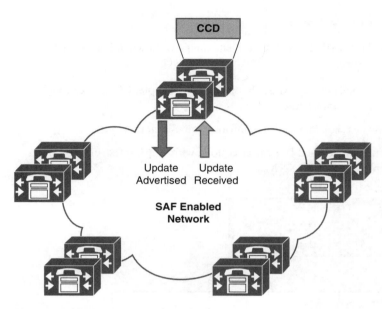

Figure 16-9 *CCD Service Pertinent to Call Controls*

CCD enables call agents to exchange dial plans, signaling protocols, corresponding PSTN numbers, and reachability information through SAF. CCD is a function of call agents and allows extending call control logic to incorporate dynamic routing based on information learned through SAF-enabled networks. CCD focuses on enterprise-owned DNs, including information on direct inward dial (DID) rules in advertisements to simplify PSTN failover (if IP routing fails). The following are the characteristics of CCD service:

- **CCD advertising service:** CCD advertising service is responsible for advertising preconfigured hosted DN ranges, PSTN failover rules, and trunk route information to the SAF network. CCD advertising service can select up to two trunks (one SAF CCD SIP trunk and one SAF-enabled H323 intercluster trunk [ICT]) and runs on the same nodes as its selected trunks. Upon any change in local configuration, the CCD advertising service sends a new advertisement to the SAF network.

- **CCD requesting service:** CCD requesting service is responsible for learning hosted DN routes from the SAF network. CCD requesting service stores learned route information locally and registers it with CUCM digit analysis. CCD requesting service performs load balancing for calls to learned routes. If a call cannot go through via the IP network, CCD requesting service will route the call via the PSTN network.

As discussed in a previous section, an SAF packet has payload (that is, SAF service data), which in this case is the CCD. The following section explains the CCD information object layout (schema).

Call Control Discovery Schema

CCD schema is based on XML 1.0 with UTF-8 encoding, as shown in Figure 16-10. The structure is as follows:

■ All CCD information is in the hosted-dn object (for example, CUCM version, enterprise parameter name, and location).

■ trunk-route provides signaling protocol information (SIP in this case).

■ dn-pattern contains advertised DN patterns along with strip/prefix rules to obtain DID (for example, +14085551XXX).

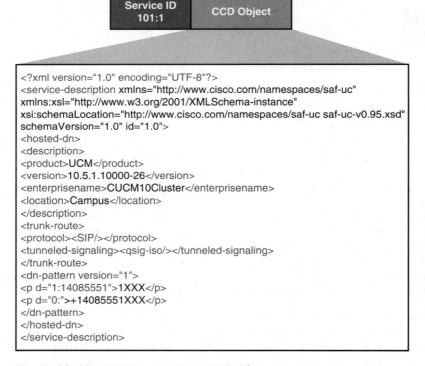

Figure 16-10   *SAF Service Data: CCD Object*

CCD Characteristics and Operation

This section details the various characteristics of CCD and the way it operates related to SAF framework.

The following are the CCD (as an SAF service) characteristics:

- CCD enables call agents to exchange dial plan, signaling protocol, and reachability information through SAF.

- CCD extends call control logic to incorporate dynamic routing based on information learned through SAF.

- CCD focuses more on enterprise-owned DNs than on PSTN egress points. However, it includes information on DID translation "rules" in advertisements to simplify PSTN failover.

Figure 16-11 illustrates CCD advertised and learned routes (dial plan) across an SAF-enabled network.

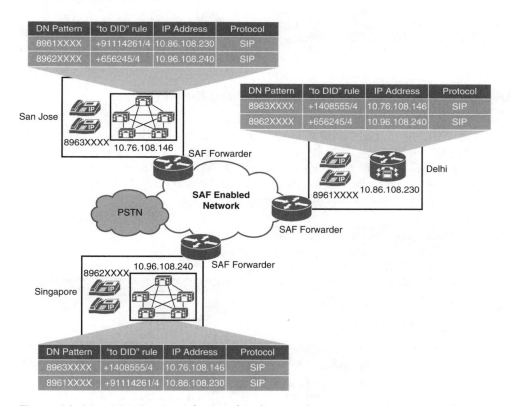

Figure 16-11 *CCD Service Advertised and Learned Routes in an SAF Network*

As shown in Figure 16-9, three locations—San Jose (CUCM cluster), Delhi (Cisco Unified CME), and Singapore (CUCM cluster)—are configured to advertise and learn their respective hosted DNs and are shown in the following steps:

Step 1.   The SAF forwarders advertise the learned route from respective SAF clients (CUCM or Cisco Unified CME) to their peers (other SAF forwarders).

Step 2.   SAF forwarders also in turn share the learned routes with the SAF clients (CUCM and Cisco Unified CME), thereby building the CCD dial plan.

Step 3.   Each SAF client builds a database, based on dynamic call routing, that has the route, PSTN prefix, remote IP address, and protocol to be used for routing calls.

Note   SAF CCD supports both H.323 and SIP protocols for dynamic DN advertisement and learning.

As shown in previous activities, there are multiple use cases that can be examined. This book discusses the most common ones.

Note   These use cases assume that DN routing is enabled in the SAF network and that the advertise/learn process is complete.

Use Case 1: Normal Calls via SAF-Enabled Network to Remote Call Control

As Shown in Figure 16-12 when Phone A in San Jose (DN=89632200) calls Phone B in Delhi (DN=89611100), CUCM CCD service intercepts the call and routes it over to SAF Forwarder which further routes it over to destination IP (that of target Cisco Unified CME). Upon receiving the invite, remote call control (Cisco Unified CME) connects the call to Phone B (going through SIP call setup).

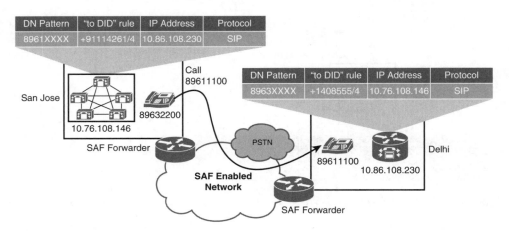

Figure 16-12   *Calls via SAF Network*

Use Case 2: Calls via PSTN When the SAF Forwarder Is Down

As shown in Figure 16-13, when Phone A in San Jose (DN=89632200) calls Phone B in Delhi (DN=89611100), CUCM CCD service intercepts the call. As a result of underlying routing updates, the SAF forwarder is aware that the remote site's SAF forwarder is down (not available) and updates CUCM. CUCM (CCD service) then routes the call over to PSTN using the +E.164 number derived from "to DID" rule. The call lands as a PSTN inbound call at the remote site, and Cisco Unified CME establishes SIP connection followed by media channel with Phone B.

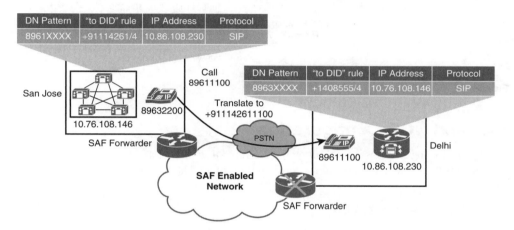

Figure 16-13 *Calls via PSTN (PSTN Failover)*

Note   When in survivable remote site telephony (SRST) mode, the remote site leverages the same mechanism (that is, uses the "to DID" rule for the CUCM cluster in San Jose). Essentially, the same procedure is followed when Phone B calls Phone A.

Use Case 3: Normal Calls via SAF-Enabled Network to CUBE

As shown in Figure 16-14, when Phone A in San Jose (DN=89632200) calls Phone C in Delhi (DN=89611101), CUCM CCD service intercepts the call and routes it over to the SAF forwarder, which further routes it over to destination IP (that of target CUBE). Upon receiving the invite, remote call control (CUBE) connects the call via the SIP/H.323 trunk to the private branch exchange (PBX), which in turn connects the call to Phone C. This call flow is very similar to use case 1, except that now calls are landing on CUBE and CUBE hands over the signaling to the connected PBX.

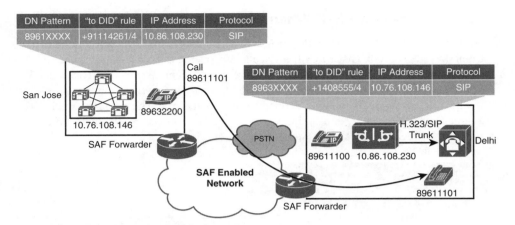

Figure 16-14   *Calls Via CUBE*

SAF and CCD Configuration

The following sections walk through the configuration of SAF and CCD on CUCM clusters (SAF clients), followed by configuration of an SAF forwarder.

SAF Client Configuration

Follow these steps to configure CUCM as an SAF client:

Step 1.   Go to **CUCM Administration GUI > Advanced Features > SAF > SAF Security Profile,** as shown in Figure 16-15, and create a new profile and enter the required information. Ensure that the username and password entered match the credentials entered in IOS external client configuration. Click **Save.**

Step 2.   Go to **Advanced Features > SAF > SAF Forwarder** and create a new forwarder. Use the external client configuration parameters for **Client Label, SAF Forwarder Address, SAF Forwarder Port (Default 5050),** and select the security profile created in the preceding step, as shown in Figure 16-16. Click **Save.**

> **Note**   The SAF client label is a unique client label in the registration message that CUCM sends to the SAF forwarder. When the SAF forwarder receives the registration message, it verifies whether the CUCM is configured with the appropriate client label. SAF forwarder address is the IP address of the IOS router.

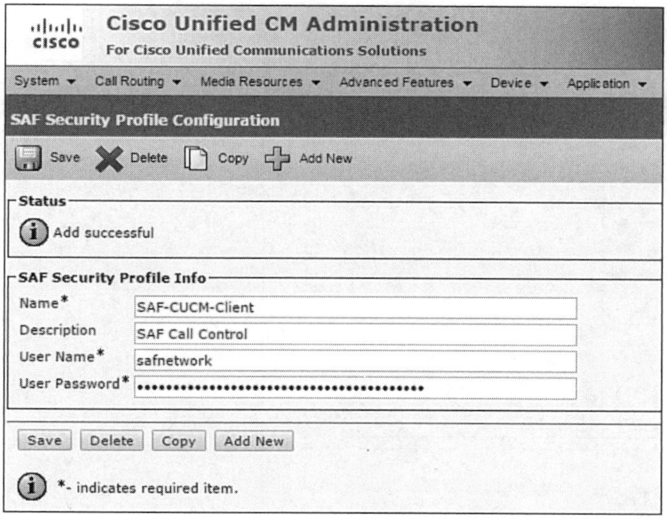

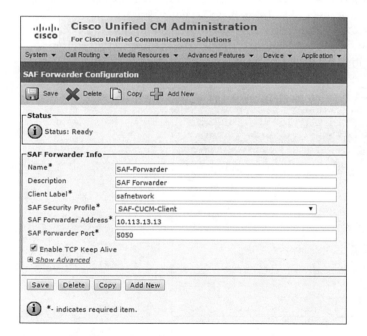

Figure 16-15   *CUCM SAF Security Profile Configuration*

Figure 16-16   *CUCM SAF Forwarder Configuration*

Step 3.   Go to **Call Routing > Call Control Discovery > Hosted DN Group**, as shown in Figure 16-17, and create a group. Enter required information and click **Save**. Hosted DN groups are a collection of hosted DN patterns (covered in Step 4) that will be advertised by CCD service.

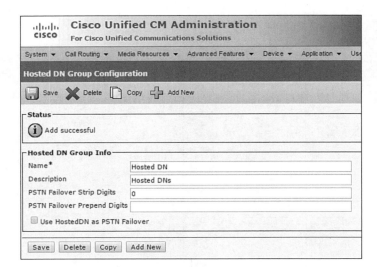

Figure 16-17   *Hosted DN Group Configuration*

Step 4.   Go to **Call Routing > Call Control Discovery > Hosted DN Pattern** and add a route pattern as you expect from the PSTN (after digit manipulation) (for example, 1XXX), as shown in Figure 16-18. Assign it to Hosted DN Group and click **Save**. One or more hosted DN patterns can be configured to be associated with the hosted DN group.

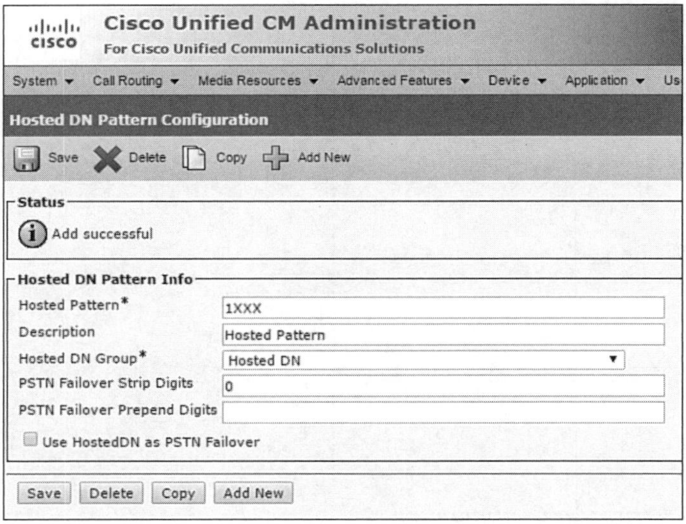

Figure 16-18   *Hosted DN Pattern Configuration*

Step 5.   Go to **Device > Trunk** and create a new SIP trunk with service type as Call Control Discovery. Configure this trunk as any other SIP trunk. Click **Save.**

Note   Alternatively, an H.323 (H.225) trunk may be configured for an SAF client that does not support SIP.

Step 6.   Go to **Call Routing > Call Control Discovery > Advertising Service** and add a new service. Select the SIP trunk (or H.323 trunk) and the hosted DN group configured earlier, as shown in Figure 16-19. Click **Save.**

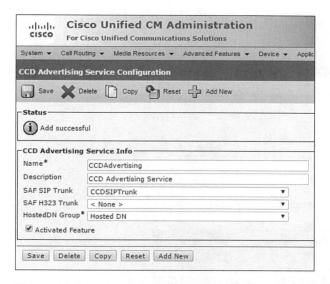

Figure 16-19   *CUCM CCD Advertising Service Configuration*

Step 7.   Go to **Call Routing > Call Control Discovery > Requesting Service** and add a new service, as illustrated in Figure 16-20. Add the SIP/H.323 trunk and click **Save.**

Step 8.   Go to **Call Routing > Call Control Discovery > Partition** and configure the default partition to which the learned routes will be placed, as shown in Figure 16-21.

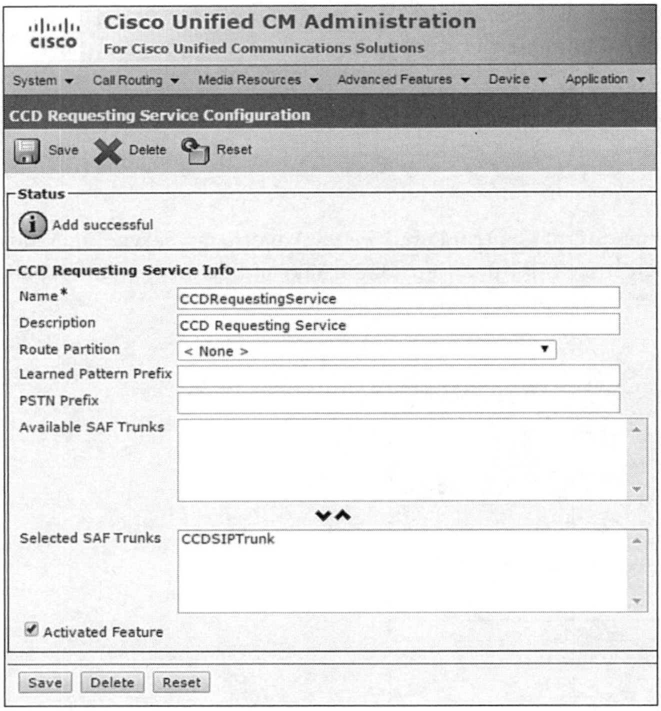

Figure 16-20   *CUCM CCD Requesting Service Configuration*

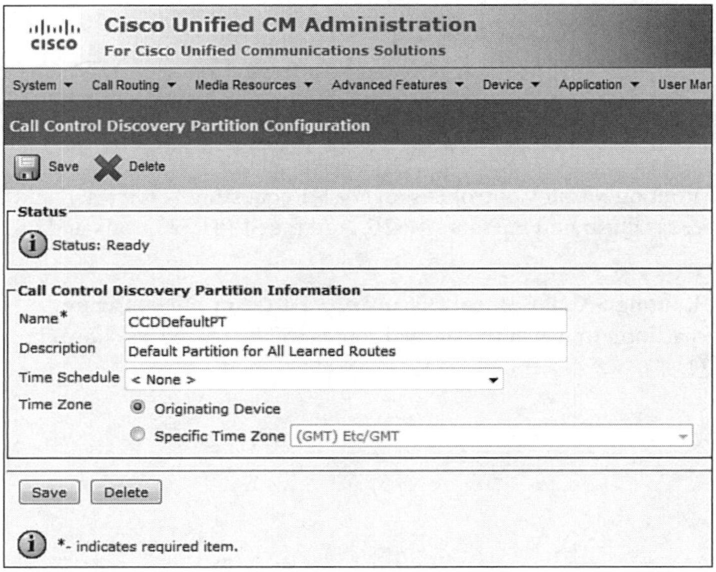

Figure 16-21   *CCD Default Partition Configuration*

In addition, the following SAF parameters can be fine-tuned by browsing to **Call Routing > Call Control Discovery > Feature Configuration:**

- **CCD Maximum Numbers of Learned Patterns:** Specifies the number of patterns that a CUCM cluster can learn from the SAF network

- **CCD Learned Pattern IP Reachable Duration:** Specifies the number of seconds that learned patterns stay active before CUCM marks them as unreachable

- **CCD PSTN Failover Duration:** Specifies number of minutes that calls to learned patterns (marked as unreachable) are routed through PSTN failover

- **Issue Alarm for Duplicate Learned Patterns:** Specifies whether to issue an alarm when it learns about duplicate patterns from different remote call control entities on the SAF network

- **CCD Stop Routing On Unallocated Unassigned Number:** Specifies whether CUCM continues to route calls to the next call control entity when the current call control entity rejects the call with cause code unallocated/unassigned number (Q.850 Cause Code = 1)

SAF Forwarder Configuration

This section explores the configuration of Cisco SAF forwarder (that is, IOS router) to support Cisco SAF clients (CUCM, Cisco Unified CME). Figure 16-22 depicts an SAF client-forwarder relationship and various protocols and entities involved in an SAF network.

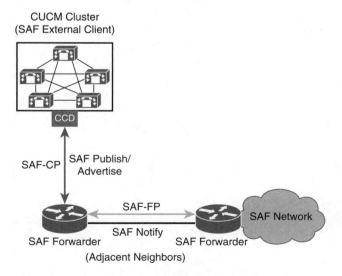

Figure 16-22 *SAF Client: Forwarder Relationship and SAF Network*

In Figure 16-22, the SAF forwarder is the medium for CUCM communicating with the SAF network. The SAF forwarder essentially ends up communicating with CUCM using SAF-CP to accept SAF advertisements and to push the learned SAF routes from the SAFE network back to CUCM. However, the SAF forwarder communicates with the SAF network using SAF-FP and notifies/gets notified of any new advertisements or routes learned.

Example 16-4 builds on Figure 16-22 for configuration of SAF and CCD on a Cisco IOS router.

Example 16-4   *SAF Forwarder Configuration*

```
SAFRouter(config)# router eigrp SAF
SAFRouter(config-router)# service-family ipv4 autonomous-system 100
SAFRouter(config-router-sf)# neighbor 10.76.108.235 loopback 0 remote 10
SAFRouter(config-router-sf)# topology base
SAFRouter(config-router-sf-topology)# external-client safnetwork
SAFRouter(config-router-sf)# sf-interface GigabitEthernet 0/0
SAFRouter(config-router-sf-interface)# no split-horizon
SAFRouter(config-router-sf-interface)# hello-interval 10
SAFRouter(config-router-sf-interface)# hold-time 20
!
SAFRouter(config)# voice service voip
SAFRouter(conf-voi-serv)# no ip address trusted authenticate
!
SAFRouter(config)# service-family external-client listen ipv4 5050
SAFRouter(config-external-client)# external-client safnetwork
SAFRouter(config-external-client-mode)# username safnetwork
SAFRouter(config-external-client-mode)# password C1sc0123456
SAFRouter(config-external-client-mode)# keepalive 5000
!
SAFRouter(config)# voice service saf
SAFRouter(conf-voi-serv-saf)# profile trunk-route 1
SAFRouter(conf-voi-serv-saf-tr)# session protocol sip interface loopback 0 transport
  udp port 5060
SAFRouter(conf-voi-serv-saf)# profile dn-block 1
SAFRouter(conf-voi-serv-saf-dnblk)# pattern 1 type global 14082221XXX
SAFRouter(conf-voi-serv-saf-dnblk)# pattern 2 type extension 1XXX
SAFRouter(conf-voi-serv-saf)# profile callcontrol 1
SAFRouter(conf-voi-serv-saf-cc)# dn-service
SAFRouter(conf-voi-serv-saf-cc-dn)# trunk-route 1
SAFRouter(conf-voi-serv-saf-cc-dn)# dn-block 1
SAFRouter(conf-voi-serv-saf)# channel 1 vrouter SAF asystem 100
SAFRouter(conf-voi-serv-saf-chan)# subscribe callcontrol wildcarded
SAFRouter(conf-voi-serv-saf-chan)# publish callcontrol 1
```

In Example 16-4, the commands used for configuring the SAF forwarder are explained as follows:

- SAF leverages **EIGRP** (virtual router) although the underlying network can use static routing or any dynamic routing protocol.

- EIGRP instance **SAF** initiates SAF-associated configuration.

- The **service-family protocol** and **AS number** are required because SAF clients must know them to register with this forwarder and neighbors will not form a neighbor relationship unless these match between forwarders.

- The **sf-interface** (default allows SAF on all interfaces) allows specifying an interface thereby limiting SAF to a particular interface.

- SAF permits multiple service **topology databases** per **service-family**, in this case defining SAF clients.

- The **service-family** allows SAF client configuration so that a client-forwarder authentication relationship can be built.

- **client-label** is unique across an AS and only one client can use a client label.

- The **username** and **password** are used in security validation with SAF client.

- **Keepalives** are used between SAF client and forwarder to ensure the forwarder is available; otherwise, the clients can reroute the calls out the alternate path (PSTN).

- The command **voice service SAF** initiates CCD service on the router.

- **trunk profile** lets other devices know of the **protocol** to contact the SAF client/ forwarder with.

- **DN-block** defines the routes to advertise.

- A **callcontrol profile** ties all of these entities together. The call control profiles need to be associated with an SAF instance as a **channel** using the EIGRP process name and autonomous system number.

- The **publish** process advertises **callcontrol profile** to the SAF network, and the **subscribe** process makes the router listen to **wildcard** (all) advertisements (could be set to listen to specific advertisements as well).

Summary

The following key points were discussed in this chapter:

- Dial plans have evolved over time and so has the need to make things easier from a management and administrative point of view (to ensure that enterprises have the flexibility to change the dial plans when required with minimal impact on ongoing operations).

- SAF-CCD is a dynamic service that allows Cisco Collaboration network administrators to leverage the underlying routing network to learn and distribute dial plans.

- SAF and CCD work together to help replicate the dial plan across an SAF-enabled network, thereby saving both administrative time and bypassing the error-prone process of manually updating the dial plan.

- SAF and CCD framework supports multiple call controls such as CUCM, Cisco Unified CME, CUBE, and Cisco IOS gateways.

References

For additional information, refer to these resources:

Cisco Systems, Inc. Cisco Collaboration Systems 10.x Solution Reference Network Designs (SRND), May 2014. http://www.cisco.com/c/en/us/td/docs/voice_ip_comm/cucm/srnd/collab10/collab10.html

Cisco Systems, Inc. Cisco Collaboration Systems 11.x Solution Reference Network Designs (SRND), July 2015. http://www.cisco.com/c/en/us/td/docs/voice_ip_comm/cucm/srnd/collab11/collab11.html

Review Questions

Use these questions to review what you have learned in this chapter. The answers appear in Appendix A, "Answers Appendix."

1. **Cisco SAF makes use of which underlying routing protocol?**

    a. IGRP

    b. RIPv2

    c. OSPF

    d. EIGRP

2. **True or False: An SAF forwarder can only interact with one SAF forwarder.**

    a. True

    b. False

3. **CCD is used by another service apart from SAF. Which other CUCM service leverages CCD?**

    a. ILS

    b. GDPR

    c. MRA

    d. EMCC

4. Which call control is not explicitly supported by SAF?

   a. CUCM

   b. SRST

   c. CUBE

   d. Cisco Unified CME

5. SAF-FP is used as a propagation protocol in an SAF network. Which SAF network entity uses SAF-FP?

   a. SAF forwarder

   b. SAF client

   c. Dark net routers

   d. SAF-CP forwarder

6. Which two services are supported by CCD?

   a. Advertising service

   b. DN advertising service

   c. DN requesting service

   d. Requesting service

7. Which protocols do the SAF trunks (in CUCM, Cisco Unified CME, CUBE) support? (Choose two.)

   a. SIP

   b. H.323

   c. MGCP

   d. ICT

Answers Appendix

Chapter 1

1. D
2. C
3. C
4. A
5. A and C
6. B
7. A
8. C
9. C

Chapter 2

1. D
2. B, C, and E
3. B and C
4. D and E
5. A
6. A
7. B and C
8. C
9. D
10. B

Chapter 3

1. C
2. B and C
3. C
4. C
5. C
6. A
7. D and E
8. A and C
9. B
10. C and D
11. D

Chapter 4

1. True
2. E
3. False
4. C and E
5. D
6. C and D
7. B
8. False
9. E
10. False

Chapter 5

1. D
2. A and B
3. B
4. False
5. A, C, and D
6. A
7. A and D
8. C, D, E, F, and G

Chapter 6

1. A, D, and E
2. A, B, D, and E
3. C
4. C
5. False
6. True
7. A
8. True

Chapter 7

1. D
2. A
3. B and D
4. B
5. A and D
6. A
7. A
8. A
9. D
10. False

Chapter 8

1. B
2. A and C
3. C
4. False
5. A and D
6. C
7. B
8. B
9. A, C, and D

Chapter 9

1. A
2. C and D
3. A, D, and E
4. A and D
5. A and B
6. A
7. D
8. B
9. C
10. True

Chapter 10

1. B and D
2. C
3. D
4. F
5. A and C
6. C
7. B and C

Chapter 11

1. False
2. C, D, and F
3. E
4. E
5. D and F
6. A and C
7. C
8. F
9. False
10. H

Chapter 12

1. A, D, and E
2. B and D
3. F
4. A, C, and G
5. A, B, E, and F
6. A, C, D, and F
7. E
8. A, B, and E
9. F

Chapter 13

1. F
2. B
3. E
4. B, C, E, and G
5. B
6. True
7. F

Chapter 14

1. D
2. A
3. B, C, and D
4. B
5. A, B, and C
6. A
7. A, B, C, and D
8. D
9. True

Chapter 15

1. C
2. A and C
3. False
4. A, B, and D
5. C
6. C
7. False
8. C
9. D
10. A and B

Chapter 16

1. D
2. False
3. B
4. B
5. A
6. A and D
7. A and B

Glossary

1080p30

An abbreviation for video transmission at 1080p resolution at 30 frames per second.

2B+D

A designation for an ISDN BRI implementation including two bearer channels and 1 data channel.

23B+D

A designation for an ISDN PRI (T1) implementation including 23 bearer channels and 1 data channel.

30B+D

A designation for an ISDN PRI (E1) implementation including 30 bearer channels and 1 data channel.

720p30

An abbreviation for video transmission at 720p resolution at 30 frames per second.

720p60

An abbreviation for video transmission at 720p resolution at 60 frames per second.

802.11a/b/g/n/ac

Standards for wireless local-area network radio transmission and associated data rates. Typically, devices with wireless network adapters will be listed as 802.11 capable followed by a list of the radios supported. In this case, 802.11a/b/g/n/ac denotes support for 802.11a, 802.11b, 802.11g, 802.11n, and 802.11ac standards.

802.3af PoE

A standard for providing a maximum of 15.4 watts of DC power to a PoE-capable device. The 802.3af standard defines only the use of Classes 0, 1, 2, and 3. Class 4 devices require 802.3at.

802.3at PoE

A standard for providing a maximum of 25.5 watts of DC power to a PoE-capable device. Class 4 devices must have 802.3at power available to function. This is also known as PoE+.

AAR

Automated alternate routing. Used to reroute calls over a traditional PSTN network if bandwidth is not available, as governed by CAC.

Active Directory

A Microsoft-based LDAP server. Microsoft AD is a common LDAP server in which CUCM end users are synchronized and authenticated against.

Ad hoc

Cisco defines an ad hoc conference as any conference where the participants joining the conference are not scheduled.

ANI

Automatic Number Identification.

API

Application programming interface. The code-line interface used by programmers and equipment administrators to issue advanced-level commands to the system.

ARJ

Admission reject; part of the H.323 RAS messaging. Sent from the gatekeeper to an endpoint confirming the call attempt failed.

ARQ

Admission request; part of the H.323 RAS messaging. Sent from an endpoint to the gatekeeper to request a call be established.

Auto attendant

A virtual reception available on MCUs. Auto attendants are used as a means for participants to choose what conference they want to join. Auto participants will hear an interactive voice response or IVR.

Auto-registration

A capability in CUCM that allows for phones to be connected to the network, register, and receive a directory number without any phone-specific administrative configuration.

AVC

Advanced Video Coding. Video compression format that is commonly used for the recording, compression, and distribution of video content.

Basic Rate Interface (BRI)

BRI is an ISDN interface to basic rate access. Basic rate access consists of a single 16-kbps D channel plus 2 64-kbps B channels for voice or data.

BAT

Bulk Administration Tool. A method inside a CUCM server for bulk administering settings into the database.

Binary Floor Control Protocol (BFCP)

Binary Floor Control Protocol is a protocol for controlling access to the media resources in a conference. It is also used in embedding shared content into a video stream during a video call or videoconference.

CA

Certification authority. Entity that issues digital certificates (especially X.509 certificates) and vouches for the binding between the data items in a certificate.

CAC

Call admission control.

Call control

A central or distributed entity that provides signaling, destination route pattern lookup, connection admission control, class of restriction, and other operations associated with call setup, state change, and teardown in telephony or video infrastructure deployments.

CAS

Channel associated signaling.

CCD

Call Control Discovery. Works along with Service Advertisement Framework (SAF) and Global Dial Plan Replication (GDPR) for dynamic dial plan distribution in a Collaboration solution.

CFA

Call Forward All. A Cisco IP phone feature to forward all calls to a user-defined number.

CFNB

Call Forward No Bandwidth. A Cisco IP phone feature.

CFQDN

Cluster fully qualified domain name.

CFUR

Call Forward Unregistered. A Cisco IP phone feature.

Cisco ASA

Cisco Adaptive Security Appliance.

CoS

Class of service. An indication of how an upper-layer protocol requires a lower-layer protocol to treat its messages. In SNA subarea routing, CoS definitions are used by subarea nodes to determine the optimal route to establish a given session. A CoS definition comprises a virtual route number and a transmission priority field. Also called ToS.

cRTP

Compressed Real-Time Transport Protocol. Specified in RFC 2508, compressed RTP provides a mechanism to reduce the IP/UDP/RTP headers from 40 bytes to 2 to 4 bytes to improve bandwidth efficiency.

CSS

Calling search space. Defines what can be called by an endpoint.

CSV

Comma-separated value. A form of data represented by columns that are separated by commas. Typically, CSV files are used to exchange information between disparate systems. CUCM uses CSV files as an import and export tool.

CTI

Computer telephony integration. Allows IP PBX to interact with computers and computer applications.

CTS

Cisco TelePresence Server. A scalable videoconferencing bridge that works with Cisco Unified Communications Manager to bring multiparty video to unified communications deployments.

CUBE

Cisco Unified Border Element.

CUCM

Cisco Unified Communications Manager is an IP PBX.

DHCP

Dynamic Host Configuration Protocol is a client/server protocol used to provide IP information to a device automatically.

DID

Direct inward dialing. Allows a user outside a company to dial an internal extension number without needing to pass through an operator or an attendant. The dialed digits are passed to the PBX, which then completes the call.

DISA

Direct inward system access.

DMZ

Demilitarized zone.

DN

Directory number.

DND

Do Not Disturb. A Cisco IP phone feature.

DNS

Domain Name System. System used on the Internet for translating names of network nodes into addresses.

DNS SRV

Domain Name System Service record. Used to identify systems that host specific services.

DSP

Digital signal processor. Used to handle analog-to-digital (IP) and vice versa conversion and to handle functions like mixing of voice streams.

DTMF

Dual-tone multifrequency. Tones generated when a button is pressed on a telephone.

E.164

An ITU-T recommendation, titled "The International Public Telecommunication Numbering Plan," that defines a numbering plan for the worldwide public switched telephone network (PSTN) and some other data networks. E.164 defines a general format for international telephone numbers.

EMCC

Cisco extension mobility cross cluster. This is a feature of Cisco Unified Communications Manager that allows a user to log in to an IP phone on a remote cluster.

Expressway-C

The internal gateway component of the Cisco Expressway (Collaboration Edge) solution. Cisco Expressway-C and Expressway-E form a secure traversal link to enable video, voice, content, and IM&P services to software clients and endpoints outside the firewall.

Expressway-E

The external gateway component of the Cisco Expressway (Collaboration Edge) solution. Cisco Expressway-C and Expressway-E form a secure traversal link to enable video, voice, content, and IM&P services to software clients and endpoints outside the firewall.

FQDN

Fully qualified domain name. FQDN is the full name of a system, rather than just its hostname. For example, abc is a hostname, and abc.local.com is an FQDN.

Gatekeeper

A call control and CAC mechanism most often associated with H.323 voice and video implementations.

GDPR

Global Dial Plan Replication. Allows numbers and patterns to be replicated dynamically using Inter-cluster Lookup Service (ILS).

H.323

H.323 allows dissimilar communication devices to communicate with each other by using a standardized communication protocol. H.323 defines a common set of codecs, call setup and negotiating procedures, and basic data transport methods.

HTTP

Hypertext Transfer Protocol is an application protocol for distributed, collaborative, hypermedia information systems. HTTP is the foundation of data communication for the World Wide Web (WWW). Hypertext is structured text that uses logical links (hyperlinks) between nodes containing text.

HTTPS

Secure Hypertext Transfer Protocol is the use of Secure Sockets Layer (SSL) or Transport Layer Security (TLS) as a sublayer under regular HTTP application layering.

ICT

Intercluster Trunk. Allows intercluster dialing between two or more clusters.

ILS

Inter-cluster Lookup Service. Dynamically networks clusters so that the directory Uniform Resource Identifier (URI) information can be replicated using an ILS network. GDPR uses this service.

IOS

Internetwork Operating System. The operating system used on most Cisco routers and Cisco network switches.

IPsec

IP Security. A framework of open standards that provides data confidentiality, data integrity, and data authentication between participating peers. IPsec provides these security services at the IP layer. IPsec uses IKE to handle the negotiation of protocols and algorithms based on local policy and to generate the encryption and authentication keys to be used by IPsec.

IPsec can protect one or more data flows between a pair of hosts, between a pair of security gateways, or between a security gateway and a host.

ISDN

Integrated Services Digital Network is a form of communication over the circuit-switched network using V.35, PRI, or BRI lines.

ITSP

Internet telephony service provider. Usually a SIP provider.

IVR

Interactive voice response. Term used to describe systems that provide information in the form of recorded messages over telephone lines in response to user input in the form of spoken words or, more commonly, DTMF signaling. Examples include banks that allow you to check your balance from any telephone and automated stock quote systems.

Jabber IM&P

A Cisco Jabber client (desktop or mobile) that is configured to provide only IM and presence services. In IM&P mode, the Jabber desktop client is still capable of providing CTI control of a Cisco collaboration desktop endpoint.

Jabber phone

A Cisco Jabber client (desktop or mobile) that is configured to provide only voice service.

LBM

Location Bandwidth Manager. Allows enhanced location-based call admission control (E-LCAC) for intercluster calls.

LCR

Least cost routing.

LDAP

Lightweight Directory Access Protocol. A protocol that provides access for management and browser applications that provide read/write interactive access to the X.500 Directory.

LRG

Local route group. A feature of Cisco Unified Communications Manager that allows using a local gateway for PSTN.

MGCP

Media Gateway Control Protocol. A merging of the IPDC and SGCP protocols.

MOH

Music on hold. Plays when a party is kept on hold by the other party. MOH can be unicast or multicast and can be played from a CUCM MOH server or an IOS router's flash.

MPLS

Multiprotocol Label Switching. A switching method that forwards IP traffic using a label. This label instructs the routers and the switches in the network where to forward the packets based on pre-established IP routing information.

MRA

Mobile and Remote Access. A core component of collaboration edge architecture (Expressway). MRA allows Cisco Jabber and other endpoints VPN-less registration, call control, provisioning, messaging, and presence services.

MRG

Media resource group. Collection of media resources such as transcoders, conferencing, and MOH resources.

MRGL

Media resource group list. Collection of MRG.

MTP

Media termination point. Helps bridge audio calls with different characteristics.

MVA

Mobile voice access.

MWI

Message waiting indicator. Indicator on a Cisco Unified IP phone to notify the user of a new voicemail.

NANP

North American Numbering Plan.

NAT

Network Address Translation. A mechanism for reducing the need for globally unique IP addresses. NAT allows an organization with addresses that are not globally unique to connect to the Internet by translating these addresses into globally routable address space. Also known as Network Address Translator.

NTP

Network Time Protocol. A protocol that is built on top of TCP that ensures accurate local timekeeping with reference to radio and atomic clocks that are located on the Internet. This protocol is capable of synchronizing distributed clocks within milliseconds over long time periods.

OOB

Out of band. One of the many ways used for management of resources in a network.

OTLD

Organization top-level domain.

PAT

Port Address Translation. Translation method that allows the user to conserve addresses in the global address pool by allowing source ports in TCP connections or UDP conversations to be translated. Different local addresses then map to the same global address, with port translation providing the necessary uniqueness. When translation is required, the new port number is picked out of the same range as the original following the convention of Berkeley Standard Distribution (BSD).

PKI

Public key infrastructure. System of CAs (and, optionally, RAs and other supporting servers and agents) that performs some set of certificate management, archive management, key management, and token management functions for a community of users in an application of asymmetric cryptography.

POTS

Plain old telephone service.

PSTN

Public switched telephone network. General term referring to the variety of telephone networks and services in places worldwide. Sometimes also called POTS.

QoS

Quality of service. Measure of performance for a transmission system that reflects its transmission quality and service availability.

RAS

Registration, Admission, Status are communication messages sent between devices and an H.323 gatekeeper.

RSIP

RestartInProgress.

RSVP

Resource Reservation Protocol. A network-control protocol that allows endpoints to request specific QoS for their data flows.

RTCP

Real-time Transport Control Protocol.

RTMT

Real-Time Monitoring Tool. A GUI-based troubleshooting tool for Cisco Unified Communications servers.

RTP

Real-time Transport Protocol. Commonly used with IP networks. RTP is designed to provide end-to-end network transport functions for applications transmitting real-time data, such as audio, video, or simulation data, over multicast or unicast network services. RTP provides such services as payload type identification, sequence numbering, timestamping, and delivery monitoring to real-time applications.

SAF

Service Advertisement Framework. Used to dynamically set up number/DN distribution in a Cisco Collaboration solution using underlying routing protocols.

SCCP

Skinny Client Control Protocol. A protocol developed by Cisco for its endpoints to communicate with CUCM and other Cisco call controls.

SDP

Session Description Protocol.

SIP

Session Initiation Protocol. Protocol developed by the IETF MMUSIC Working Group as an alternative to H.323. SIP features are compliant with IETF RFC 2543, published in March 1999. SIP equips platforms to signal the setup of voice and multimedia calls over IP networks.

SNAP

Simple Network-enabled Auto Provision.

SRST

Survivable remote site telephony.

SRTP

Secure Real-time Transport Protocol.

SSH

Secure Shell Protocol. Protocol that provides a secure remote connection to a router through a TCP application.

SVC

Switched virtual circuit. A virtual circuit that is dynamically established on demand and is torn down when transmission is complete. SVCs are used in situations where data transmission is sporadic. Called a switched virtual connection in ATM terminology.

TDM

Time-division multiplexing. Technique in which information from multiple channels can be allocated bandwidth on a single wire based on preassigned time slots. Bandwidth is allocated to each channel regardless of whether the station has data to transmit.

TEHO

Tail-end hop-off. For LCR of toll calls via IP until the penultimate hop and then hand off to PSTN.

TLS

Transport Layer Security. An IETF protocol.

TMS

TelePresence Management Suite.

TTL

Time To Live. A mechanism that limits the lifespan or lifetime of data in a computer or network.

TUI

Telephone user interface.

TURN

Traversal Using Relay NAT.

UCL

User Connect Licensing.

UDS

User Data Services. A REST-based set of operations that provide authenticated access to user resources and entities such as users' devices, subscribed services, speed dials, and much more from the Unified Communications configuration database. UDS is available with CUCM 10.0 and later.

URI

Uniform Resource Identifier. Type of formatted identifier that encapsulates the name of an Internet object and labels it with an identification of the name space, thus producing a member of the universal set of names in registered name spaces and of addresses referring to registered protocols or name spaces [RFC 1630].

UWL

Unified Workspace Licensing.

VCS

[Cisco TelePresence] Video Communications Server.

VPN

Virtual private network. Enables IP traffic to travel securely over a public TCP/IP network by encrypting all traffic from one network to another. A VPN uses tunneling to encrypt all information at the IP level.

VXML

Voice XML.

XMPP

Extensible Messaging and Presence Protocol.

Index

Symbols

. (dot), 82

! calling-party transformation pattern, 109

Numbers

112, emergency dialing, EU, 111-112

888, emergency dialing, Australia, 110

999, emergency dialing, UK, 111-112

A

AAR (automated alternate routing), 38, 54

  availability, 59

  CAC (Call Admission Control), 202-204

AAR Destination Mask setting, 204

access codes, 18

  digit manipulation requirements, 82-83

  requirements, for centralized call-processing deployments, 83-84

access control lists. *See* ACLs (access control lists)

access list functions, Cisco Unified Mobility, 274

  time-of-day access control, 274-275

access lists, Cisco Unified Mobility, 270

ACLs (access control lists), 7

addressing

  blended addressing, 127-128

  URI endpoint addressing, overview, 123-125

admin policies, VCS (Video Communications Server), dial plans, 338-339

Advanced Voice BusyOut (AVBO), 206

advantages of QoS (quality of service), multisite deployment solutions, 37-39

advertised patterns configuration, GDPR (Global Dial Plan Replication), 389

Aggregation Services Routers (ASRs), 29

Allow List, registration restriction policy, 318

alternate patterns, GDPR (Global Dial Plan Replication), 387

ANI (automatic number identification), 16, 82

annunciators, 40

disabling for remote branches, 43

appliance-based hardware versions, VCS (Video Communications Server), 296

area codes, 18-19

ASRs (Aggregation Services Routers), 29

associate ccm, 176

associate profile (number), 176, 179

audio-only calls, E-LCAC (enhanced location-based CAC), 194

AudioStream, RSVP (Reservation Protocol), 197

Australia, emergency dialing, 111

authenticating endpoints, 316-317

authentication methods, VCS (Video Communications Server), 317-318

automated alternate routing (AAR). *See* AAR (automated alternate routing), 38, 54

automatic number identification (ANI), 16, 82

availability

multisite deployment issues, 10-11

multisite deployment solutions, 53-54

AAR (automated alternate routing), 59

CFNB (Call Forward on No Bandwidth), 59

CFUR (Call Forward Unregistered), 36, 58-59

fallback for IP phones, 56-57

MGCP fallback, 55-56

PSTN backup, 55

AVBO (Advanced Voice BusyOut), 206

B

B2B (business-to-business), 4

B2B video communications, 294

B2C (business-to-consumer) video communications, 294

backups, PSTN backup, 53-55

dial plan challenges, 22-23

bandwidth

codec selection, 162-166

codecs, 161-162

issues with, 7-10

local conference bridges, 172-176

management options, 159-161

managing

CAC (Call Admission Control), 160

QoS (quality of service), 160

MRG (Media Resource Group), 166-168

MRGL (media resource group list), 166-168

multicast MOH, 168-171

multisite deployment solutions, 39-41

codec configuration in CUCM, 42-43

disabling annunciators for remote branches, 43

local versus remote conference bridges, 44

low-bandwidth codecs and RTP header compression, 41-42

mixed conference bridges, 46-47

multicast MOH from branch router flash, 47-49

transcoder design, 45-46

transcoders, 44-45

provisioning, 183

transcoders, 176-179

bandwidth restrictions, pipes, 328-329

Base DN for Accounts parameter, 314

Base DN for Groups, 314

BAT (Bulk Administration Tool), 124

blended addressing, 127-128

branch router flash

multicast MOH, 49-51

multicast MOH from branch router flash, Cisco IOS configuration, 51

bridges, local conference bridges, 172-176

bulk administration, URI endpoint addressing, 124-125

Bulk Administration Tool (BAT), 124

business-to-business (B2B), 4

C

CAC (Call Admission Control), 3, 184

AAR (automated alternate routing), 202-204

bandwidth, 160

characteristics of, 184

E-LCAC (enhanced location-based CAC), 187-196

LBM (Locations Bandwidth Manager), 187-189

local CAC, 184, 204-205

location-based CAC (LCAC), 185-187

measurement-based CAC, 184, 206

preventing, too many calls, 52-53

reservation-based CAC, 184, 205-206

RSVP (Reservation Protocol), 196-197

configurations, 198-199

SIP preconditions, 199-201

topology unaware CAC, 184

topology-aware CAC, 184

Call Control Discovery (CCD). *See* CCD (Call Control Discovery)

trunks, 79

call egress, 107-109

gateways, 102

globalized call routing, 99

localized call egress

gateways, 105-107

phones, 107-109

phones, 102

call flow

between, CUCM and VCS-C, 335-336

Mobile Connect, 264-266

MVA (mobile voice access), 266-267

with MGCP or SCCP gateway access, 272

RSVP (Reservation Protocol), 197

SDP (Session Description Protocol), 199

Call Forward on No Bandwidth (CFNB), 54

Call Forward Unregistered (CFUR), 54, 58-59

call ingress
  globalized call routing, 99
  normalization of localized call ingress
    on gateways, 102-104
    on phones, 104-105

Call Manager Express (CME), 54

Call Processing Language (CPL), 323

call routing
  CFUR (Call Forward Unregistered), 154-155
  device mobility, 226-228
  globalized call routing. See globalized call routing
  numeric URI call routing process, 134
    routing URI calls over SIP trunks, 134-136
  URI call routing, 129-132

call routing logic, directory URIs, 126

call routing non-numerical URI call routing process, 132-134

call sources, URI call sources, 126-127

called-party transformation patterns, 106-107

caller IDs
  limiting, 274
  time-of-day access control, 275

calling search space (CSS), 4, 60
  Cisco Extension Mobility, 247-248
  URI partitions, 125-126

calls, preventing by CAC, 52-53

CAPEX (capital expenditure), 37

capital expenditure (CAPEX), 37

CBWFQ (class-based weighted fair queue), 37

CCD (Call Control Discovery), 27, 397, 400
  characteristics of, 408-410
  overview, 406-407
  SAF (Service Advertisement Framework), 401
  schemas, 408
  trunks, 79
  use cases, 410-411

CCD advertising service, 407

CCD requesting service, 407

CEF (Cisco Express Forwarding), 5

centralized administration, TMS (TelePresence Management Suite), 319

centralized call processing
  access and site codes, 83-84
  CUCM (Cisco Unified Communications Manager), 10

centralized conference resources, 9

CFB (conference bridges), 8

CFNB (Call Forward on No Bandwidth), 54
  availability, 59

CFQDN (cluster fully qualified name), 127

CFUR (Call Forward Unregistered), 54, 154
  availability, 36, 58-59

challenges
  of dial plan implementation, 397-399
  of roaming, 210-211
    solving with device mobility, 212-213
  of roaming users, 241-242

characteristics of
  CAC (Call Admission Control), 184
  CCD (Call Control Discovery),
    408-410
  extension mobility, 243-244
Cisco Business Edition, 143
Cisco Express Forwarding. *See*
  CEF (Cisco Express Forwarding)
Cisco Expressway C, 63-64
Cisco Expressway Core and Edge
  solution, 71
Cisco Expressway E, 63-64
Cisco Expressway series. *See*
  Expressway series
Cisco Extension Mobility
  configurations, 252-256
  CSS (calling search space) and, 247-248
  device profiles, 248-249
    configurations, 254
    feature safe, 251-252
    *relationships between
      configuration elements,
      249-251*
  phone service, configurations,
    253-254
Cisco IOS configuration, multicast
  MOH from branch router flash, 51
Cisco IOS gateway protocol,
  functions review, 72-73
Cisco Jabber, 289
  troubleshooting, MRA (Mobile
    Remote Access), 373
Cisco Jabber for TelePresence (Jabber
  Movi), 291-292
Cisco Jabber Guest, 291
Cisco Jabber Video for TelePresence,
  deploying, 320
Cisco Mobile Remote Access. *See*
  MRA (Mobile Remote Access)

Cisco Service Advertisement
  Framework (SAF), 386
Cisco TelePresence Management
  Suite. *See* TMS (TelePresence
  Management Suite)
Cisco Unified Border Elements
  (CUBEs), 4
  ILS calls, 383-385
Cisco Unified CME (Communications
  Manager Express), 141-142
Cisco Unified Communications
  Manager (CUCM), 4
Cisco Unified Mobility, 60-61, 261
  access list functions, 274
    *time-of-day access control,
      274-275*
  call flow
    Mobile Connect, 264-266
    *MVA (mobile voice access),
      266-267*
    *MVA (mobile voice access) with
      MGCP or SCCP gateway
      access, 272*
  configuration elements, 268-271
  CSS (calling search space), 273
  implementing, 267-268
  MGCP (Media Gateway Control
    Protocol), 271
  Mobile Connect, 262-263
    configuring, 275-281
  MVA (mobile voice access), 263
    characteristics of, 263-264
    configuring, 281-284
    CSS (calling search space), 273
  overview, 262-263
  SCCP (Skinny Client Control
    Protocol), 271
Cisco Unified Workspace Licensing
  (CUWL), 297

Cisco Validated Designs (CVDs), 9

class-based weighted fair queue (CBWFQ), 37

cluster fully qualified name (CFQDN), 127

clustering
  deployment, 300
  Expressway series, 298-300
  VCS (Video Communications Server), 298-300

CME (Call Manager Express), 54, 141-142

codec configuration in CUCM, 42-43

codec preferences, 164-165

codec selection, bandwidth, 162-166

codecs, 161
  factory default lossy codecs, 163
  factory default low loss options, 162
  G.711, 161
  G.722, 161
  G.729, 161
  H.264, 161
  iLBC (Internet Low Bitrate Codec), 161
  iSAC (Internet Speech Audio Codec), 162
  LATM (Low Overhead Audio Transport Multiplex), 162

collaboration edge, 290

commands
  associate ccm, 176
  associate profile (number), 176, 179
  dn template, 147
  dsp services dsp-farm under voice-card, 175
  dspfarm profile (number), 176, 179
  ephone-dns, 147
  ephone-template, 147
  ip multicast-routing, 171

ip pim sparse-dense-mode, 171
ip source-address, 146
max-dn, 148
max-ephones, 146
maximum conference participants, 176
maximum sessions, 179
max-pool, 148
moh, 147
sccp ccm, 175
sccp ccm group (number), 176
sccp enables, 175
sccp local, 175
secondary-dialtone, 146
srst, 147
switchback, 176
telephony-service, 147

comparing VCS (Video Communications Server) and Expressway series, 297-298

components of MRA, 351-352

conference bridges (CFB), 8
  local versus remote, 44
  mixed conference bridges, 46-47

configuration elements
  Cisco Unified Mobility, 268-271
  device mobility, 217-220

configuration parameters, extension mobility, 244

configurations
  CUBE configurations for ILS calls, 384
  CUCM (Cisco Unified Communications Manager), for MRA, 358-363
  CUCM SRST configuration, 149-150
  device mobility, 233-236
  Expressway series, for MRA, 366-372

extension mobility, 252-256

   phone service, 253-254

FindMe feature, VCS (Video Communications Server), 341-344

GDPR (Global Dial Plan Replication), 388-391

hardware conference bridge configuration, 175

ILS networking, 380-381

IM&P for MRA, 363-366

interconnection, CUCM and VCS, 340-341

IOS transcoder configurations, 179

LDAP authentication configuration example, 313-314

MGCP fallback configuration, 154

mobile access at the Cisco IOS gateway, 283

multicast configuration for IOS routers supporting multicast MOH, 171

RSVP (Reservation Protocol), 198-199

SAF clients, 412-417

SAF forwarder, 417-419

SAF forwarder L2 adjacent configuration, 404

SAF forwarder nonadjacent peer configuration, 405

SIP E-SRST, 147-148

SRST (survivable remote site telephony), 146-147

SRST router multicast configuration, 152-153

VCS (Video Communications Server), 301-306

configuring

AAR (automated alternate routing), 203

IP phones, PSTN gateways, 88-89

Mobile Connect, 275-281

MRG (Media Resource Group), 168

MRGL (media resource group list), 168

MVA (mobile voice access), 281-284

transcoders, 178

congestion-management tools, 36

connection options, CUCM (Cisco Unified Communications Manager), overview, 71

connections, multisite connection options. *See* **multisite connection options**

considerations for, clustering, 299-300

country codes, 18

CPL (Call Processing Language), 323

cRTP, 42

CSS (calling search space), 4, 245

Cisco Extension Mobility, 247-248

Cisco Unified Mobility, 273

device CSS, 225

device mobility, 225-226

line CSS, 225

MVA (mobile voice access), 273

CUBEs (Cisco Unified Border Elements), 4, 62

flow-through mode, 62-63

ILS calls, 383-385

SAF (Service Advertisement Framework), 411

CUCM (Cisco Unified Communications Manager), 4

availability challenges, 10-11

CAC (Call Admission Control), 184

configurations, for MRA, 358-363

connection options, overview, 71

with Expressway series, 289

interconnection

with VCS, 340-341

with VCS-C, 334

regions relationships, transcoder design, 46

VCS-C (Video Communications Server Control), 290

call flow, 335-336

interconnection, 295

CUCM Cluster A, 74

CUCM extension mobility, 60

CUCM SRST configuration, 149-150

CUWL (Cisco Unified Workspace Licensing), 297

CVDs (Cisco Validated Designs), 9

D

data applications, 8

default device profiles, feature safe, Extension Mobility, 251-252

default subzones, VCS (Video Communications Server), 322-323

Deny List, registration restriction policy, 318

deploying Cisco Jabber Video for TelePresence, 320

deployment

clustering, 300

Expressway series, 293-294

multisite deployment issues, overview, 2-5

VCS (Video Communications Server), 292-293

VCS Expressway series, 292-293

design, transcoders, 45-46

detection of end of dialing invariable-length plans, dial plan challenges, 20-21

device CSS, device mobility, 225

device mobility, 60

call routing paths, 226-228

configuration elements, 217-220

configurations, 233-236

considerations for, 224

CSS (calling search space), 225-226

device CSS, 225

dynamic configuration by location-dependent device pools, 216-217

dynamic phone configuration parameters, 213-216

globalized call routing, 228-229

examples, 230-233

local route groups, 229

line CSS, 225

operations, 220-221

flowcharts, 221-223

overview, 213

roaming, 210

issues with, 210-211

solving issues, 212-213

device mobility group (DMG), 217

call routing, 226-228

device mobility info (DMI), 217

device pool (DP), 217

device profiles, Extension Mobility, 248-249

configurations, 254

feature safe, 251-252

relationships between configuration elements, 249-251

dial emergency dialing, 24

dial plan challenges, 12

complex dial plan implementation, 397-399

detection of end of dialing invariable-length plans, 20-21

DID (Direct Inward Dialing) ranges, 14-15

different methods of PSTN dialing, 24-25

E.164 addressing, 14-15

fixed versus variable-length numbering plans, 17-20

nonconsecutive numbers, 13

optimized call routing, 15-16, 22-23

overlapping numbers, 12

PSTN backup, 22-23

PSTN requirements, 16, 23-24

scalability, 17

variable-length numbering, 13-14

dial plan overlap, 4

dial plan requirements, for multisite deployments, with distributed call processing, 79-80

dial plan scalability issues, 26-27

dial plan solutions, 61-62

dial plans

global dial plan catalogs, 391-392

SRST IOS dial plan, 148-149

VCS (Video Communications Server), 337-338

admin policies, 338-339

FindMe feature, 339

search rules, 340

transforms, 338

dial presentation, 23

dial rules for called-party numbers, 23

dial while, 26

dialing

emergency dialing. *See* emergency dialing

URI dialing, overview, 120-123

DID (direct inward dial), 60-61

DID (Direct Inward Dialing) ranges, 14-15, 92

solutions, 61

different methods of PSTN dialing, 24-25

digit manipulation requirements

access and site codes, 82-83

for PSTN backup, on-net intersite calls, 90-92

digital signal processors (DSPs), 10

direct inward dial (DID), 60-61

Direct Inward Dialing (DID) ranges, 14-15

directory numbers (DNs), 4

directory URIs

dial ring interpretations, 130

FQDNs (Fully Qualified Domain Names), 128-129

disabling, annunciators, for remote branches, 43

distributed call processing

CUCM (Cisco Unified Communications Manager), 10

dial plan requirements, multisite deployment, 79-80

DMG (device mobility group), 217

call routing, 226-228

DMI (device mobility info), 217

dn template, 147

DNs (directory numbers), 4

internal DNs, 15

DNS resolution

Expressway-C, 355

Expressway-E, 355

DNS SRV (Domain Name System service records), 349

Expressway-C, 355

Expressway-E, 354-355

MRA (Mobile Remote Access), 354-355

dot (.), 82

DP (device pool), 217

dsp services dsp-farm under voice-card, 175

dspfarm profile (number), 176, 179

DSPs (digital signal processors), 10, 40, 161

DTMF (dual-tone multifrequency), 73

dynamic configuration by location-dependent device pools, device mobility, 216-217

dynamic phone configuration

by device profiles, extension mobility, 245

extension mobility, 244

dynamic phone configuration parameters, device mobility, 213-216

E

E.164, normalized call routing, 101

E.164 (number), globalized call routing, 99

E.164 addressing, 14-15, 24

+E.164

globalized call routing, 97, 101

ILS (Intercluster Lookup Service), 385

E911, 110

easy contacts management, TMS (TelePresence Management Suite), 320

E-LCAC (enhanced location-based CAC), 52, 187, 189-196

audio-only calls, 194

immersive calls, 195

intercluster E-LCAC, 191

intracluster E-LCAC, 190

network modeling, 189

overview, 196

shadow locations, 193

shared locations, 194

video calls, 194

EMCC (extension mobility cross clusters), 79, 162

emergency dialing, 24

globalized call routing, 109-112

United Kingdom, 112

encapsulation, Layer 2 encapsulation, 8

End User Configuration page, URI endpoint addressing, 124

End User License Agreement (EULA), 51

end users, Cisco Unified Mobility, 268-269

endpoint authentication, 316-317

endpoint registration, 314-316

endpoints, blended addressing, 128

end-user home cluster discovery, 385

enhanced LCAC, 160, 187, 189-196

audio-only calls, 194

immersive calls, 195

intercluster E-LCAC, 191

intracluster E-LCAC, 190

network modeling, 189

overview, 196

shadow locations, 193

shared locations, 194

video calls, 194

enhanced SRST (E-SRST), 144

Enterprise Alternate, ILS (Intercluster Lookup Service), 385

enterprise alternate number, 84

ephone-dns, 147

ephone-template, 147

E-SRST (enhanced SRST), 144

SCCP based E-SRST, 147

SIP E-SRST, 147-148

EULA (End User License Agreement), 51

European Union, emergency dialing, 111

Expressway series, 291-293

clustering, 298-300

comparing to, VCS (Video Communications Server), 297-298

configurations, 301-306

for MRA, 366-372

CUCM (Cisco Unified Communications Manager), 289

deployment, 293-294

licensing, 297

overview, 288-289

platforms, 296

remote access, 306

Expressway-C, 350-352

DNS resolution, 355

DNS SRV (Domain Name System service records), 355

Expressway-E, 350-351

DNS resolution, 355

DNS SRV (Domain Name System service records), 354-355

HTTPS reverse proxy, 354

Extension Mobility, 60, 241

configuration parameters, 244

configurations, 252-256

CSS (calling search space) and, 247-248

device profiles, 248-249

configurations, 254

feature safe, 251-252

dynamic phone configuration by device profiles, 245

login process, 245

operations, 245-247

overview, 243-244

phone service, configurations, 253-254

roaming users, 241-243

user logins, 244

extension mobility cross clusters (EMCC), 79, 162

external callers, globalized call routing, 100

F

factory default lossy codecs, 163

factory default low loss options, codecs, 162

fallback

for IP phones, 56-57

for IP phones with SRST, 54

MGCP (Media Gateway Control Protocol), 53-54, 153-154

MGCP fallback, 55-56

fallback mode, 151

feature safe, device profiles (Extension Mobility), 251-252

FIFO (first-in, first-out), 6

FindMe feature, VCS (Video Communications Server), 339

configurations, 341-344

firewall traversal, MRA (Mobile Remote Access), 352-353

firewalls

inside firewalls, 351

MRA (Mobile Remote Access), firewall traversal, 352-353

outside firewalls, 351

first-in, first-out. *See* FIFO (first-in, first-out)

fixed numbering plans, versus variable-length numbering plans, 17-20

flexible scheduling, TMS (TelePresence Management Suite), 319

flow-through mode, CUBEs (Cisco Unified Border Elements), 62-63

flowcharts, operations (device mobility), 221-223

foreign exchange office (FXO), 71

forwarders, SAF (Service Advertisement Framework), 403

FQDNs (Fully Qualified Domain Names), directory URIs, 128-129

functions review, Cisco IOS gateway protocol, 72-73

FXO (foreign exchange office), 71

G

G.711, 7-8, 41, 161

G.722, 42, 161

G.729, 8, 39, 42, 45, 161

MOH (music on hold), 52

G.729r8, 160

gadget overload, 261

gatekeeper-based CAC, 205

gatekeeper-based RSVP, 205-206

gatekeeper-controlled ICT, 75-78

gatekeepers, 53

gateways

call egress, 102

localized call egress, 105-107

normalization of localized call ingress, 102-104

GDPR (Global Dial Plan Replication), 27, 80

configurations, 388-391

imported global dial plan, 387

learned global dial plans, 387

overview, 386-388

partitions, 388

PSTN failover number, 387

German roaming users, 224

global dial plan catalogs, 391-392

Global Dial Plan Replication (GDPR). *See* GDPR

globalized call routing, 61-62

+E.164, 97

device mobility, 228-229

   examples, 230-233

   local route groups, 229

emergency dialing, 109-112

implementing, 96-98

interdependencies, 112-113

number formats, 98-102

TEHO (tail-end hop-off)

   examples, 113-114

   interdependencies, 113

H

H.225, 71

trunks, 77-78

H.225 trunk, 76

H.264, 161

H.323, 72

trunks, 74-76

H.323 gatekeepers, 26, 71

H.323 gateway, 272

HD (high-density), 35

Hookflash transfers, H.323, 75

hot desking, 241

hoteling, 241

HTTPS reverse proxy, MRA (Mobile Remote Access), 354

hubs, LBM (Locations Bandwidth Manager), 191-193

I

ICT, digit manipulation requirements, 91

iLBC (Internet Low Bitrate Codec), 161

ILS (Intercluster Lookup Service), 27

+E.164, 385

Enterprise Alternate, 385

networking, overview, 378-379

networking configurations, 380-381

ILS calls via SIP trunk and Cisco Unified Border Element, 383-385

ILS-based SIP URI dialing/ routing, 381-383

overview, 378

trunks, 79

URIs, 385

ILS configuration for GDPR, 388

ILS-based SIP URI dialing/routing, 381-383

ILS-learned URIs, 133

IM (instant messaging), 4

IM&P, configurations for MRA, 363-366

immersive calls, E-LCAC (enhanced location-based CAC), 195

implementing

Cisco Unified Mobility, 267-268

dial plans, challenges of, 397-399

globalized call routing, 96-98

PSTN access, in Cisco IOS gateways, 84-88

PSTN backup, for on-net intersite calls, 90

selective PSTN breakout, 88-89

site codes for on-net calls, 81

TEHO (tail-end hop-off), 92-93

imported global dial plan, GDPR (Global Dial Plan Replication), 387

In&Out bandwidth restrictions, VCS-C (Video Communications Server Control), 325-326

incoming calls

ISDN TON, 85-86

from PSTN, 84

incoming PSTN call, globalized call routing, 99

inside firewalls, MRA (Mobile Remote Access), 351

instant messaging (IM), 4

Integrated Services Routers (ISRs), 29

Intelligent WAN. *See* IWAN (Intelligent WAN)

interactive voice response (IVR), 267

intercluster E-LCAC, 191

Intercluster Lookup Service (ILS). *See* ILS (Intercluster Lookup Service)

intercluster trunk, 71

interconnection

CUCM and VCS, configurations, 340-341

CUCM and VCS-C

call flow, 335-336

overview, 334

VCS-C (Video Communications Server Control), CUCM (Cisco Unified Communications Manager), 295

interdependencies, globalized call routing, 112-113

TEHO (tail-end hop-off), 113

interdigit timeout, 20

internal callers, globalized call routing, 101

internal DNs, 15

international prefix, 18, 19

international TONs, 86

Internet Low Bitrate Codec (iLBC), 161

Internet Speech Audio Codec (iSAC), 162

Internet telephony service providers (ITSPs), 4

intracluster E-LCAC, 190

IOS dial plans, SRST (survivable remote site telephony), 148-149

IP addresses, 4

multicast MOH, 172

ip multicast-routing, 171

IP phones

Cisco Unified Mobility, 269

configuring, PSTN gateways, 88-89

ip pim sparse-dense-mode, 171

ip source-address, 146

IP WAN

availability issues, 11

broken connections, 11

CUCM, availability challenges, 10-11

voice packets, 8

iSAC (Internet Speech Audio Codec), 162

ISDN TON

examples, 87-88

incoming calls, 85-86

ISRs (Integrated Services Routers), 29, 34

issues, multisite deployment issues

availability, 10-11

bandwidth, 7-10

dial plan challenges. *See* dial plan challenges

dial plan scalability issues, 26-27

NAT (Network Address Translation), 27-29

overview, 2-5

security, 27-29

voice and video call quality, 5-6

ITSPs (Internet telephony service providers), 4

IVR (interactive voice response), 267

IWAN (Intelligent WAN), 7

J-K

Jabber, troubleshooting MRA (Mobile Remote Access), 373

Jabber clients, registering remote Jabber clients, 355-357

Jabber Guest, 291

Jabber Movi, 291-292

jitter, 6

L

LATM (Low Overhead Audio Transport Multiplex), 162

Layer 2 encapsulation, 8

LBM (Locations Bandwidth Manager), 187-189

hubs, 191-193

LBM intercluster replication network, 191

LCAC (location-based CAC), 160, 185-187

LCR (least cost routing), 25

LDAP (Lightweight Directory Access Protocol), 311

URI endpoint addressing, 124

VCS, user authentication options, 312-314

learned global dial plans, GDPR, 387

least cost routing (LCR), 25

legacy video, terminology, 290-292

licensing

Expressway series, 297

VCS (Video Communications Server), 297

limiting caller IDs, 274

line CSS, device mobility, 225

links, VCS-C (Video Communications Server Control), 324

zone bandwidth restrictions, 325-327

LLQ (low-latency queueing), 37

local CAC, 184, 204-205

local conference bridges, 40, 44

bandwidth, 172-176

local route groups

device mobility, 229

selective PSTN breakout, 88

TEHO (tail-end hop-off), 95-96

Local Voice BusyOut (LVBO), 205

local zones, VCS (Video Communications Server), 321-322

local-dependent device pools, device mobility, dynamic configuration, 216-217

localized call egress

gateways, 105-107

globalized call routing, 98

at phones, 107-109

localized call ingress

normalization of, on gateways, 102-104

normalization of localized call ingress, on phones, 104-105

localized E.164 (number), globalized call routing, 99

location-based CAC (LCAC), 160, 185-187

Locations Bandwidth Manager (LBM), 187-189

login process, extension mobility, 245

Low Overhead Audio Transport Multiplex (LATM), 162

low-bandwidth codecs, 39-40

RTP header compression, 41-42

low-latency queueing (LLQ), 37

LVBO (Local Voice BusyOut), 205

M

MAN (metropolitan-area network), 7

managing bandwidth, 159-161

CAC (Call Admission Control), 160

max-dn, 148

max-ephones, 146

max-hops parameter, 169

maximum conference participants, 176

maximum sessions, 179

maximum transmission unit. See MTU (maximum transmission unit)

max-pool, 148

measurement-based CAC, 184, 206

media exchange, CUCM, 10

Media Gateway Control Protocol (MGCP), 4

Media Resource Group (MRG), 152
  bandwidth, 166-168

media resource group list (MRGL), 48
  bandwidth, 166-168

media termination points (MTPs), 40

Medianet, 3

message waiting indicator. *See* MWI (message waiting indicator)

messages, SAF (Service Advertisement Framework), 406

metropolitan-area network (MAN), 7

MGCP (Media Gateway Control Protocol), 4, 39, 72
  Cisco Unified Mobility, 271
  fallback, 53-56, 153-154

mixed conference bridges, 46-47

Mobile Connect, 262-263
  call flow, 264-266
  configuring, 275-281

Mobile Remote Access (MRA), 291

mobile voice access. *See* MVA (mobile voice access)

mobility solutions, 54, 60-61

MOH (music on hold), 8, 41
  G.729, 52
  multicast MOH
    bandwidth, 168-171
    remote site router flash, 52
    SRST (survivable remote site telephony), 150-153
  multicast MOH from branch router flash, 47-49
    Cisco IOS configuration, 51
    examples, 49-51

moh command, 147

MPLS (Multiprotocol Label Switching), 29, 159

MRA (Mobile Remote Access), 291, 349
  components, 351-352
  configurations
    CUCM, 358-363
    Expressway series, 366-372
    IM&P, 363-366
  DNS SRV (Domain Name System service records), setup, 354-355
  firewall traversal, 352-353
  HTTPS reverse proxy, 354
  overview, 349-351
  troubleshooting, 373

MRG (Media Resource Group), 152
  bandwidth, 166-168
  configuring, 168

MRGL (media resource group list), 48
  bandwidth, 166-168
  configuring, 168

MTPs (media termination points), 40

MTU (maximum transmission unit), 8

multicast MOH
  bandwidth, 168-171
  IP address and port considerations, 172
  remote site router flash, alternatives, 52
  SRST (survivable remote site telephony), 150-153

multicast MOH from branch router flash, 47-49
  Cisco IOS configuration, 51
  examples, 49-51

multicast traffic, 50

Multiprotocol Label Switching (MPLS), 29

multisite connection options,
overview, 70-71

multisite deployment issues
availability, 10-11
bandwidth, 7-10
dial plan challenges, 12
*detection of end of dialing
invariable-length plans,
20-21*
*DID (Direct Inward Dialing)
ranges, 14-15*
*different methods of PSTN
dialing, 24-25*
E.164 addressing, 14-15
*fixed versus variable-length
numbering plans, 17-20*
nonconsecutive numbers, 13
*optimized call routing, 15-16,
22-23*
overlapping numbers, 12
PSTN backup, 22-23
PSTN requirements, 16, 23-24
scalability, 17
*variable-length numbering,
13-14*
dial plan scalability issues, 26-27
NAT (Network Address Translation),
27-29
overview, 2-5
security, 27-29
voice and video call quality, 5-6

multisite deployment solutions
availability, 53-54
*AAR (automated alternate
routing), 59*
*CFNB (Call Forward on No
Bandwidth), 59*
*CFUR (Call Forward
Unregistered), 36, 58-59*

fallback for IP phones, 56-57
MGCP fallback, 55-56
PSTN backup, 55
bandwidth, 39-41
*codec configuration in CUCM,
42-43*
*disabling annunicators for
remote branches, 43*
*local versus remote conference
bridges, 44*
*low-bandwidth codecs and RTP
header compression,
41-42*
*mixed conference bridges,
46-47*
*multicast MOH from branch
router flash, 47-49*
transcoder design, 45-46
transcoders, 44-45
dial plan requirements, distributed
call processing, 79-80
dial plan solutions, 61-62
mobility solutions, 60-61
NAT (Network Address Translation),
62-64
overview, 34-36
QoS (quality of service), 36
advantages of, 37-39

music on hold (MOH), 8

MVA (mobile voice access),
262-263
call flow, 266-267
characteristics of, 263-264
configuring, 281-284
CSS (calling search space), 273

MVA media resource, Cisco Unified
Mobility, 270

MWI (message waiting indicator),
241

N

NANP (North American Numbering Plan), 13, 86

NAT (Network Address Translation), 5

issues, 27-29

solutions, 62-64

national TONs, 86

natural user experience, TMS (TelePresence Management Suite), 320

Network Address Translation. *See* NAT (Network Address Translation)

network modeling, E-LCAC, 189

networking, ILS (Intercluster Lookup Service), 378-379

networking configurations, ILS (Intercluster Lookup Service), 380-381

ILS calls via SIP trunk and Cisco Unified Border Element, 383-385

ILS-based SIP URI dialing/routing, 381-383

non-gatekeeper-controlled ICT, 75-77

non-numerical URI call routing process, 132-134

nonconsecutive numbers

dial plan challenges, 13

solutions, 61

nonfallback mode, 151

normalized call routing, E.164, 101

North American Numbering Plan (NANP), 13, 86

number formats, globalized call routing, 98-102

number globalization, globalized call routing, 99

number localization, globalized call routing, 99

number normalization, globalized call routing, 99

numeric URI call routing process, 134

O

on-net calls, site codes, 81

on-net intersite calls

digit manipulation requirements, PSTN backup, 90-92

implementing, PSTN backup, 90

operational expenditure (OPEX), 37

operations, device mobility, 220-221

flowcharts, 221-223

OPEX (operational expenditure), 37

optimized call routing, 15-16

dial plan challenges, 22-23

organization top-level domain (OTLD), 127

OTLD (organization top-level domain), 127

outgoing calls, PSTN, 84

outgoing PSTN calls, globalized call routing, 99

outside firewalls, MRA (Mobile Remote Access), 351

overlapping numbers

dial plan challenges, 12

solutions, 61

P

packet buffers, 6

packet drop, 6

packet voice data modules (PVDMs), 9

Packet Voice DSP Module (PVDM), 160

packets, 6

parameters, dynamic phone configuration parameters, device mobility, 213-216

partitions, 4

  configurations, GDPR (Global Dial Plan Replication), 390

  GDPR (Global Dial Plan Replication), 388

  URI partitions, calling search space (CSS), 125-126

patterns, alternate patterns, GDPR (Global Dial Plan Replication), 387

percent encoding, 122

phone buttons, configuration parameters, 244

phone service, extension mobility, configurations, 253-254

phones

  call egress, 102

  localized call egress, 107-109

  normalization of localized call ingress, 104-105

physical location (PL), 217

pipes, VCS-C (Video Communications Server Control), 327-329

PL (physical location), 217

plain old telephone system (POTS), 4

platforms

  Expressway series, 296

  VCS (Video Communications Server), 296

Policy Service, registration restriction policy, 318

ports, multicast MOH, 172

POTS (plain old telephone system), 4

PQ (priority queue), 37

preconditions-based call processing, RSVP (Reservation Protocol), 200-201

Presence-enabled call lists, 126

preventing calls by CAC, 52-53

priority queue (PQ), 37

protocols

  LDAP (Lightweight Directory Access Protocol), 311

  MGCP (Media Gateway Control Protocol), fallback, 153-154

  RSVP (Reservation Protocol). See RSVP (Reservation Protocol)

  RTP (Real-time Transport Protocol), 3

  SAF-CP (SAF Client Protocol), 403

  SAF-FP (SAF Forwarding Protocol), 400, 405

  SCCP (Skinny Client Control Protocol), 176

  TCP (Transmission Control Protocol), 5

  UDP (User Datagram Protocol), 5

provisioning

  bandwidth, 183

  TMS (TelePresence Management Suite), 319-320

PSTN (public switched telephone network), 4

  incoming calls, 84

  outgoing calls, 84

  selective PSTN breakout, implementing, 88-89

PSTN access

  examples, 85

  implementing for on-net calls in Cisco IOS gateways, 84-88

PSTN ANI, 92

PSTN backup, 53, 55

  dial plan challenges, 22-23

digit manipulation requirements, on-net intersite calls, 90-92

implementing for on-net intersite calls, 90

PSTN dialing, methods of, 24-25

PSTN failover number, GDPR (Global Dial Plan Replication), 387

PSTN Format, globalized call routing, 99

PSTN gateways, IP phones, configuring, 88-89

PSTN requirements, dial plan challenges, 16, 23-24

public switched telephone network (PSTN), 4

PVDM (Packet Voice DSP Module), 160

PVDMs (packet voice data modules), 9, 40

Q

QoS (quality of service), 3, 6

advantages of, 37-39

bandwidth, managing, 160

multisite deployment solutions, 36

VoIP packet prioritization, 38

queue-management tools, 36

queueing, 6

R

RDP (remote destination profile), Cisco Unified Mobility, 269

Real-Time Monitoring Tool (RTMT), 401

troubleshooting, MRA, 373

Real-time Transport Protocol. *See* RTP (Real-Time Transport Protocol)

region Remote-Site, 165

regions relationships, CUCM (Cisco Unified Communications Manager), transcoder design, 46

registering

endpoints, 314-316

remote Jabber clients, with CUCM, 355-357

registration restriction policy, 318-319

relationships, between Extension Mobility configuration elements, 249-251

remote access, Expressway series, 306

remote branches, disabling, 43

remote Cisco Jabber users, 294

remote conference bridges, 44

remote destination, Cisco Unified Mobility, 270

remote destination profile (RDP), Cisco Unified Mobility, 269-271

remote Jabber clients, registering, 355-357

remote site router flash, multicast MOH (alternatives), 52

Reservation Protocol, 3

reservation-based CAC, 184, 205-206

Resource Reservation Protocol. *See* RSVP (Reservation Protocol)

right to use (RUU), 51

roaming, 25, 210

issues with, 210-211

solving with device mobility, 212-213

roaming users, 241-242

extension mobility, 243

roaming-sensitive settings

device mobility, 215

globalized call routing, device mobility, 228

routing
    optimized call routing, 15-16
        dial plan challenges, 22-23
    URI calls over SIP trunks, 134-136
RSVP (Reservation Protocol), 3,
    52-53, 160, 196-197
    call flow, 197
    configurations, 198-199
    preconditions-based call processing,
        200-201
    SIP preconditions, 199-201
RTMT (Real-Time Monitoring
    Tool), 401
RTP (Real-Time Transport Protocol),
    3-5
RTP header compression, 40
    low-bandwidth codecs, 41-42
RUU (right to use), 51

S

SAF (message) advertisement, 400
SAF (Service Advertisement
    Framework), 386, 397
    architecture, 399-401
    CCD (Call Control Discovery), 401
    clients, 402
    messages, 406
    overview, 399
        architecture, 399-401
    SAF-CP (SAF Client Protocol), 403
SAF Client Protocol (SAF-CP), 400
SAF clients, 400, 402
    configurations, 412-417
SAF forwarder, 400, 403-405,
    409-410
    configurations, 417-419
SAF forwarder protocol, 405

SAF forwarding nodes, 403-405
SAF Forwarding Protocol
    (SAF-FP), 400
SAF unaware (node) routers, 400
SAF-CP (SAF Client Protocol),
    400, 403
SAF-FP (SAF Forwarding Protocol),
    400, 405
sampling rates, 8
scalability
    dial plan challenges, 17
    dial plan scalability issues, 26-27
scalable provisioning, TMS
    (TelePresence Management
    Suite), 319
SCCP (Skinny Client Control
    Protocol), 4, 39, 176
    Cisco Unified Mobility, 271
SCCP based E-SRST, 147
sccp ccm, 175
sccp ccm group (number), 176
sccp enables, 175
sccp local, 175
schemas, CCD (Call Control
    Discovery), 408
SDP (Session Description Protocol), 199
    call flow, 199
search rules, VCS (Video
    Communications Server), 340
secondary-dialtone, 146
security
    issues, 27-29
    solutions, 62-64
selective PSTN breakout,
    implementing, 88-89
Service Advertisement Framework.
    See SAF (Service Advertisement
    Framework)
Session Description Protocol (SDP), 199

Session Initiation Protocol (SIP), 4

Session Management Edition (SME), 191

shadow locations, E-LCAC, 193

shared locations, E-LCAC, 194

single number reach (SNR), 60-61

SIP (Session Initiation Protocol), 4, 53, 71-73

    preconditions, RSVP (Reservation Protocol), 199-201

    trunk characteristics, 73

    trunks, 76-77

SIP E-SRST, 147-148

SIP INVITE, 127

SIP profiles

    dial string interpretations, 130

    FQDNs (Fully Qualified Domain Names), directory URIs, 129

SIP trunks, 334

    routing URI calls, 134-136

site codes

    digit manipulation requirements, 82-83

    implementing for on-net calls, 81

    requirements, for centralized call-processing deployments, 83-84

Skinny Client Control Protocol (SCCP), 4, 176

small office/home office. See SOHO (small office/home office)

small to medium-size businesses. See SMBs (small to medium-size businesses)

SMBs (small to medium-size businesses), 141

    Cisco Business Edition, 143

    Cisco Unified CME (Communications Manager Express), 141-142

CUCM SRST configuration, 149-150

SRST (survivable remote site telephony), 144-146

SRST IOS dial plan, 148-149

SME (Session Management Edition), 191, 381

SNR (single number reach), 60-61, 262-264, 320

SNR fatigue, 274

software-based CAC, 41

SOHO (small office/home office), 141

Solutions Reference Network Designs (SRNDs), 9

SRNDs (Solutions Reference Network Designs), 9, 33

SRST (survivable remote site telephony), 4, 141, 144-146

    CFUR (Call Forward Unregistered), 154

    configurations, 146-147

    fallback, 56-57

    fallback for IP phones, 54

    multicast MOH, 150-153

srst command, 147

SRST IOS dial plan, 148-149

SRST reference, 149

storing, international telephone numbers of contacts, 25

stub forwarders, SAF (Service Advertisement Framework), 404

subscriber numbers, 18-19

subscriber TONs, 86

subzones, VCS (Video Communications Server), 323

survivability features, MGCP (Media Gateway Control Protocol), 80

survivable remote site telephony (SRST), 4, 144-146

switchback, 176

T

T.302 timer, 14

tail drop, 6

tail-end hop-off (TEHO), 16, 22-23

Tandberg Company, 289

TCP (Transmission Control Protocol), 5

TEHO (tail-end hop-off), 16, 22-23

   call routing, device mobility, 226-228

   examples

      with local route groups, 95-96

      without local route groups, 93-95

   globalized call routing

      examples, 113-114

      interdependencies, 113

   implementing, 92-93

telephony-service, 147

TelePresence Management Suite. *See* TMS (TelePresence Management Suite)

TelePresence Management Suite Provisioning Extension (TMSPE), 320

terminology for video/legacy video, 290-292

time-of-day access control, access list functions, 274-275

TMS (TelePresence Management Suite), provisioning, 319-320

TMSPE (TelePresence Management Suite Provisioning Extension), 320

toll bypass, 22

TON (type of number), 16

   incoming calls, ISDN TON, 85-86

   international TONs, 86

   ISDN TON, 87-88

   national TONs, 86

   subscriber TONs, 86

tools

   congestion-management tools, 36

   queue-management tools, 36

topology unaware CAC, 184

topology-aware CAC, 184

Total bandwidth restriction, VCS-C (Video Communications Server Control), 326-327

transcoders, 44-45

   bandwidth, 176-179

   configuring, 178

   design, 45-46

   guidelines for, 177-178

transforms, VCS (dial plans), 338

transitive forwarders, SAF (Service Advertisement Framework), 404

Transmission Control Protocol. *See* TCP (Transmission Control Protocol)

traversal subzones, VCS (Video Communications Server), 321-323

troubleshooting, MRA (Mobile Remote Access), 373

trunk prefix, 18-19

trunks, 71

   Call Control Discovery (CCD), 79

   EMCC (extension mobility cross clusters), 79

   gatekeeper-controlled ICT, 77-78

   H.225, 76-78

   H.323, 74-76

   ILS (Intercluster Lookup Service), 79

   non-gatekeeper-controlled ICT, 76-77

SIP (Session Initiation Protocol), 73, 76-77

VCS (Video Communications Server), 79

type of number (TON), 16

U

UC (Unified Communications), 1

UCCX (Unity Connection and Unity Contact Center Express), 9

UDP (User Datagram Protocol), 5

Unified Communications (UC), 1

Unified IP phones, transcoders, 178

Unified Mobility. *See* Cisco Unified Mobility

United Kingdom, emergency dialing, 111-112

Unity Connection and Unity Contact Center Express (UCCX), 9

Universal Resource Indicators. *See* URIs (Universal Resource Indicators)

URI call routing, 129-132

URI call sources, overview, 126-127

URI calls over SIP trunks, routing over SIP trunks, 134-136

URI dialing, overview, 120-123

URI endpoint addressing, overview, 123-125

URI partitions, calling search spaces (CSSs), 125-126

URIs (Universal Resource Indicators), 377

directory URIs, FQDNs (Fully Qualified Domain Names), 128-129

global dial plan catalogs, 391-392

ILS (Intercluster Lookup Service), 378, 385

ILS (Intercluster Lookup Service) networking, 378-379

ILS-based SIP URI dialing/routing, 381-383

ILS-learned URIs, 133

numeric URI call routing process, 134

URIs non-numerical URI call routing process, 132-134

URL encoding, 122

use cases, CCD (Call Control Discovery), 410-411

user authentication options, VCS (Video Communications Server), 312-313

LDAP (Lightweight Directory Access Protocol), 313-314

User Datagram Protocol. *See* UDP (User Datagram Protocol)

user logins, extension mobility, 244

user-specific device-level parameters, extension mobility, 244

V

variable-length numbering

detection of end of dialing invariable-length plans, 20-21

dial plan challenges, 13-14

versus fixed numbering plans, 17-20

solutions, 61

VCS (Video Communications Server), 287, 290-291

authentication methods, 317-318

clustering, 298-300

comparing to, Expressway series, 297-298

configurations, 301-306

deployment, 292-293

dial plans, 337-338
    admin policies, 338-339
    FindMe feature, 339
    search rules, 340
    transforms, 338
endpoint authentication, 316-317
endpoint registration, 314-316
FindMe feature, configurations, 341-344
interconnection, with CUCM, 340-341
licensing, 297
overview, 288-289
platforms, 296
registration restriction policy, 318-319
trunks, 79
user authentication options, 312-313
    LDAP (Lightweight Directory Access Protocol), 313-314
zones, 320-321
    default subzones, 322-323
    local zones, 321-322
    subzones, 323
    traversal subzones, 321-323
VCS Expressway Core, 288
VCS Expressway Edge, 288
VCS Expressway series
deployment, 292-293
VCS-C (Video Communications Server Control), 290
VCS-C (Video Communications Server Control), 311
CUCM (Cisco Unified Communications Manager), 290
    call flow, 335-336
interconnection, 295
    with CUCM, 334

links, 324
    zone bandwidth restrictions, 325-327
pipes, 327-329
VCS Expressway series, 290
VICs (virtual interface cards), 4, 35
video, terminology, 290-292
video call quality, 5-6
video calls, E-LCAC (enhanced location-based CAC), 194
Video Communications Server. *See* VCS
Video Communications Server Control. *See* VCS-C
VideoStream, RSVP (Reservation Protocol), 197
virtual interface cards (VICs), 4
virtual routing and forwarding (VRF), 399
VMware-based, VCS (Video Communications Server), 296
VoFR (Voice over Frame Relay), 205
voice call quality, 5-6
Voice Extensible Markup Language (VXML), 267
Voice over Frame Relay (VoFR), 205
Voice over IP (VoIP), 161
voice packets
bandwidth, 8
IP WAN, 8
voice RTP streams, 7
VoIP (Voice over IP), 161
VoIP packet prioritization, QoS (quality of service), 38
VPN tunnels, 28
VRF (virtual routing and forwarding), 399
VXML (Voice Extensible Markup Language), 267

W-X-Y

WAAS (Wide Area Application Services), 7

WAN failure, CFUR (Call Forward Unregistered), 58

WAN links, bandwidth, 7-8

Wide Area Application Services (WAAS), 7

Within bandwidth restrictions, VCS-C, 325

Z

zone bandwidth restrictions, VCS-C (Video Communications Server Control)

In&Out, 325-326

Total bandwidth restriction, 326-327

Within, 325

zones, VCS (Video Communications Server), 320-321

default subzones, 322-323

local zones, 321-322

subzones, 323

traversal subzones, 323

CISCO

Connect, Engage, Collaborate

The Award Winning Cisco Support Community

Attend and Participate in Events

Ask the Experts
Live Webcasts

Knowledge Sharing

Documents
Blogs
Videos

Top Contributor Programs

Cisco Designated VIP
Hall of Fame
Spotlight Awards

Multi-Language Support

https://supportforums.cisco.com